Rough Computing:
Theories, Technologies, and Applications

Aboul Ella Hassanien
Kuwait University, Kuwait

Zbigniew Suraj
University of Information Technology and Management, Poland

Dominik Ślęzak
University of Regina, Canada

Pawan Lingras
Saint Mary's University, Canada

INFORMATION SCIENCE REFERENCE
Hershey · New York

Acquisitions Editor:	Kristin Klinger
Development Editor:	Kristin Roth
Senior Managing Editor:	Jennifer Neidig
Managing Editor:	Sara Reed
Copy Editor:	Angela Thor
Typesetter:	Jamie Snavely
Cover Design:	Lisa Tosheff
Printed at:	Yurchak Printing Inc.

Published in the United States of America by
Information Science Reference (an imprint of IGI Global)
701 E. Chocolate Avenue, Suite 200
Hershey PA 17033
Tel: 717-533-8845
Fax: 717-533-8661
E-mail: cust@igi-pub.com
Web site: http://www.igi-global.com/reference

and in the United Kingdom by
Information Science Reference (an imprint of IGI Global)
3 Henrietta Street
Covent Garden
London WC2E 8LU
Tel: 44 20 7240 0856
Fax: 44 20 7379 0609
Web site: http://www.eurospanonline.com

Copyright © 2008 by IGI Global. All rights reserved. No part of this publication may be reproduced, stored or distributed in any form or by any means, electronic or mechanical, including photocopying, without written permission from the publisher.

Product or company names used in this set are for identification purposes only. Inclusion of the names of the products or companies does not indicate a claim of ownership by IGI Global of the trademark or registered trademark.

Library of Congress Cataloging-in-Publication Data

Rough computing : theories, technologies and applications / Aboul Ella Hassanien ... [et al.], eds.

   p. cm.

  Summary: "This book offers the most comprehensive coverage of key rough computing research, surveying a full range of topics from granular computing to pansystems theory. With its unique coverage of the defining issues of the field, it provides libraries with a single, authoritative reference to this highly advanced technological topic"--Provided by publisher.

  Includes bibliographical references and index.

  ISBN-13: 978-1-59904-552-8 (hardcover)

  ISBN-13: 978-1-59904-554-2 (ebook)

  1. Soft computing. 2. Rough sets. I. Hassanien, Aboul Ella.

  QA76.9.S63R66 2008

  006.3--dc22

                                    2007016855

British Cataloguing in Publication Data
A Cataloguing in Publication record for this book is available from the British Library.

All work contributed to this book set is new, previously-unpublished material. The views expressed in this book are those of the authors, but not necessarily of the publisher.

In Memoriam

This book is dedicated to Professor Zdzisław Pawlak, a father of rough sets, who passed away on April 7, 2006 in Warsaw, Poland.

Table of Contents

Preface x

Acknowledgment xiii

Section I
Foundations of Rough Sets

Chapter I
Foundations of Rough Sets from Vagueness Perspective / *Piotr Wasilewski and Dominik Ślęzak* 1

Chapter II
Rough Sets and Boolean Reasoning / *Hung Son Nguyen* 38

Chapter III
Rough Set-Based Feature Selection: A Review / *Richard Jensen and Qiang Shen* 70

Chapter IV
Rough Set Analysis and Formal Concept Analysis / *Yiyu Yao and Yaohua Chen* 108

Section II
Current Trends and Models

Chapter V
Rough Sets: A Versatile Theory for Approaches to Uncertainty Management in Databases /
Theresa Beaubouef and Frederick E. Petry 129

Chapter VI
Current Trends in Rough Set Flow Graphs / *Cory J. Butz and Wen Yan* 152

Chapter VII
Probabilistic Indices of Quality of Approximation / *Annibal Parracho Sant'Anna* 162

Chapter VIII
Extended Action Rule Discovery Based on Single Classification Rules and Reducts /
Zbigniew W. Ras and Elzbieta M. Wyrzykowska 175

Section III
Rough Sets and Hybrid Systems

Chapter IX
Monocular Vision System that Learns with Approximation Spaces /
James F. Peters, Maciej Borkowski, Christopher Henry, and Dan Lockery 186

Chapter X
Hybridization of Rough Sets and Multi-Objective Evolutionary Algorithms for
Classificatory Signal Decomposition / *Tomasz G. Smolinski, Astrid A. Prinz,
and Jacek M. Zurada* 204

Chapter XI
Two Rough Set Approaches to Mining Hop Extraction Data / *Jerzy W. Grzymala-Busse,
Zdzislaw S. Hippe, Teresa Mroczek, Edward Roj, and Bolesław Skowroński* 228

Chapter XII
Rough Sets for Discovering Concurrent System Models from Data Tables /
Krzysztof Pancerz and Zbigniew Suraj 239

Compilation of References 269

About the Contributors 290

Index 297

Detailed Table of Contents

Preface x

Acknowledgment xiii

Section I
Foundations of Rough Sets

Chapter I
Foundations of Rough Sets from Vagueness Perspective / *Piotr Wasilewski and Dominik Ślęzak* 1

This chapter presents the algebraic foundations behind the rough-set theory. Algebraic structures arise from various types of data-based knowledge and information. The authors examine those structures with respect to satisfying or violating different versions of the excluded middle principle. Besides some novel results, the chapter may be also treated as a survey on the rough-set-based methodologies of representing and analyzing the information systems.

Chapter II
Rough Sets and Boolean Reasoning / *Hung Son Nguyen* 38

This chapter presents the approximate Boolean reasoning approach to problem solving. It is based on general framework for the concept approximation. It combines classical Boolean reasoning with modern methods in machine learning and data mining. The author shows advantages of approximate Boolean reasoning using examples of some representative challenges of KDD, emphasizing ability of balancing between the quality of the designed solution and its computational cost.

Chapter III
Rough Set-Based Feature Selection: A Review / *Richard Jensen and Qiang Shen* 70

This chapter introduces the fundamental ideas behind rough-set-based approaches and reviews the related feature selection methods. The authors discuss extensions to the traditional rough set approach, including recent feature selection methods based on tolerance rough sets, variable precision rough sets, and fuzzy-rough sets. The chapter reports also the latest developments in the search methods supporting the rough-set-based feature selection search methods, including hill-climbing, genetic algorithms, and ant colony optimization.

Chapter IV
Rough Set Analysis and Formal Concept Analysis / *Yiyu Yao and Yaohua Chen* 108

This chapter reviews the existing studies on the comparisons and combinations of rough-set analysis and formal concept analysis. Such unified framework provides a better understanding of the current data analysis challenges. The authors also report some new results, important for both the theoretical foundations and applications.

Section II
Current Trends and Models

Chapter V
Rough Sets: A Versatile Theory for Approaches to Uncertainty Management in Databases /
Theresa Beaubouef and Frederick E. Petry 129

This chapter discusses how rough sets can enhance databases by allowing for management of uncertainty. The authors discuss the rough relational database and rough object-oriented database models, as well as their fuzzy and intuitionistic extensions. Benefits of those various methods are discussed, illustrating usefulness and versatility of rough sets for the database extensions.

Chapter VI
Current Trends in Rough Set Flow Graphs / *Cory J. Butz and Wen Yan* 152

The chapter reviews a recently developed framework for reasoning from data, called the rough set flow graphs (RSFG). The authors examine two methods for conducting inference in RSFG. They further show how the order of variable elimination affects the amount of computation. The culminating result is the incorporation of an algorithm for obtaining a good ordering into the RSFG inference.

Chapter VII
Probabilistic Indices of Quality of Approximation / *Annibal Parracho Sant'Anna* 162

The chapter presents a new index of quality of approximation. It measures the mutual information between the relations respectively determined by conditional and decision attributes. It is based on comparison of two graphs, each one representing a set of attributes. Applications in the context of indiscernibility and dominance relations are considered. Combination with the idea of transformation into probabilities of classes being a preferred option is explored as well. The algorithmic procedure to select the most important attributes is outlined.

Chapter VIII
Extended Action Rule Discovery Based on Single Classification Rules and Reducts /
Zbigniew W. Ras and Elzbieta M. Wyrzykowska 175

This Chapter focuses on a novel strategy of construction of action rules, directly from single classification rules instead of the pairs of classification rules. This way, not only the simplicity of the algorithm, but also its reduced-time complexity is achieved. The chapter also presents a modified tree-based strategy for constructing action rules, with comparative analysis included.

Section III
Rough Sets and Hybrid Systems

Chapter IX
Monocular Vision System that Learns with Approximation Spaces /
James F. Peters, Maciej Borkowski, Christopher Henry, and Dan Lockery 186

The chapter discusses the monocular vision system that learns to control the pan and tilt operations of a digital camera that tracks a moving target. The authors consider various forms of the actor critic learning methods. Rough set approximation spaces are applied to handle degrees of overlapping of the behavior patterns. The conventional actor critic methods are successfully extended with the built-in run-and-twiddle rough-set-based control strategy mechanism.

Chapter X
Hybridization of Rough Sets and Multi-Objective Evolutionary Algorithms for
Classificatory Signal Decomposition / *Tomasz G. Smolinski, Astrid A. Prinz,
and Jacek M. Zurada* 204

This chapter proposes hybridization of rough-set theory and multiobjective evolutionary algorithms to perform the task of signal decomposition in the light of the underlying classification problem itself. The authors present results for several variants of implementation of the introduced methodology.

Chapter XI
Two Rough Set Approaches to Mining Hop Extraction Data / *Jerzy W. Grzymala-Busse,
Zdzislaw S. Hippe, Teresa Mroczek, Edward Roj, and Bolesław Skowroński* 228

This chapter presents application of two rough-set approaches to mining the data sets related to bed caking during the hop extraction process. The authors use direct rule induction by the MLEM2 algorithm; they also generate belief networks conversed into the rule sets within the BeliefSEEKER system. Statistics for the rule sets are presented. Six rule sets are also ranked by the expert. The overall results show that both approaches are of approximately the same quality.

Chapter XII
Rough Sets for Discovering Concurrent System Models from Data Tables /
Krzysztof Pancerz and Zbigniew Suraj 239

This chapter constitutes the continuation of research trend binding rough-set theory with concurrency. Automatic methods of discovering concurrent system models from data tables are presented. Data tables are created on the basis of observations or specifications of the process behaviors in the modeled systems. The proposed methods are based on rough set theory and colored Petri nets.

Compilation of References 269

About the Contributors 290

Index 297

Preface

Since introduction of rough set methodology by Zdzisław Pawlak in the early eighties, we have witnessed its great advances in both theory and applications. There is a growing research interest in foundations of rough sets, with some relationships to other already established methodologies. As a result, rough sets are linked with decision support and intelligent systems, soft and granular computing, data mining and KDD, as well as pattern recognition and machine learning. A wide range of applications of rough sets, alone or combined with other techniques, have been proposed in bioinformatics and medicine, business and finances, environmental and social studies, multimedia data mining and processing, and many others.

The objective of this book is to provide a representative and authoritative source of current research results and methodology in the area of rough set methods and applications. The book consists of 12 chapters organized in three section. Section I (four chapters) presents the foundations of rough sets, including comparison with other methodologies and discussion on future perspectives. This part will be useful for the beginners and students who are interested in carrying out his/her research projects on the rough sets.

Section II (four chapters) continues with current trends of extending, combining, as well as applying rough set techniques. Section III (four chapters) discusses hybrid intelligent systems involving the elements of rough set techniques, as well as illustrates the place for rough sets in real-life applications. These two parts will be useful for the professionals who are interested in catching a new idea or using the book as a reference work.

Chapter I, by **Piotr Wasilewski** and **Dominik Ślęzak**, presents the algebraic foundations behind the rough set theory. Algebraic structures arise from various types of data-based knowledge and information. The authors examine those structures with respect to satisfying or violating different versions of the excluded middle principle. Besides some novel results, the chapter may be also treated as a survey on the rough-set-based methodologies of representing and analyzing the information systems.

Chapter II, by **Hung Son Nguyen**, presents the approximate Boolean reasoning approach to problem solving. It is based on the general framework for the concept approximation. It combines classical Boolean reasoning with modern methods in machine learning and data mining. The author shows advantages of approximate Boolean reasoning using examples of some representative challenges of KDD, emphasizing ability of balancing between the quality of the designed solution and its computational cost.

Chapter III, by **Richard Jensen** and **Qiang Shen**, introduces the fundamental ideas behind rough-set-based approaches and reviews the related feature selection methods. The authors discuss extensions to the traditional rough set approach, including recent feature selection methods based on tolerance rough sets, variable precision rough sets and fuzzy rough sets. The chapter reports also the latest developments in the search methods supporting the rough-set-based feature selection search methods, including hill climbing, genetic algorithms, and ant colony optimization.

Chapter IV, by **Yiyu Yao** and **Yaohua Chen**, reviews the existing studies on the comparisons and combinations of rough set analysis and formal concept analysis. Such unified framework provides a better understanding of the current data analysis challenges. The authors also report some new results important for both the theoretical foundations and applications.

Chapter V, by **Theresa Beaubouef** and **Frederick E. Petry**, discusses how rough sets can enhance databases by allowing for management of uncertainty. The authors discuss the rough relational database and rough object-oriented database models, as well as their fuzzy and intuitionistic extensions. Benefits of those various methods are discussed, illustrating usefulness and versatility of rough sets for the database extensions.

Chapter VI, by **Cory J. Butz** and **Wen Yan**, reviews a recently developed framework for reasoning from data, called the rough set flow graphs (RSFG). The authors examine two methods for conducting inference in RSFG. They further show how the order of variable elimination affects the amount of computation. The culminating result is the incorporation of an algorithm for obtaining a good ordering into the RSFG inference.

Chapter VII, by **Annibal Parracho Sant'Anna**, presents a new index of quality of approximation. It measures the mutual information between the relations respectively determined by conditional and decision attributes. It is based on comparison of two graphs, each one representing a set of attributes. Applications in the context of indiscernibility and dominance relations are considered. Combination with the idea of transformation into probabilities of classes being a preferred option is explored as well. The algorithmic procedure to select the most important attributes is outlined.

Chapter VIII, by **Zbigniew W. Ras** and **Elzbieta M. Wyrzykowska**, focuses on a novel strategy of construction of action rules, directly from single classification rules instead of the pairs of classification rules. This way, not only the simplicity of the algorithm, but also its reduced time complexity is achieved. The chapter also presents a modified tree-based strategy for constructing action rules, with comparative analysis included.

Chapter IX, by **James F. Peters, Maciej Borkowski, Christopher Henry**, and **Dan Lockery**, discusses the monocular vision system that learns to control the pan and tilt operations of a digital camera that tracks a moving target. The authors consider various forms of the actor critic learning methods. Rough set approximation spaces are applied to handle degrees of overlapping of the behavior patterns. The conventional actor critic methods are successfully extended with the built-in run-and-twiddle rough-set-based control strategy mechanism.

Chapter X, by **Tomasz G. Smolinski, Astrid A. Prinz**, and **Jacek M. Zurada**, proposes hybridization of rough set theory and multiobjective evolutionary algorithms to perform the task of signal decomposition in the light of the underlying classification problem itself. The authors present results for several variants of implementation of the introduced methodology.

Chapter XI, by **Jerzy W. Grzymala-Busse, Zdzisław S. Hippe, Teresa Mroczek, Edward Roj**, and **Boleslaw Skowronski**, presents application of two rough set approaches to mining the data sets related to bed caking during the hop extraction process. The authors use direct rule induction by the MLEM2 algorithm; they also generate belief networks conversed into the rule sets within the BeliefSEEKER system. Statistics for the rule sets are presented. Six rule sets are also ranked by the expert. The overall results show that both approaches are of approximately the same quality.

Chapter XII, by **Krzysztof Pancerz** and **Zbigniew Suraj**, constitutes the continuation of research trend binding rough set theory with concurrency. Automatic methods of discovering concurrent system models from data tables are presented. Data tables are created on the basis of observations or specifications of the process behaviors in the modeled systems. The proposed methods are based on rough set theory and colored Petri nets.

Overall, this book provides a critical analysis of many important issues in rough sets and applications, including introductory material and research results, as well as hybrid intelligent systems involving the elements of rough set techniques.

Acknowledgment

The editors would like to thank all people involved in the development and review process of the book, especially Professor Guoyin Wang, the Chairman of the Rough Set and Soft Computation Society, Chinese Association for Artificial Intelligence, Dean of the College of Computer Science and Technology, Chongqing University of Posts and Telecommunications, and Professor Sheela Ramanna, Dept. of Applied Computer Science, University of Winnipeg. Most of the authors of chapters included in this book also served as referees for chapters written by other authors. Thanks go to all those who provided constructive and comprehensive reviews.

We would also like to thank all the authors for their contributions. Special thanks go to Kristin Roth and Meg Stocking at IGI Global for their great support. Special thanks also go to the rest of the publishing team at IGI Global, whose contributions throughout the whole process from inception of the initial idea to final publication have been invaluable. In particular, to Kristin Roth and Meg Stocking, who continuously prodded via e-mail for keeping the project on schedule, and to Jan Travers, whose enthusiasm motivated me to initially accept his invitation for taking on this project.

Section I
Foundations of Rough Sets

Chapter I
Foundations of Rough Sets from Vagueness Perspective

Piotr Wasilewski
Warsaw University, Poland

Dominik Ślęzak
Infobright Inc., Canada

ABSTRACT

We present three types of knowledge that can be specified according to the rough set theory. Then, we present three corresponding types of algebraic structures appearing in the rough set theory. This leads to three following types of vagueness: crispness, classical vagueness, and a new concept of "intermediate" vagueness. We also propose two classifications of information systems and approximation spaces. Based on them, we differentiate between information and knowledge.

INTRODUCTION

Handling vagueness was one of the motivations for proposing the rough set theory (Pawlak, 2004, see also Pawlak, 1982, 1991, 2003). In this chapter, we present algebraic foundations of the rough set theory from that perspective. Vagueness is understood according to the *tertium non datur* principle from traditional logic; the contemporary version of this principle is called the law of excluded middle (Frege, 1903; Hempel, 1939; Pawlak, 2004; Russell, 1923). Concepts in the rough set theory are represented by subsets of a universe of discourse. Each concept, X, can be represented by two subsets: that consisting of examples of X (i.e., objects, which can be certainly classified as belonging to X) called a positive region of X, and that consisting of objects that certainly do not belong to X, called a negative region of X. Positive region of a given concept is represented by its lower approximation. Negative region is represented as a complementation of an upper approximation of the set. A complementation of the union of these two sets is called a boundary. Thus, boundary consists of objects that cannot be certainly classified as belonging to a concept, or as not belonging. A concept is vague if its boundary is nonempty, otherwise it is crisp (Pawlak, 2004). Vague concepts are represented in the rough set theory as rough sets. Crisp concepts satisfy the *tertium non datur* principle, their boundaries are empty, there are no objects of third type or, in other words, the middles between their positive and negative regions are excluded. Rough sets (representing vague concepts) violate this principle. However, classical rough sets, based on equivalence relations, satisfy the weak law of excluded middle. This law states that a pseudo-complement of a set X with a pseudocomplement of a pseudocomplement of a set X is equal to whole space or, in more general version it states that a pseudocomplement of an element x of a pseudo-complemented lattice L with a pseudocomplement of a pseudocomplement of this element x is equal to unit of the lattice L. This observation gives a reason for differentiating two types of vagueness: strong vagueness (or simply vagueness) and weak vagueness. Weakly vague concepts satisfy the weak law of excluded middle, while strongly vague concepts violate this law. Concepts of both types violate the law of excluded middle.

We show that an incompleteness of information is a source of these two types of vagueness.

Incomplete information, represented by deterministic incomplete information systems, can be analyzed by means of classical approximation spaces. Such spaces consist of a universe and an indiscernibility relation, which is always equivalence relation. Knowledge granules determined by these spaces are equivalence classes. Rough sets within classical approximation spaces violate the law of excluded middle, but they satisfy the weak law of excluded middle. Thus, such rough sets represent vague concepts in an "intermediate" sense.

Nondeterministic information systems gave a reason for introducing many information relations (Orłowska, 1993, 1997, 1998). Most of these relations are tolerance relations, which are not necessarily transitive, that is are not equivalence relations. Therefore, incomplete information, represented by nondeterministic information systems, can be analyzed by tolerance spaces that consist of a universe and a tolerance relation. Tolerance spaces admit various ways of granulations of the universe. We present a type of knowledge granulations of the tolerance spaces such that if a tolerance relation is not transitive, then rough sets determined by this granulation violate even the weak law of excluded middle. Thus, such rough sets represent vague concepts in the strong sense.

We propose a differentiation between information and knowledge. It follows from foundations of the rough set theory. Starting point is reflected by two suggested classifications of information systems and approximation spaces. We point out that properties of incompleteness and determinacy of information systems are independent and, particularly, that there are complete indeterministic information systems. Information, represented in both complete deterministic and complete indeterministic information systems, enables us to discern between two arbitrary objects from the universe. So, indiscernibility relations determined by information systems of these types are identity relations. Since, according to Pawlak (1982, 1991, 2004), knowledge is based on ability to discern between objects, then complete information systems are bases for constructing complete knowledge, which is represented by identity relations. So, knowledge has more general nature than information. Complete knowledge is necessarily exact; thus, there is no complete inexact knowledge. It follows that properties of completeness and exactness of knowledge are not independent. This underlies the difference between information and knowledge, since there are only three types of knowledge. In terms of knowledge, we show which types of knowledge are responsible for crispness or vagueness of concepts. Firstly, we show that rough sets, based on complete knowledge, are crisp, so they represent crisp concepts. Secondly, we show that rough sets, based on incomplete exact knowledge, are not crisp, but they satisfy the weak law of excluded middle, so they represent concepts that are vague in an "intermediate" sense. Lastly, we show that rough sets based on incomplete inexact knowledge are not crisp, and they also violate the weak law of excluded middle, so they represent strongly vague concepts.

We present algebraic foundations of rough sets from the perspective of vagueness. From various types of rough set algebraic structures, we choose those connected with algebraic versions of the law of excluded middle: Boolean algebras, Stone algebras, and Heyting algebras. We present algebras of definable sets and rough sets, determined by different types of knowledge, and we discuss informational representations of abstract Boolean, regular double Stone algebras by means of these algebras. We point out how those types coincide with crispness, and the two introduced types of vagueness of concepts.

The chapter consists of five sections, followed by conclusions. In the first section, we describe how information is represented by information systems, and we present different types of information systems together with their classification. In the next section, we present some of information relations, which are used within analysis of information. The third section is devoted to basics of the rough set theory. We discuss a representing of knowledge within the rough set theory. We introduce a classification of types of knowledge, and we propose some rough set approach to vagueness of concepts. We present different types of approximation spaces based on information relations, various approximation operators connected with that spaces. In the fourth section, we present some algebraic structures that appeared in the rough set theory, and we discuss connections of these structures with different types of knowledge connected with rough sets. In the last section, we introduce some Heyting algebras of tolerance rough sets, and we show that these algebras are connected with incomplete and inexact knowledge.

REPRESENTATION OF INFORMATION

Nowadays, there is no agreement on what information is. Instead of a precise definition of the notion of information, researchers acknowledge various approaches to information representation. Information systems are one of the main methods here. Nondeterministic information systems were introduced by Lipski (1976). Deterministic information systems were developed independently by Pawlak in 1980. He also introduced one of the main tools for analyzing information systems: the rough set theory. The field of applications of rough set methods, from analysis of information systems to handling incomplete information, has been vigorously developing (e.g., Demri & Orłowska, 2002; Greco, Matarazzo, & Słowiński, 2002; Grzymała-Busse, 1988; Grzymała-Busse, Stefanowski, & Ziarko, 1996; Inuiguchi, Hirano, & Tsumoto, 2003; Lin & Cercone, 1997; Nakata & Sakai, 2005; Orłowska, 1998a; Orłowska & Pawlak, 1984; Pal, Polkowski, & Skowron, 2004; Pawlak, 1982, 1991, 2003, 2004; Polkowski, 2002; Polkowski & Skowron, 1996, 1998a, 1998b; Sakai & Nakata, 2005; Skowron, 2001, 2003; Skowron & Rauszer, 1991; Skowron & Stepaniuk, 1996, 2003; Słowiński, 1992; Słowiński & Stefanowski, 1989, 1996; Stefanowski, 1998; Ziarko, 1993). Information systems handle information tables, where rows are labeled by names of objects, and columns by names of attributes. Table 1 illustrates an example of an information table. The attribute Degrees assigns to each person all of university degrees that he/she possesses, while the attribute Languages assigns all languages that he/she knows.

Concerning objects, for example, the person P_4 is a woman with green eyes, 169 cm high, 58 kg weight, Bachelor and Master of Arts, speaking English and French.

Formally, attributes can be treated as functions that map objects to some value domains, for example, temperature measured on the Calvin scale can be represented as a function from a set of objects to the set of nonnegative real numbers. A given information table determines the set of objects and the set of attributes. Each attribute possesses a value domain, the particular set of its possible values. Sometimes it is convenient to assign to a given object a set of values (which is a subset of its value domain) instead of a single value. Thus, generally, there are two kinds of attributes, namely those mapping objects to single values, and those mapping objects to sets of values. In Table 1, Eyes, Gender, Height, and Weight are attributes of the first type while Degrees and Languages are of the second kind. To handle attributes of both kinds, one can treat them as functions from an object domain into the family of subsets of the set of all attribute values.

Concerning applications of information systems within real data representation, we should point out that one of the main problems is missing attribute values. There are different approaches to dealing with situations of that type, in particular, based on the rough set theory-based approach to imcomplete information (Demri & Orłowska, 2002; Düntsch & Gediga, 1997, 2000; Düntsch & Orłowska, 2001; Grzymała-Busse, 1988; Järvinen, 1999; Nakata & Sakai, 2005; Orłowska, 1993, 1997, 1998; Orłowska & Pawlak, 1984; Pagliani, 1998; Pawlak, 1982, 1991, 2004; Słowiński & Stefanowski, 1989, 1994; Vakarelov, 1989, 1998). One of the possibilities is introducing a special value placed instead of the missing ones. In order to define information systems, formally, we will use that approach. Instead of *Unknown* and *None*, we put the set of all natural numbers and the empty set, respectively.

Definition 1

An information system is a triple $S = \langle U, At, \{Val_a\}_{a \in At} \rangle$ where U is a set of objects called universe of the information system, At is a set of attributes, Val_a is a value domain of an attribute a and each attribute $a \in At$ is a function $a: \longrightarrow \wp(\bigcup_{a \in At} Val_a)$ such that $a(x) \subseteq Val_a$. If $a(x) \neq \varnothing$ for all $x \in U$ and $a \in At$, then S is total.

Example 1

In the case of the information presented in Table 1, an information system $S = \langle U, At, \{Val_a\}_{a \in At} \rangle$ connected with that table has the object set $U = \{P_1, P_2, P_3, P_4\}$, the attribute set At = {Eyes, Gender, Height, Weight, Degree, Language} and the following value domains for particular attributes: Val_{Eyes} = {Grey, Blue, Brown, Green}, Val_{Gender} = {Male, Female}, $Val_{Height} = \Omega$, where Ω denotes the set of natural numbers, $Val_{Weight} = \Omega \cup \{\Omega\}$,

Table 1.

Attributes Objects	Eyes	Gender	Height (cm)	Weight (kg)	Degrees	Languages
P_1	Grey	Male	185	93	{BSc, MSc, PhD}	{English, Polish}
P_2	Blue	Female	174	Unknown	{BSc, BA, MA, PhD}	{English, German, Polish}
P_3	Brown	Male	169	75	None	{English, Polish}
P_4	Green	Female	169	58	{BA, MA}	{English, French}

$Val_{Degrees} = \{BSc, BA, MSc, MA, PhD\}$, $Val_{Languages} = \{English, French German, Polish\}$.

Definition 2

An information system $S = \langle U, At, \{Val_a\}_{a \in At} \rangle$ is *deterministic* if $card(a(x))=1$ for every $a \in At$ and for every $a \in U$, otherwise the system S is called *nondeterministic*. If $card(a(x))=1$ for every $x \in U$, then an attribute a is called *deterministic*. In the case when $a(x)$ is a singleton, for example, $\{v\}$, we often omit parentheses and we write $a(x) = v$. If $card(a(x)) > 1$ for some $x \in U$ then a is called a nondeterministic attribute.

Example 2 (Deterministic Information System)

Let $S = \langle U, At, \{Val_a\}_{a \in At} \rangle$ be an information system such that At = {Eyes, Gender, Height} where attributes are taken from Example 1.

Information represented by nondeterministic attributes can be interpreted generally in two ways (see Jarvinen, 1999). Let $S = \langle U, At, \{Val_a\}_{a \in At} \rangle$ be a nondeterministic information system and let $a \in At$, $x \in U$. If for every $v \in a(x)$, v is an actual value of a for an object x, then an attribute a is said to be *multivalued*. If the unique value of an attribute $a \in At$ for the object $x \in U$ is assumed to be in the set $a(x)$, the attribute a is called *approximate*. Such an interpretation reflects a lack of knowledge within a particular domain. In such a case, for $a \in At$ and $x \in U$ a complete ignorance can be denoted by $a(x) = Val_a$. Let us provide some examples: In the case of empirical measurements, one can assign a particular value of measured property to each object, but because of a measurement error, it may be convenient to assign small sets of values supposed to contain exact attribute values. Further numerical or ordered attributes (i.e., deterministic attributes with ordered value domains) are often changed into approximate attributes (Nguyen, 1998; Nguyen & Nguyen, 1996; Nguyen & Skowron, 1997). Finally, in the rough set methodology, *decision attributes* are often replaced by multivalued *generalized decision attributes* (Greco, Pawlak, & Słowiński, 2002; Skowron & Rauszer, 1991; Ślęzak, 1999).

With respect to those two interpretations of nondeterministic attributes, two kinds of nondeterministic information systems can be formed: *multivalued systems*, where attributes are multivalued, and *approximate systems*, where attributes are approximate. It should be also noted that these two types of attributes are based just on an interpretation of information; they do not follow from formal properties of nondeterministic information systems.

Example 3 (Many-Valued Information System)

Let $S = \langle U, At, \{Val_a\}_{a \in At} \rangle$ be information system such that At = {Degrees, Language} while $Val_{Degrees}$ = {BSc, BA, MSc, MA, PhD} and $Val_{Languages}$ = {English, French German, Polish} and the values of the attributes are defined within Table 1. Let us note that, in the case of the attribute Degrees, instead of *None*, the empty set is used. Such approach to dealing with missing values seems to be natural for many-valued information systems.

Table 2.

Attributes Objects	Gender	Height (cm)	Weight (kg)
P_1	{Male}	{183, ..., 187}	{93, ..., 99}
P_2	{Male, Female}	{172, ..., 176}	Ω
P_3	{Male, Female}	{169, ..., 171}	{72, ..., 78}
P_4	{Female}	{169, ..., 171}	{72, ..., 78}

Example 4. (Approximate Information System)
Let $S = \langle U, At, \{Val_a\}_{a \in At} \rangle$ be information system such that At = {Gender, Height, Weight} and values of appropriate attributes are presented in Table 2.

In the column of the attribute Weight instead of Unknown, the set of all natural numbers is placed. Such an approach to dealing with missing values seems to be natural for approximate information systems. The attribute values now are only approximated. For example, the height of the person P_2 is between 172 and 176, and his/her Gender and Weight are not known; instead of unknown values the whole value domains of appropriate attributes are placed.

Information Relations Determined by Information Systems

Information presented in information systems can be analyzed in many ways. One of the main reasons for inventing the rough set theory was introducing new formal methods for analyzing that type of information. Rough set methods of information analysis are based on information relations that reflect different kinds of discernibility between objects. These relations give a bridge between information and abstract, formal methods of its analyzing. Here we present some basic notions from the set theory that will be used in the sequel.

R is *binary relation on* a set U if $R \subseteq U \times U$; in such a case, the set U is said to be a *universe* or a *domain* of the relation R. Instead of $(x, y) \in R$, we may also write xRy. We use the following notation: $\Delta_U := \{(x, y) \in U^2 : x = y\}$ for *diagonal* or *identity relation on* U, and $\nabla_U = U \times U$ for *full relation on* U. Further, R is *tolerance relation* (shortly: *tolerance*) if it is reflexive and symmetric, and R is *equivalence relation* (or shortly: *equivalence*) if R is reflexive, symmetric, and transitive. $Tol(U)$ and $Eq(U)$ denote the families of all tolerance and equivalence relations, respectively. Within $Eq(U)$ and $Tol(U)$, Δ_U and ∇_U are the least and the greatest elements, respectively. For $X \subseteq U$, we put $R(X) := \{y \in U : \exists x \in U, xRy\}$ and $R(X)^{-1} := \{y \in U : \exists x \in U, yRx\}$, as image and counterimage of X, respectively. Instead of $R(\{x\})$ and $R^{-1}(\{x\})$, we simply write $R(x)$ and $R^{-1}(x)$. If R is an equivalence, then $R(x)$ is denoted by $x_{/R}$ and called equivalence class of the relation R.

Generally, by information relations we mean relations defined on sets of objects of information systems by means of attributes and values. These relations are aimed in analysis of information represented within information systems. There are many information relations discussed in literature (see Demri & Orłowska, 2002; Orłowska, 1989, 1993, 1997; Vakarelov, 1989, 1991, 1998). We present some of the most frequently used information relations, namely indiscernibility and similarity relations.

Definition 3
Let $S = \langle U, At, \{Val_a\}_{a \in At} \rangle$ be an information system, let $B \subseteq At$ and $x, y, \in U$.

1. The *strong indiscernibility relation* $ind(B)$ is a relation such that $(x, y) \in ind(B) :\Leftrightarrow a(x) = a(y)$, for all $a \in B$.
2. The *weak indiscernibility relation* $wind(B)$ is a relation such that $(x, y) \in wind(B) :\Leftrightarrow a(x) = a(y)$, for some $a \in B$.
3. The *strong similarity relation* $sim(B)$ is a relation such that $(x,y) \in sim(B) :\Leftrightarrow a(x) \cap a(y) \neq \emptyset$, for all $a \in B$.
4. The *weak similarity relation* $wsim(B)$ is a relation such that $(x,y) \in wsim(B) :\Leftrightarrow a(x) \cap a(y) \neq \emptyset$, for some $a \in B$.

The strong indiscernibility relation was introduced in Konrad, Orłowska, and Pawlak (1981), while the other three by Ewa Orłowska (1995, 1997, 1998). The most popular information relation is strong indiscernibility. Usually, it is called an *indiscernibility relation*. (In the sequel, we use that name referring to the strong indiscernibility relation). This relation plays a fundamental role in the rough set theory, though the other ones have been considered too. Let us note that similarity relations have been considered mostly for nondeterministic systems. On the other hand, indiscernibility relations can be applied to both deterministic and nondeterministic information systems, although they seem to better reflect deterministic case.

Example 5

Let $S = \langle U, At, \{Val_a\}_{a \in At} \rangle$ be an information system presented in Example 1. In that system, $P_1, P_3 \in ind(\{Gender, Languages\})$, since both P_1 and P_2 are men who speak English and Polish while $P_1, P_3 \in wind(At)$ since they have the same height. Persons P_1 and P_2 are in the relation $sim(\{Gender, Languages\})$ because they are Bachelors of Science who speak Polish while $wsim(At) = \nabla_U$ that is, all of persons P_1 - P_4 are in the relation $wsim(At)$ since all of them speak English.

In the following proposition, we present some basic properties of indiscernibility and similarity relations (see Demri & Orłowska, 2002; Jarvinen, 1999, 2006; Orłowska, 1995, 1997, 1998).

Proposition 1

Let $S = \langle U, At, \{Val_a\}_{a \in At} \rangle$ be an information system, let $B \subseteq At$ and $a \in B$, then:

1. $ind(B)$ is an equivalence,
2. $wind(B)$, $sim(B)$ and $wsim(B)$ are tolerances,
3. $ind(\{a\}) = wind(\{a\})$ and $sim(\{a\}) = wsim(\{a\})$,
4. $ind(B) \subseteq wind(B)$ and $sim(B) \subseteq wsim(B)$,
5. $ind(B) \subseteq sim(B)$ and $wind(B) \subseteq wsim(B)$,
6. $ind(\emptyset) = sim(\emptyset) = \nabla_U$ and $wind(\emptyset) = wsim(\emptyset) = \emptyset$.

Example 6

Let $S = \langle U, At, \{Val_a\}_{a \in At} \rangle$ be the information system presented in Example 1. We can see that $wind(At)$, $sim(\{Degrees, Languages\})$, and $wsim(At \setminus sim(\{Languages\}))$ are not transitive.

Thus, generally, the weak indiscernibility relations and similarity relations of both kinds are not necessarily transitive; so in light of conditions 1 and 2 from Proposition 1 and Example 6, we can see that the strong indiscernibility relations are distinguished between indiscernibility and similarity information relations: they are always equivalence relations. The strong indiscernibility relation is a basic element in the construction of the rough set theory. The indiscernibility relations can be interpreted in such a way that they exemplify our knowledge about the universe of objects (Pawlak 1991, 2004). In such an interpretation, knowledge is based on an ability to discern between objects. Thus, if for a given information system its strong indiscernibility relation is equal to the diagonal of a universe set of that system, then this relation represents complete knowledge about objects of the universe of that information system: we are able to discern between any two distinct elements from the universe of the information system using attributes of the system. It gives a reason for division of class of information systems into two groups.

Definition 4

An information system $S = \langle U, At, \{Val_a\}_{a \in At} \rangle$ is *complete* if $ind(At) = \Delta_U$. Otherwise S is *incomplete*.

Example 7

1. Let $S = \langle U, At, \{Val_a\}_{a \in At} \rangle$ be a deterministic information system from Example 2. Since $ind(At) = \Delta_U$, thus S is a complete deterministic information system.
2. Let $S_1 = \langle U, At, \{Val_a\}_{a \in At} \rangle$ be an information system such that $At = \{Gender, Height, Weight\}$ and values of appropriate attributes are presented in Table 3.

Table 3.

Attributes Objects	Gender	Highest Academic Degree	First speaking language
P_1	Male	PhD	Polish
P_2	Female	PhD	German
P_3	Male	BSc	English
P_4	Female	MA	French
P_5	Female	PhD	German
P_6	Female	MA	French

Since $(P_2, P_5) \in ind(At)$, so $ind(At) \neq \Delta_U$. Thus, S_1 is an incomplete deterministic information system.

3. Let $S_2 = \langle U, At, \{Val_a\}_{a \in At} \rangle$ be an information system such that At = {Height, Weight} and values of appropriate attributes are presented in Table 2. Since $(P_3, P_4) \in ind(At)$, we have $ind(At) \neq \Delta_U$. Thus, S_2 is an incomplete nondeterministic information system.

4. Let $S = \langle U, At, \{Val_a\}_{a \in At} \rangle$ be a nondeterministic information system from Example 4. Since $ind(At) = \Delta_U$, thus S is a complete nondeterministic information system.

Example 7 shows that properties of completeness and determinacy of information systems are independent, that is,. all four possibilities hold. On the basis of these properties, we can propose the following classification of information systems.

The most interesting class in a classification presenting in Table 7 is that consisting of complete nondeterministic information systems. At first sight, it is surprising, but later it appears natural. It shows that determinacy of information has a local nature while completeness, global. It reflects a situation when we are not able to assign to particular objects specified attribute values, but we are still able to discern between any two objects from the universe. Information systems represent known information. Knowledge appears on the basis of information systems, and chosen information relations and types of knowledge are represented by formal properties of those information relations. These formal properties depend, as it is shown in Example 6, on information presented in information systems, and on the nature (definitions) of information relations.

Table 4. Classification of information systems with respect to the determinacy of attributes and the completeness of knowledge

Information Systems	Complete	Incomplete
deterministic	complete deterministic information systems	incomplete deterministic information systems
nondeterministic	complete nondeterministic information systems	incomplete nondeterministic information systems

APPROXIMATION SPACES BASED ON INFORMATION SYSTEMS

The notion of approximation space is the basic notion of the rough set theory. Approximation spaces are derived from information systems, and they represent knowledge about objects from a given universe (Pawlak, 1991). It reflects a supposition that knowledge is not knowledge about everything, but it is always focused on a particular domain of discourse. This is a natural assumption within data analysis, where data are assumed to be collected from more or less specified domains.

Definition 5

Let U be a set of objects and R be a relation on U. A pair (U, R) is an *approximation space*.

At the beginning of the rough set theory, a relation R was assumed to be an equivalence relation (Pawlak, 1982, 1991). In the middle of the nineties, it was generalized to be a tolerance relation, thus not necessarily transitive (Skowron & Stepaniuk, 1996). Nowadays, it is allowed to be any binary relation on a given set (Pawlak, 2004; Yao, 1998b).

Knowledge Representation in Rough Set Theory

According to Zdzisław Pawlak, within an approximation space (U, R), a relation R reflects our knowledge about objects of the universe. Thus, we will identify knowledge with particular classes of indiscernibility relations characterized by some formal conditions, or, more strictly, with particular types of approximation spaces.

Definition 6

If $R = \Delta_U$, then an approximation space (U, R) is *complete*, otherwise it is *incomplete*. If a space (U, R) is complete (incomplete), then we say also that knowledge represented by a space (U, R) is complete (incomplete) or simply that knowledge R is complete (incomplete).

Complete approximation spaces reflect knowledge that is complete in the sense that we are able to discern between any two objects from the universe. Thus, if (U, R) is incomplete, that is $R \neq \Delta_U$, then it reflects a lack of our knowledge about objects from universe U, in such a sense that we are not able to discern between any two of them. With each approximation space (U, R), there is connected a family of *granules of knowledge* or *knowledge granules*. Granules of knowledge are subsets of U determined by the relation R. Usually, granules are sets of the form $R(x), x \in U$. However, in the spirit of the late Pawlak's approach (2004), other ways of defining knowledge granules also can be admitted, as we will show in the sequel. Such plurality of types of knowledge granules seems to be very natural from the perspective of granular computing, which is vigorously developing now and was inspired, among others, by the rough set theory (Lin, 1998b; Pawlak, 1991, 2003, 2004; Skowron, 2001; Skowron & Stepaniuk, 1996, 2003; Yao, 2001). The family of knowledge granules of (U, R) is denoted by $G_R(U)$. Thus, if $R = \Delta_U$ and $G_R(U) := \{R(x) : x \in U\}$, then we obtain $G_R(U) = \{\{x\} : x \in U\}$. Let us note that if $R \in Eq(U)$, then $\{R(x) : x \in U\}$ is a partition of U, denoted by $U_{/R}$. Thus, if $R \in Eq(U)$ and $G_R(U) := \{R(x) : x \in U\}$, then $G_R(U) = U_{/R}$.

If an approximation space (U, R) represents incomplete knowledge (i.e., $R \neq \Delta_U$), it can be interpreted in such a way that we are not able to perceive particular objects from a universe U (see Pawlak, 2004). We are able only to perceive of and to deal with granules of knowledge. Thus, there is another important factor connected with knowledge, namely, it is a relation between granules: whether they are mutually disjoint or not, that is there exist at least two granules that are not disjoint.

Definition 7

Let (U, R) be an approximation space. If the family of knowledge granules $G_R(U)$ is a partition of the set U, then a space (U, R) is *exact*, otherwise, it is *inexact*. If a space (U, R) is exact (inexact), then we say also that knowledge represented by a space (U, R) is exact (inexact) or simply that knowledge R is exact (inexact).

Let us note that if knowledge is represented by tolerance relations, then the following proposition holds:

Proposition 2

If $R \in Tol(U)$ and $G_R(U) := \{R(x) : x \in U\}$, then the following conditions are equivalent:

1. Knowledge represented in an approximation space (U, R) is exact
2. R is transitive
3. $R \in Eq(U)$

Thus, if knowledge is complete, then it is exact. So there are two kinds of exact knowledge: complete and incomplete. There is no inexact knowledge that is complete. It holds because if knowledge is inexact, that is, it is represented by an approximation space (U, R), where R is a tolerance relation that is not transitive, then $R \neq \Delta_U$. Completeness and exactness are not independent: there are only three types of knowledge with respect to these properties, as it is shown in Table 5.

Inexactness of knowledge can be viewed as analogical to nondeterminacy of information systems. Nonexistence of complete inexact knowledge underlines a difference between information and knowledge that emerges from information by application of chosen information relations. If a given nondeterministic information systems is complete, then we are able to discern between any two objects of the universe; thus, knowledge is also complete and exact. This is true if we acknowledge one of fundamental assumptions of the rough set theory, that is, that knowledge reflects an ability of discerning between any two objects of the universe of discourse. This assumption determines also which granulations of the universe should be admitted. If granules are understood as sets consisting of indiscernible objects, then, in the case of complete nondeterministic information systems, granules used in the analysis should be singletons. Otherwise, if granules are not understood as sets consisting of indiscernible objects, then, concerning complete nondeterministic information systems, families of granules at least should include all singletons of objects of the universe. This is necessary, if

Table 5. Types of knowledge according to properties of completeness and exactness

Knowledge	Complete	Incomplete
Exact	complete exact knowledge (approximation spaces (U, R), where $R = \Delta_U$)	incomplete exact knowledge (approximation spaces (U, R), where $R \in Eq(U)$ and $R \neq \Delta_U$)
Inexact		incomplete exact knowledge (approximation spaces (U, R), where $R \in Tol(U) \setminus Eq(U)$)

knowledge is understood as reflecting ability of discerning between objects. Or in other words, we should take into account only such granulations of universe that reflect a possibility of discerning between any two objects in the case of complete nondeterministic information systems.

Rough Set Theory Approach to Vagueness

Rough sets were designed to handle vague concepts (Pawlak, 2004). Generally, with each concept (not necessary vague), C there are connected in a natural way two subsets of the universe: that of objects that belong to the concept, and that of objects that do not. The first set can be viewed as a *positive region* of the concept C, denoted by *Pos*(C), while the second, a *negative region* C (since it consists of negative examples) denoted by *Neg*(C). Objects that do not belong to both of these sets create a *boundary* of C. Boundary is not necessarily empty. Concepts with an empty boundary are called *crisp*, otherwise they are called *vague*. This idea is due to Frege (1903), who noticed that most of the concepts used by people are vague, while concepts needed in mathematics should be clarified enough to be crisp or exact (see also Pawlak, 2004). We follow this idea, although one should be aware that there are other approaches to vagueness too (Keefe, 2000). Vagueness can be treated in different formal ways (Dubois, Godo, Prade, & Esteva, 2005; Peters, 2007; Smith, 2005) also using rough set methods (Bazan, Skowron & Swiniarski, 2006; Skowron, 2004, 2005; Skowron, Stepaniuk, Peters, & Swiniarski, 2006). Going back to the basic idea that we follow, the objects can be described with respect to a given concept C as:

1. the objects that *surely belong* to C,
2. the objects that *possibly belong* to C,
3. the objects that *surely do not belong* to C.

Within this approach, it is assumed that every object surely belonging to concept C also possibly belongs to C. Thus, the set of objects of the first type is contained in the set of objects of the second type. It is also assumed that there are no objects that both possibly belong to C and surely do not belong to C, that is, the set of objects of the second type is disjoint with that of the third type. Within this approach, a boundary of C can be defined as a set of those objects that can be neither classified as surely in C nor as surely outside C. Thus, a boundary of C consists of those objects possibly belonging to C that do not belong to C surely. As we can see, both approaches allow us for defining the boundary region of the concept. Frege also noticed that vague concepts violate *tertium non datur* principle from traditional logic (nowadays also called the principle of excluded middle) since their boundaries are nonempty; thus, a "middle" between positive and negative examples of a given concept are not excluded. In a mathematical logic, the *tertium non datur* principle has its algebraic and logical counterparts. In the sequel, we present algebraic counterparts and some consequences of their violations.

Concepts are represented within a given approximation space as subsets of the universe U. When knowledge is incomplete, it is interpreted in such a way that concepts can be described only by means of knowledge granules, not elements of the universe. Boundaries, positive, and negative regions can be determined with usage of knowledge granules, not objects. Thus, knowledge granules are basic building blocks from which abstract knowledge is constructed.

Classical (Equivalence) Approximation Spaces

Definition 8

An approximation space (U, R) is *classical* if and only if $R \in Eq(U)$. In the sequel, if it does not lead to misunderstandings, we will refer to classical approximation spaces as approximation spaces.

Let (U, R) be an approximation space and $G_R(U) := \{R(x) : x \in U\} = U_{/R}$. For any $X \subseteq U$ we define $R_* : \wp(U) \longrightarrow \wp(U)$, $R^* : \wp(U) \longrightarrow \wp(U)$, and $BN_R : \wp(U) \longrightarrow \wp(U)$ as follows:

$R_*(X) := \bigcup \{A \in U_{/R} : A \subseteq X\}$
$R^*(X) := \bigcup \{A \in U_{/R} : A \cap X \neq \emptyset\}$
$BN_R(X) := R^*(X) \setminus R_*(X)$

R_* and R^* are *lower* and *upper* approximation operator respectively, and BN_R is a *boundary* operator. For any $X \subseteq U$, the sets $R_*(X)$ and $R^*(X)$ are called *lower* and *upper approximations of X* respectively. $BN_R(X)$ is the *boundary of X*. X is *definable in* (U, R) *on the basis of* $G_R(U)$ if and only if X is a union of knowledge granules from $G_R(U)$, that is, there is a family $A \subseteq G_R(U)$ such that $X = \bigcup A$. The family of all definable sets in (U, R) on the basis of $G_R(U)$ is denoted by $Def(G_R(U))$. $X \subseteq U$ is *rough* if and only if $BN_R(X) \neq \emptyset$.

Notice that $\emptyset, U \in Def(G_R(U))$ since $\emptyset = \bigcup \emptyset$ and $U = \bigcup G_R(U)$. Definition 8 reflects an idea that granules of knowledge are the basic building blocks of knowledge. In the case of classical approximation spaces, knowledge granules are always equivalence classes of the indiscernibility relation.

For any $X \subseteq U$, the elements of $R_*(X)$ can be viewed as surely belonging to X, $R^*(X)$, possibly belonging to X, while $U \setminus R^*(X)$, surely not belonging to X. In the sequel, we will use a set theoretical complement denoted by " ' ". $R_*(X)$ and $R^*(X)$' can be interpreted as a positive and a negative region of X, respectively.

Let us note that a set $X \subseteq U$ is definable if $X = R_*(X)$ if $X = R^*(X)$ if $R_*(X) = R^*(X)$ if $BN_R(X) = \emptyset$. Thus, each set is rough if and only if it is not definable. Let us also note that $BN_R(X) = \bigcup \{A \in U_{/R} : A \cap X \neq \emptyset \ \& \ A \cap X' \neq \emptyset\}$. Thus, both approximations and a boundary of any set are definable sets, that is, they are unions of knowledge granules. One can see that $R_*(X)$

Table 6.

1a. $R_*(X) \subseteq X$	1b. $X \subseteq R^*(X)$
2a. $X \subseteq Y \Rightarrow R_*(X) \subseteq R_*(Y)$	2b. $X \subseteq Y \Rightarrow R^*(X) \subseteq R^*(Y)$
3a. $R_*(\emptyset) = \emptyset$	3b. $R^*(\emptyset) = \emptyset$
4a. $R_*(U) = U$	4b. $R^*(U) = U$
5a. $R_* R_*(X) = R_*(X)$	5b. $R^* R^*(X) = R^*(X)$
6a. $R_*(X \cap Y) = R_*(X) \cap R_*(Y)$	6b. $R^*(X \cap Y) \subseteq R^*(X) \cap R^*(Y)$
7a. $R_*(X) \cup R_*(Y) \subseteq R_*(X \cup Y)$	7b. $R^*(X) \cup R^*(Y) = R^*(X \cup Y)$
8a. $R_*(X) = R^*(R_*(X))$	8b. $R_*(R^*(X)) = R^*(X)$
9. $R_*(X) \subseteq R^*(X)$	
10. $R_*(X)' = R^*(X')$	
11. $R^*(X)' = R_*(X')$	
12. X is definable $\Leftrightarrow R_*(X) = X \Leftrightarrow R^*(X) = X \Leftrightarrow R_*(X) = R^*(X)$	
13. If X or Y are definable, then $R_*(X) \cup R_*(Y) = R_*(X \cup Y)$ and $R^*(X \cap Y) = R^*(X) \cap R^*(Y)$	

is the greatest definable set contained in X, and $R^*(X)$ is the least definable set containing X. Notice $R^*(X)' = R_*(X')$. Thus, the negative region of X is the greatest definable set disjoint with X. Let us note also that the family $Def(G_R(U))$ is a topology on a set U, and $R_*(X)$ and $R^*(X)$ are topological interior and closure operators, respectively, that are connected with a topological space $(U, Def(G_R(U)))$. Summing up, for any sets $X, Y \subseteq U$ the following formulas hold:

Now we can face the problem how to represent rough sets. We are following the intuition that in the case of incomplete knowledge, we are only able to handle knowledge granules. So, we are not able to discern between two sets that have the same lower and upper approximations. This interpretation gives reason for defining a new relation $\approx_{(U,R)}$ on the power set $\wp(U)$, such that for any $X, Y \subseteq U$:

$$X \approx_{(U,R)} Y :\Leftrightarrow R_*(X) = R_*(Y) \;\&\; R^*(X) = R^*(Y).$$

This relation is called a *rough equality of sets*. Notice that $\approx_{(U,R)} \in Eq(\wp(U))$. Because of a lack of knowledge, we are not able to discern between arbitrary sets by means of knowledge granules, and instead of dealing with subsets of U, we handle equivalence classes of the rough equality. Let us note that a set X is rough if and only if $card(X_{/\approx_{(U,R)}}) > 1$. In order to distinguish between rough and definable sets, we still have to be able to discern between any two subsets of U. However, according to the acknowledged interpretation, we can not do this by means of granules of incomplete knowledge. Rough equality suggests a different way of representing rough sets. Each set can be represented as a pair consisting of its lower and upper approximation that is settled by the following mapping $X \mapsto (R_*(X), R^{**}(X))$. Now, a set X is rough if $R_*(X) \neq R^*(X)$, otherwise X is not rough, which means it is definable. This representation is acceptable from the point of view of the interpretation discussed previously. It is so that approximations are definable sets, that is, they are constructed from knowledge granules that are known; thus, we are able to discern between any two of them.

The pairs of the form $(R_*(X), R^*(X))$ often are called rough sets straightly (see e.g., Banerjee & Chakraborty, 2003; Iwiński, 1987; Demri & Orlowska, 2002). The family of all such pairs will be denoted by $\Re_R(U)$. It contains also pairs representing definable sets: if a set $X \subseteq U$ is definable, then $(R_*(X), R^*(X)) = (X, X)$. Calling such pairs "rough sets" seems to be a little bit counterintuitive. However, it is convenient and widely accepted, so we follow it in the sequel. Note that for an approximation space (U, R), any pair of the form $(R_*(X), R^*(X))$, where $X \subseteq U$, is a pair of definable sets $A, B \in Def(G_R(U))$, such that $A \subseteq B$. Therefore, on the abstract level of knowledge, we can regard rough sets as the pairs of definable sets $A, B \in Def(G_R(U))$, such that $A \subseteq B$. Such pairs, in the sequel of this chapter, will be called the *abstract rough sets* of the space (U, R), and the family of all abstract rough sets will be denoted by $Ab(G_R(U))$. That way of representing rough sets was introduced in Iwiński (1987). He considered also a mapping $X \mapsto (R_*(X), R^*(X))$ as an algebraic representation of rough sets; however, he still called the elements of $Ab(G_R(U))$ as rough sets. Concerning such representation of rough sets, a question arises: Does any abstract rough set represent a rough set? In other words: whether for each pair $(A, B) \in Ab(G_R(U))$ there is a set $X \subseteq U$, such that $R_*(X) = A$ and $R^*(X) = B$? The answer is negative and the following proposition holds:

Proposition 3

Let (U, R) be an approximation space. Then the following conditions are equivalent:

1. For any pair $(A, B) \in Ab(G_R(U))$ there is a set $X \subseteq U$, such that $R_*(X) = A$ and $R^*(X) = B$
2. There are no one-element equivalence classes of R.

Let us note that $\mathfrak{R}_R(U) \subseteq Ab(G_R(U))$ always holds, but the proposition implies that if knowledge is locally complete, that is, there are one-element granules, then $\mathfrak{R}_R(U) \neq Ab(G_R(U))$. Oppositely, if knowledge is absolutely incomplete, that is, there are no one-element equivalence classes of R, then abstract rough sets are exactly rough sets, that is, $Ab(G_R(U)) = \mathfrak{R}_R(U)$.

Tolerance Approximation Spaces

Tolerance approximation spaces were introduced in Skowron and Stepaniuk (1996). They are more complex objects than classical equivalence approximation spaces and, in some sense, they are generalizations of approximation spaces. In Skowron and Stepaniuk (1996), if the given tolerance relation is transitive (i.e., it is equivalence), then the classical Pawlak's approximation operators are particular examples of the tolerance approximation operators. Operators presented here will be defined for tolerance spaces that are simpler objects than those in Skowron and Stepaniuk (1996). These operators are straight generalizations of equivalence approximation operators, and are particular examples of the ones stated previously.

All equivalence approximation spaces are tolerance spaces, but there are also tolerance spaces such that their tolerance relations are not transitive. Thus, tolerance spaces are natural generalizations of classical ones. Knowledge is represented by tolerance relations; families of knowledge granules are determined by these relations in some way. As we will see, this way is not unique. Approximation operators are defined on the basis of families of knowledge granules. As we will show, for tolerances, there are many ways of defining approximation operators that are all equal to classical approximation operators for transitive tolerances.

Definition 9

Let U be a set of objects and τ be a tolerance relation on U. A pair (U, τ) is a *tolerance approximation space*.

We put $G_\tau(U) := \{\tau(x) : x \in U\}$. The set $\tau(x)$ is called the neighborhood of x. We define operators $\tau_* : \wp(U) \longrightarrow \wp(U)$, $\tau^* : \wp(U) \longrightarrow \wp(U)$, and $BN_\tau : \wp(U) \longrightarrow \wp(U)$ such that, for any $X \subseteq U$:

$\tau_*(X) := \{x : \tau(x) \subseteq X\}$,
$\tau^*(X) := \{x : \tau(x) \cap X \neq \emptyset\}$,
$BN_\tau(X) := \tau^*(X) \setminus \tau_*(X)$.

The sets $\tau_*(X)$, $\tau^*(X)$ are called τ-*lower* and τ-*upper approximations* of X respectively, while $BN_\tau(X)$ is τ-*boundary* of X.

Although operators τ_* and τ^* are based on the knowledge granules, they do not summarize them; instead, they only pick up the elements such that their neighborhoods are in some relation with the approximated set. The sets do not need to be approximated as the unions of granules. Nevertheless, if a tolerance relation is transitive, then approximations are definable sets, as the following proposition shows:

Proposition 4

Let (U, τ) be a tolerance space. The following conditions are equivalent:

1. A tolerance τ is transitive, and so $\tau \in Eq(U)$
2. $\{\tau(x) : x \in U\}$ is a partition of U
3. $\tau_*(X) = \bigcup\{\tau(x) : \tau(x) \subseteq X\}$ and $\tau^*(X) = \bigcup\{\tau(x) : \tau(x) \cap X \neq \emptyset\}$

Operators τ_* and τ^* are natural generalizations equivalence approximation operators introduced by Pawlak. Other properties of these operators are presented in Table 7.

If tolerance τ is not transitive, then τ-lower and τ-upper approximations are not unions of

Table 7.

1a. $\tau_*(X) \subseteq X$	1b. $X \subseteq \tau^*(X)$
2a. $X \subseteq Y \Rightarrow \tau_*(X) \subseteq \tau_*(Y)$	2b. $X \subseteq Y \Rightarrow \tau^*(X) \subseteq \tau^*(Y)$
3a. $\tau_*(\emptyset) = \emptyset$	3b. $\tau^*(\emptyset) = \emptyset$
4a. $\tau_*(U) = U$	4b. $\tau^*(U) = U$
5a. $\tau_*\tau_*(X) \subseteq \tau_*(X)$	5b. $\tau^*(X) \subseteq \tau^*\tau^*(X)$
6a. $\tau_*(X \cap Y) = \tau_*(X) \cap \tau_*(Y)$	6b. $\tau^*(X \cap Y) \subseteq \tau^*(X) \cap \tau^*(Y)$
7a. $\tau_*(X) \cup \tau_*(Y) \subseteq \tau_*(X \cup Y)$	7b. $\tau^*(X) \cup \tau^*(Y) = \tau^*(X \cup Y)$
8. $\tau_*(X) \subseteq \tau^*(X)$	
9. $\tau_*(X)' = \tau^*(X')$	
10. $\tau^*(X)' = \tau_*(X')$	

knowledge granules, which is not compliant with Pawlak's idea of granular construction of knowledge. On the other hand, properties 9 and 10 in Table 10 show that for any $X \subseteq U$, there is always $\tau^*(X) \cap \tau_*(X') = \emptyset$; thus, there is no paradox of elements that possibly belong and simultaneously surely do not belong to X.

The next version of approximations follows the Pawlak's idea more closely.

Definition 10

Let (U,τ) be a tolerance space, $G_\tau(U) := \{\tau(x) : x \in U\}$ and $Def(G_\tau(U)) := \{\bigcup A : A \subseteq G_\tau(U)\}$. We define $P_\tau : \wp(U) \longrightarrow \wp(U)$, $P^\tau : \wp(U) \longrightarrow \wp(U)$ and $BN_P : \wp(U) \longrightarrow \wp(U)$ such that, for any $X \subseteq U$:

$P_\tau(X) := \bigcup\{\tau(x) : \tau(x) \subseteq X\}$,
$P^\tau(X) := \bigcup\{\tau(x) : \tau(x) \cap X \neq \emptyset\}$,
$BN_P(X) := P^\tau(X) \setminus P_\tau(X)$.

We call $P_\tau(X)$ and $P^\tau(X)$ as *P-lower* and *P-upper approximations* of X respectively, while $BN_P(X)$ is called *P-boundary* of X.

Notice that a set is definable if it is a union of knowledge granules. From Proposition 4 it follows that if τ is transitive, then $P_\tau(X) = \bigcup\{\tau(x) : \tau(x) \subseteq X\} = \tau_*(X)$ and $P^\tau(X) = \bigcup\{\tau(x) : \tau(x) \cap X \neq \emptyset\} = \tau^*(X)$. Thus operators P_τ and P^τ are natural generalizations of classical equivalence approximations. Let us also note that for any set X, both P-lower and P-upper approximations are definable sets, that is, unions of knowledge granules. Properties of operators P_τ and P^τ are presented in Table 8.

Any set X is definable if $P_\tau(X)=X$, so X is a union of all knowledge granules contained in it. However, properties 9 and 10 of operators P_τ and P^τ do not protect this model from the previously mentioned paradox of objects that possibly belong, and simultaneously surely do not belong, to the approximated set.

Example 8

Let $U := \{a,b,c\}$ and $\tau := \Delta_U \cup \{(a,b), (b,a), (b,c), (c,b)\}$. There is $\tau \in Tol(U)$, $\tau(a)=\{a, b\}$, $\tau(b)= \{a, b, c\}$, $\tau(c)=\{b, c\}$. Let $X:=\{a\}$. Then $P^\tau(X)=\{a,b\}$, $P_\tau(X') = P_\tau(\{b,c\}) = \{b,c\}$. Hence, $b \in P^\tau(X)$ and $b \in P_\tau(X')$, so $P^\tau(X) \cap P_\tau(X') \neq \emptyset$.

Proposition 5

Let (U,τ) be a tolerance space. The following conditions are equivalent:

Table 8.

1a. $P_\tau(X) \subseteq X$	1b. $X \subseteq P^\tau(X)$
2a. $X \subseteq Y \Rightarrow P_\tau(X) \subseteq P_\tau(Y)$	2b. $X \subseteq Y \Rightarrow P^\tau(X) \subseteq P^\tau(Y)$
3a. $P_\tau(\emptyset) = \emptyset$	3b. $P^\tau(\emptyset) = \emptyset$
4a. $P_\tau(U) = U$	4b. $P^\tau(U) = U$
5a. $P_\tau P_\tau(X) = P_\tau(X)$	5b. $P^\tau(X) \subseteq P^\tau P^\tau(X)$
6a. $P_\tau(X \cap Y) \subseteq P_\tau(X) \cap P_\tau(Y)$	6b. $P^\tau(X \cap Y) \subseteq P^\tau(X) \cap P^\tau(Y)$
7a. $P_\tau(X) \cup P_\tau(Y) \subseteq P_\tau(X \cup Y)$	7b. $P^\tau(X) \cup P^\tau(Y) = P^\tau(X \cup Y)$
8. $P_\tau(X) \subseteq P^\tau(X)$	
9. $P_\tau(X)' \subseteq P^\tau(X')$	
10. $P^\tau(X)' \subseteq P_\tau(X')$	
11. X is definable $\Leftrightarrow P_\tau(X) = X$	
12. If X and Y are definable, then $P_\tau(X) \cup P_\tau(Y) = P_\tau(X \cup Y)$	

1. τ is not transitive
2. There are $x, y \in U$ such that $(x,y) \notin \tau$ but $\tau(x) \cap \tau(x) \neq \emptyset$
3. There is a set $X \subseteq U$ such that $P^\tau(X) \cap P_\tau(X') \neq \emptyset$

Let us note that if $R \in Eq(U)$, then any two elements of each equivalence class are indiscernible with respect to R, that is, for each $a \in U$, if $x, y \in a_{/R}$, then $(x, y) \in R$. It reflects intuition that equivalence classes, as knowledge granules, are "compact": all of their elements are indiscernible. But this is not the case for a tolerance $\tau \in Tol(U)$ that is not transitive: for such τ there is at least one element $a \in U$ such that there are $x, y \in \tau(a)$ and $(x, y) \notin \tau$. Thus, tolerance neighborhoods of elements, as knowledge granules, are not "compact." Still, even for tolerance relations there are granules reflecting "compactness."

Definition 11

Let (U, τ) be a tolerance space. A set $A \subseteq U$ is a *preclass* of tolerance τ, shortly a *preclass*, if for any $x, y \in U$, if $x, y \in A$, then $(x, y) \in \tau$. Maximal preclasses with respect to inclusions are called *classes*. Sometimes preclasses and classes are called *cliques* and *maximal cliques* respectively. The families of all preclasses and classes of τ will be denoted by PH_τ and H_τ respectively. For $x \in U$ we put $H_\tau(x) := \{A \in H_\tau : x \in A\}$.

From Definition 11, one can see that preclasses are "compact" in the sense that any of their elements are indiscernible with respect to a given tolerance. Thus, all classes can be regarded by their maximality as generalizations of equivalence classes. A question arises whether maximal preclasses always exist, that is, if for any $\tau \in Tol(U)$, $H_\tau \neq \emptyset$. If a set U is finite or a tolerance $\tau \in Tol(U)$ contains finitely many pairs (x, y) such that $x \neq y$, then $H_\tau \neq \emptyset$. In a general case, a statement that for a given tolerance $\tau \in Tol(U)$ maximal preclasses exist is equivalent to the Axiom of Choice (see Wasilewski, 2004). Thus, in the sequel we assume the Axiom of Choice.

If $\tau \in Tol(U)$ is not transitive, then there are at least two classes $A, B \in H_\tau$ such that $A \cap B \neq \emptyset$. Thus, tolerance classes taken as knowledge granules also reflect inexactness of knowledge. Let us note that if $\tau \in Tol(U)$ is not transitive, then there is $x \in U$ such that $card(H_\tau(x)) > 1$. It means that in the case of inexact knowledge, when classes are regarded as knowledge granules, there is at least one object such that it does not determine a unique knowledge granule. It is different from the situation when knowledge granules are regarded as neighborhoods, that is, for each granule $A \in G_\tau(U) = \{\tau(x) : x \in U\}$ there is at least one $a \in U$ such that $A = \tau(a)$. Still, in that case each object can belong to many granules. Now we present new tolerance approximation operators (Ślęzak & Wasilewski, 2007) that are based on classes as knowledge granules and that avoid the paradox considered previously.

Definition 12

Let (U, τ) be a tolerance space, $G_\tau(U) := H_\tau$ and $Def(G_\tau(U)) := \{\bigcup A : A \subseteq G_\tau(U)\}$. We define operators $K_\tau : \wp(U) \longrightarrow \wp(U), K^\tau : \wp(U) \longrightarrow \wp(U)$ and $BN_K : \wp(U) \longrightarrow \wp(U)$ such that, for any $X \subseteq U$:

$K_\tau(X) := \bigcup\{A \in H_\tau : A \subseteq X\}$,
$K^\tau(X) := \Sigma^\tau(X) \setminus K_\tau(X')$,
$BN_K(X) := K^\tau(X) \setminus K_\tau(X)$,

where $\Sigma^\tau(X) := \bigcup\{A \in H_\tau : A \cap X \neq \emptyset\}$. $K_\tau(X)$ and $K^\tau(X)$ are called *K-lower* and *K-upper approximations* of X respectively, while $BN_K(X)$ is *K-boundary* of X.

One can say that the *K*-upper approximation is defined by "biting" of union of granules having nonempty intersection with an approximated set by its negative region. Notice also that $BN_K(X) = U \setminus K_\tau(X) \cup K_\tau(X')$. A result analogous to Proposition 5 can be proved:

Proposition 6

Let (U, τ) be a tolerance space. The following conditions are equivalent:

1. τ is not transitive
2. There are $A, B \in H_\tau$ such that $A \cap B \neq \emptyset$ and $A \neq B$
3. There is $X \subseteq U$ such that $\Sigma^\tau(X) \cap K_\tau(X') \neq \emptyset$

One can also easily show that:

Proposition 7

If (U, τ) is a tolerance space, then for any $X \subseteq U$ the following conditions hold:

1. $K^\tau(X)' = K_\tau(X')$
2. $K_\tau(X)' = K^\tau(X')$
3. $K^\tau(X) \cap K_\tau(X') = \emptyset$

Proposition 7 shows that approximation operators defined above avoid the paradox discussed earlier, while Proposition 6 shows the role of "biting" in avoiding that paradox.

ALGEBRAIC STRUCTURES RELATED TO ROUGH SETS

In this section, we are going to present some algebraic structures that appear in the rough set theory. These structures are connected with approximation spaces: they are based on various operations defined within families of definable sets and families of rough sets of various approximation spaces. We do not intend to present all algebraic structures that appeared in the rough set theory (see e.g., Banerjee & Chakraborty 1993, 1996; Cattaneo 1997, 1998; Chuchro, 1994; Comer, 1991, 1995; Demri & Orłowska, 2002; Düntsch, 1997; Düntsch & Gediga, 1997; Düntsch & Orłowska, 2001; Gehrke & Walker, 1992; Iwiński, 1987; Järvinen, 1999, 2006; Obtułowicz, 1987; Orłowska, 1995; Pagliani, 1996, 1998a, 1998b; Pomykała & Pomykała, 1988; Wasilewska, 1997; Wasilewska & Banerjee, 1995; Wasilewska & Vigneron, 1995; Wasilewski, 2004, 2005; Yao, 1998). For more details, we refer to survey works (Banerjee & Chakraborty, 2003; Pagliani, 1998) or to monographs about the rough set theory (Demri & Orłowska, 2002; Pawlak, 1991; Polkowski, 2002).

We are focusing on algebraical foundations of the rough set theory from the perspective of its approach to vagueness. In the field of algebra, this topic relates to an algebraic version of the *tertium non datur* law (the law of excluded middle). Thus, we discuss Boolean algebras, Stone algebras, and Heyting algebras, since these classes of algebras model different cases of satisfaction and violation of this law. Boolean algebras satisfy the law of excluded middle (which is often used as an axiom for them), while both Stone algebras and Heyting algebras violate the law in the sense that it does not hold for all algebras within those classes. On the other hand, Stone algebras satisfy the Stone identity, which is an algebraic version of the weak law of excluded middle.

Informational Representation of Algebraic Structures

One of the main directions in a development of mathematics is a generalization of different structures appearing in mathematical considerations. Generalization should lead to defining new mathematical structures in an axiomatic way. It can be done for various structures such as algebras, relational structures, topological structures, and so forth. Groups can serve as examples. They appeared as a generalization of some structures from the number theory and geometry. Such axiomatic definitions introduce abstract structures that can be applied to the problems other than originally considered. Such new abstract structures can also be often directly applied to practical issues, as it was done with Boolean algebras. They were introduced in the second half of the XIX century to describe formal rules of so-called calculus of propositions. In fact, they reflected formal properties that hold both for logical operations on truth values and for some operations on extensions of general names. Afterwards, it appeared that Boolean algebras describe the properties of operations on sets (application in mathematics). In the middle of the XX century, Boolean algebras became one of the main tools in computer science, and became successfully applied in electrical engineering and the electronics.

Axiomatic definitions give a reason for introducing an informal division of mathematical structures in two groups: abstract and "concrete." Groups of linear transformations of vector spaces or fields of sets can serve as examples of "concrete" structures, while groups in general, defined by well-known axioms, as abstract structures. Abstract structures from a given class C are intended to be formal patterns of some "concrete" structures from the class C. If so, a natural question arises: whether for every abstract structure s from a class C there is a "concrete" structure a that belongs to the class C such that a is isomorphic to s in a suitable sense (or more informal: a has the same structure as s). Positive answers to questions of such type are called *representation theorems*. The first such theorem was proved by Peter Stone in 1936:

Theorem 1 (Stone Representation Theorem)
Every abstract Boolean algebra is isomorphic to some field of sets.

Here we are going to present some informational representation theorems. In such representation theorems, the concrete structures have to be derived from information systems. Many information representation theorems have been proved (e.g.,: Banerjee & Chakraborty, 1996; Cattaneo, 1997, 1998; Comer, 1991, 1995; Demri & Orłowska, 1998; Järvinen, 1999, 2006; Orłowska, 1993b; Pagliani, 1996, 1998a, 1998b; Vakarelov, 1989, 1991, 1998; Wasilewska, 1997; Wasilewska & Banerjee, 1995; Wasilewska & Vigneron, 1995). The notion of informational representability was proposed in Orłowska (1993b). In this section, we recall informational representation theorems for Boolean and Stone algebras.

Boolean Algebras and Rough Sets

Boolean algebras are typical examples of abstract structures defined axiomatically. Many axiomatic systems for them have been proposed. One of the main results of that kind was presenting Boolean algebras as lattices of special kind (Burris & Sankapanavar, 1981; Gratzer, 1971, 1978; Kopelberg, 1989). Boolean algebras correspond to the complemented distributive lattices, that is, every Boolean algebra is a complemented distributive lattice, and every complemented distributive lattice is a Boolean algebra. Next, we define Boolean algebras using axiomatics presenting them as complemented distributive lattices. Contemporary, this is one of the most popular axiomatics for Boolean algebras.

Definition 13

An algebra $\langle B, \wedge, \vee, \neg, 0, 1 \rangle$ with two binary operations $\vee$ and $\wedge$ a unary operation $\neg$, and two nullary operations 0 and 1 is a *Boolean algebra* if it satisfies, for any $x, y, z \in B$, (see Box 1).

Let us note that axiom B5 is an algebraic counterpart of the *tertium non datur* principle. If a given Boolean algebra is a field of sets on a given set U, then that law has the form of $A \cup A' = U$ for any $A \subseteq U$.

Because fields of sets are subalgebras of the power set algebras, the Stone Representation Theorem can be rephrased in the following way: Every abstract Boolean algebra is embeddable into some power set algebra or into some complete field of sets.

Theorem 2

The family of definable sets of every equivalence approximation space is a complete and atomic field of sets, and so it is a complete and atomic Boolean algebra.

Proof. Let (U, R) be an equivalence approximation space. The family of definable sets $Def(G_R(U))$ by its definition (definable sets are unions of granules) is closed for arbitrary unions of sets. Since all granules are disjoint, then $Def(G_R(U))$ is also closed for complement of sets, thus, through infinite De Morgan law it is closed for arbitrary unions, arbitrary intersections and complements. Notice that $\emptyset, U \in Def(G_R(U))$. Thus, $Def(G_R(U))$ together with set-theoretical operations, the null set, and the universe is a complete field of sets. Since every complete field of sets is atomic, also $Def(G_R(U))$ is atomic. From the definition of $Def(G_R(U))$ it follows that $G_R(U)$ is a family of its atoms. Thus, the structure $\langle Def(G_R(U)), \cup, \cap, ', \emptyset, U \rangle$ is a complete and atomic Boolean algebra. *q.e.d.*

In terms of knowledge, this theorem can be rephrased as follows: if knowledge is exact, then a family of definable sets (unions of knowledge granules) is a complete and atomic Boolean algebra. Thus, since the law of excluded middle holds in any Boolean algebra, **if knowledge is exact, then all definable sets are crisp** or in other words, **exact knowledge makes definable sets crisp**. It can be also shown (Wasilewski, 2004) that:

Theorem 3 (Informational Representation of Complete Fields of Sets)

Every complete field of sets is an algebra of definable sets of some approximation space.

Since every field of sets is embeddable into some complete field of sets, this theorem, together with Stone theorem, implies the following:

Box 1.

(associativity)	B1: $x \vee y = y \vee x$	B1': $x \wedge y = y \wedge z$
(commutativity)	B2: $x \vee (y \vee z) = (x \vee y) \vee z$	B2': $x \wedge (y \wedge z) = (x \wedge y) \wedge z$
(absorption)	B3: $x \vee (y \wedge x) = x$	B3': $x \vee (y \wedge x) = x$
(distributivity)	B4: $x \vee (y \wedge z) = (x \vee y) \wedge (x \vee z)$	B4': $x \vee (y \wedge z) = (x \vee y) \wedge (x \vee z)$
(complementation)	B5: $x \vee \neg x = 1$	B5': $x \wedge \neg x = 0$

Theorem 4 (Informational Representation of Boolean Algebras I)

Every abstract Boolean algebra is embeddable into an algebra of definable sets in some approximation space.

Now, let us see what follows from assumption that knowledge is complete. Let (U, R) be an equivalence approximation space and let $R = \Delta_U$. Notice that in this case, every subset of U is definable, that is, $Def(G_R(U)) = \wp(U)$. For every $X \subseteq U, (R_*(X), R^*(X)) = (X, X)$, so the function $\kappa : X \mapsto (R_*(X), R^*(X))$ is one-to-one mapping of $\wp(U)$ onto $\Re_R(U)$. Therefore, $\Re_R(U) = \Delta_{\wp(U)}$ where we put $\Delta_{\wp(U)} := \{(X, X) : X \subseteq U\}$. We have mentioned before that calling the pairs representing definable sets as "rough sets" is problematic. Hence, in case all pairs of the form $(R_*(X), R^*(X))$ represent definable sets, it will be more convenient to call them *approximated sets*. On $\Delta_{\wp(U)}$ we define the following operations, for any $X, Y \subseteq U$:

$(X,X) \wedge (Y,Y) := (X \cap Y, X \cap Y),$
$(X,X) \vee (Y,Y) := (X \cup Y, X \cup Y),$
$\neg (X,X) := (X', X').$

Notice that the family $\Delta_{\wp(U)}$ is closed for the operations defined. On $\Delta_{\wp(U)}$ we can also define relation $\leq_{\Delta_{\wp(U)}}$ as follows: $(X,X) \leq_{\Delta_{\wp(U)}} (Y,Y) :\Leftrightarrow X \subseteq Y$. Relation $\leq_{\Delta_{\wp(U)}}$ is a partial order on $\Delta_{\wp(U)}$, with $(\emptyset, \emptyset), (U, U) \in \Delta_{\wp(U)}$ as its least and greatest element, respectively. Straight calculations show that the structure is a Boolean algebra and function $\kappa : \wp(U) \longrightarrow \Re_R(U)$ is a Boolean isomorphism. Thus, a Boolean algebra of approximated sets $\langle \Re_R(U), \wedge, \vee, \neg, (\emptyset, \emptyset), (U, U) \rangle$ is complete and atomic. Consequently:

Theorem 5

Let (U, R) be an equivalence approximation space and let $R = \Delta_U$. The structure of approximated sets $\langle \Re_R(U), \wedge, \vee, \neg, (\emptyset, \emptyset), (U, U) \rangle$ is a complete and atomic Boolean algebra that is isomorphic to a power set algebra $\langle \wp(U), \cap, \cup, ', \emptyset, U \rangle$ and to an algebra of definable sets $\langle Def(G_R(U)), \cap, \cup, ', \emptyset, U \rangle$. From Theorem 5 it follows that if $R = \Delta_U$, then the law of excluded middle holds in algebra of the approximated sets $\langle \Re_R(U), \wedge, \vee, \neg, (\emptyset, \emptyset), (U, U) \rangle$. In knowledge terms, this fact can be rephrased as follows: **if knowledge is complete, then all rough sets are crisp** or, in other words, **complete knowledge makes all concepts crisp**. In the case of mathematical concepts, this is quite intuitive: mathematical concepts are usually defined with usage of sets that are crisp. It is consistent with Frege's postulate that all concepts in mathematics should be crisp.

Since every field of sets is subalgebra of some power set algebra, this theorem and the Stone Representation Theorem imply the following fact:

Theorem 6 (Informational Representation of Boolean Algebras II)

Every abstract Boolean algebra is embeddable into an algebra of approximated sets in some approximation space with complete knowledge.

Stone Algebras and Rough Sets

Stone algebras are another example of algebraic structures appearing in the rough set theory. We choose them because they are defined by the weak law of excluded middle and they reflect a rough set approach to vagueness. Stone algebras were introduced into the rough set theory in the work of J. Pomykała and J.A. Pomykała (1988) shown that the family of rough sets of any approximation space is a complete and atomic Stone algebra. Let us recall now a definition of Stone algebras.

Definition 14

An algebra $\langle S, \wedge, \vee, \neg, 0, 1 \rangle$ with two binary operations $\vee$ and $\wedge$ a unary operation $\neg$, and two nullary operations 0 and 1 is a *Stone algebra* if it satisfies, for any $x \in S$:

S1: $\langle S, \wedge, \vee \rangle$ is a distributive lattice
S2: $x \wedge 0 = 0; x \vee 1 = 1$
S3: $x \wedge \neg x = 0$
S4: $x \wedge y = 0$ implies $y \leq \neg x$
S5: $\neg x \vee \neg \neg x = 1$

An operation $\neg$ is called a *pseudocomplementation*.

It follows from this definition that an element $\neg x$ is a pseudocomplement of x, that is $\neg x$ is the greatest element in the set $\{y : x \wedge y = 0\}$. Thus, each Stone algebra is a distributive lattice with pseudocomplementation (Gratzer, 1971). In this definition, the axiom S5 is an algebraic version of the weak *tertium non datur* principle.

In Pomykała and Pomykała (1988) it is shown that the family of rough sets of any approximation space is a complete and atomic Stone algebra defining rough sets as equivalence classes of a rough equality of sets. It was further widened in Comer (1995), where the family $\Re_R(U)$ of rough sets is extended by an operation of dual pseudocomplementation. It was proved that in such a case, the family of rough sets of any approximation space is a regular double Stone algebra. Let us recall the definition of regular double Stone algebras.

Definition 15
An algebra $\langle S, \wedge, \vee, \neg, \circ, 0, 1 \rangle$ is a *double Stone algebra* if $\langle S, \wedge, \vee, \neg, 0, 1 \rangle$ is a Stone algebra and, additionally, $\circ$ is a unary operation that satisfies for any $x \in S$:

S6: $x \vee \circ x = 1$
S7: $x \vee y = 1$ implies $\circ x \leq y$
S8: $\circ x \wedge \circ \circ x = 0$

Operation $\circ$ is called a *dual pseudocomplementation*.

A double Stone algebra $\langle S, \vee, \wedge, \neg, \circ, 0, 1 \rangle$ is *regular* if for any $x, y \in S$ the following implication holds:

S9: if $\neg x = \neg y$ and $\circ x = \circ y$, then $x = y$

It follows from this definition that $\circ x$ is a dual pseudocomplement of x, that is $\circ x$ is the least element in the set $\{y : x \vee y = 1\}$. The following is shown in Comer (1995).

Theorem 7
Let (U, R) be an equivalence approximation space, and let the family of rough sets $\Re_R(U)$ be equipped with the operations $\wedge, \vee, \neg, \circ$ defined in the following way:

$(R_*(X), R^*(X)) \wedge (R_*(Y), R^*(Y))$
$:= (R_*(X) \cap R_*(Y), R^*(X) \cap R^*(Y))$,

$(R_*(X), R^*(X)) \vee (R_*(Y), R^*(Y))$
$:= (R_*(X) \cup R_*(Y), R^*(X) \cup R^*(Y))$,

$\neg(R_*(X), R^*(X)) := (R^*(X)', R^*(X)')$,

$\circ(R_*(X), R^*(X)) := (R_*(X)', R_*(X)')$.

Then the structure $\langle \Re_R(U), \wedge, \vee, \neg, \circ, (\emptyset, \emptyset), (U, U) \rangle$ is a complete and atomic regular double Stone algebra in which infinite operations have the following form:

$$\bigwedge_{i \in I}(R_*(X), R^*(X)) = \left(\bigcap_{x \in I} R_*(X), \bigcap_{i \in I} R^*(X) \right),$$

$$\bigvee_{i \in I}(R_*(X), R^*(X)) = \left(\bigcup_{i \in I} R_*(X), \bigcup_{i \in I} R^*(X) \right),$$

where $\{(R_*(X), R^*(X))\}_{i \in I} \subseteq \Re_R(U)$.

In the sequel, we will refer to the structure $\langle \Re_R(U), \wedge, \vee, \neg, \circ, (\emptyset, \emptyset), (U, U) \rangle$ as an algebra of rough sets of an approximation space (U, R) or simply a rough set algebra.

Since $R^*(X)' = R_*(X')$, we have $\neg(R_*(X), R^*(X)) = (R^*(X)', R^*(X)')$. Thus, pseudocomplementation of a given rough set is taken by its negative region. It shows that this algebraic representation of rough sets is consistent with a basic idea of rough sets, and it means that a complementation of a given rough set consists of elements that surely do not belong to it. Since $R_*(X)' = R^*(X')$, we have $\circ(R_*(X), R^*(X)) = (R_*(X)', R_*(X)')$. Thus, similarly, we can interpret a dual pseudocomplementation of a given rough set as the set consisting of the elements that possibly do not belong to it.

If the given set $X \subseteq U$ is definable in an approximation space (U, R) ($X \in Def(G_R(U))$), then $(R_*(X), R^*(X)) = (X, X)$ as well as $\neg(R_*(X), R^*(X)) = (R^*(X)', R^*(X)') = (X', X')$ So, $(R_*(X), R^*(X)) \vee \neg(R_*(X), R^*(X)) = (X, X) \vee (X', X') = (X \cup X', X \cup X') = (U, U)$, therefore $(R_*(X), R^*(X))$ satisfies the law of excluded middle and it is a crisps set. In knowledge terms, Theorem 7 can be rephrased as follows: **if knowledge is incomplete but exact, then sets that are not definable are not crisp**. However, all rough sets, both crisp and not crisp, satisfy the Stone identity, that is, the weak law of excluded middle. In the case of crisp (definable) sets, it holds since the Stone identity follows from the law of excluded middle. So, there is an interesting situation concerning the sets that are not crisp. Since they violate the law of excluded middle, they are vague, but they still preserve the weak law of excluded middle. This can be interpreted in such a way that they are vague in a weak sense. Such interpretation gives a reason for introducing a new type of vagueness. Sets are *weakly vague* if they are not crisp but they satisfy the weak law of excluded middle. Thus, the last theorem can be rephrased in knowledge terms in the following way **if knowledge is incomplete but exact, then nondefinable sets are weakly vague**. In other words, **incomplete and exact knowledge causes weak vagueness of concepts**. It is easy to note also that

Proposition 8

Let $\langle \Re_R(U), \wedge, \vee, \neg, \circ, (\emptyset, \emptyset), (U, U) \rangle$ be an algebra of rough sets of an approximation space (U, R). Then for any $X \subseteq U$ the following holds:

$X \in Def(G_R(U))$ if and only if $\neg(R_*(X), R^*(X)) = \circ(R_*(X), R^*(X))$.

□

Theorem 8 (Comer Representation Theorem)

Every abstract regular double Stone algebra is embeddable into rough sets algebra of some equivalence approximation space.

Taking into account this theorem for Stone algebras together with the previous results, we can say that regular Stone algebras are algebraic representations of incomplete and exact knowledge that is expressible in weakly vague concepts.

HEYTING ALGEBRAS AND TOLERANCE ROUGH SETS

We have shown that if knowledge is exact, then algebras of definable sets are Boolean and so definable sets are crisp. In the case of rough sets, knowledge characteristics are a little bit more compound. If knowledge is complete, and so it is exact, then algebras of rough sets are Boolean algebras, so all rough sets are crisp. Thus, complete knowledge is expressible by crisp sets. If knowledge is incomplete but exact, then algebras of rough sets are Stone algebras and rough sets that are not definable, are weakly vague. Thus, incomplete exact knowledge is expressible in weakly vague concepts. We can ask now what is an algebraic characteristic of incomplete inexact knowledge? Does inexactness of knowledge make definable sets not crisp? If the knowledge is incomplete and inexact, are nondefinable rough sets weakly vague, or do they even violate the weak law of excluded middle? Examples of algebraic structures that do not satisfy the weak law of excluded middle (and so the law of excluded middle) are Heyting algebras; so, they can represent incomplete and inexact knowledge. Thus, we can better specify our questions: Does inexactness of knowledge make algebras of definable sets Heyting algebras violating the weak law of excluded middle? And, in the case of incomplete inexact knowledge, are algebras of rough sets Heyting algebras violating the weak law of excluded middle?

Heyting algebras, introduced by Birkhoff in 1940 under the name *Brouwerian algebras* (Birkhoff, 1940, 1967), are intended to be an algebraic counterpart of the intuitionistic logic. This logic was built on the rejection of nonconstructive methods of proofs. Nondirect proofs methods are examples of such methods. Nondirect proofs are based on the law of excluded middle. Thus, the intuitionistic logic was constructed in such a way that it denies the logical version of *tertium non datur* principle or even the logical version of the weak law of excluded middle. Heyting algebras were involved in research within the field of the rough set theory (Obtułowicz 1987; Pagliani 1998). We present the definition of Heyting algebras following Burris and Sankappanavar (1981):

Definition 16
An algebra $\langle H, \wedge, \vee, \rightarrow, 0, 1 \rangle$ with three binary operations $\vee, \wedge, \rightarrow$ and two nullary operations 0, and 1 is a *Heyting algebra*, if it satisfies, for any $x, y, z \in H$:

H1: $\langle H, \wedge, \vee \rangle$ is a distributive lattice
H2: $x \wedge 0 = 0; x \vee 1 = 1$
H3: $x \rightarrow x = 1$
H4: $(x \rightarrow y) \wedge y = y; x \wedge (x \rightarrow y) = x \rightarrow y$
H5: $x \rightarrow (y \wedge z) = (x \rightarrow y) \wedge (x \rightarrow z);$
$(x \vee y) \rightarrow z = (x \rightarrow z) \wedge (y \rightarrow z)$

In each Heyting algebra $\langle H, \wedge, \vee, \rightarrow, 0, 1 \rangle$ the element $x \rightarrow y$ is a relative pseudocomplement of x to y. Moreover, Heyting algebras are exactly relatively pseudocomplemented lattices with the least element. Thus, the following holds (Birkhoff 1940; Rasiowa & Sikorski, 1970):

Lemma 1
Let $\langle H, \wedge, \vee, 0 \rangle$ be a lattice with the least element 0. Then the following conditions are equivalent:

1. $\langle H, \wedge, \vee, \rightarrow, 0, 1 \rangle$ is a Heyting algebra,
2. $z \leq x \rightarrow y \Leftrightarrow x \wedge z \leq y,$
3. $x \rightarrow y$ is the greatest element a, such that $x \wedge a \leq y$ for all $x, y \in H$.

Heyting algebras are the pseudocomplemented lattices, where $\neg x := x \rightarrow 0$ is a pseudocomplementation of $x \in H$. However, not all pseudocomplemented lattices are Heyting algebras. Double Heyting algebras were introduced by Cecylia Rauszer (1974) under the name of semi-Boolean algebras. They were used in an algebraic treatment of intuitionistic logic with two additional connectives that are dual to the intuitionistic implication and the intuitionistic negation.

Definition 17
An algebra $\langle H, \wedge, \vee, \rightarrow, \div, 0, 1 \rangle$ is a *double Heyting algebra*, if $\langle H, \wedge, \vee, \rightarrow, 0, 1 \rangle$ is a Heyting algebra and additionally $\div$ is a binary operation that satisfies, for any $x, y, z \in H$:

H6: $x \div x = 0$
H7: $x \vee (x \div y) = x; (x \div y) \vee y = x \vee y$
H8: $(x \vee y) \div z = (x \div z) \vee (y \div z);$
$z \div (x \wedge y) = (z \div x) \vee (z \div y)$

Operation $\div$ is called *pseudodifference* and $x \div y$ is said to be pseudodifference of x and y.

If an algebra $\langle H, \wedge, \vee, \rightarrow, \div, 0, 1 \rangle$ is a double Heyting algebra, then $\langle H, \vee, \wedge, \div, 0, 1 \rangle$ is a *Brouwerian algebra*, that is it satisfies the axioms H1, H2, and H6, H7, H8.

A double Heyting algebra $\langle H, \wedge, \vee, \rightarrow, \div, 0, 1 \rangle$ is *regular* if for any $x, y \in H$ the following implication holds:

H9: if $x \rightarrow 0 = y \rightarrow 0$ and $1 \div x = 1 \div y$, then $x = y$

Let us note that $1 \div x$ is the least element a such that $x \vee a = 1$. Thus, the operation $\circ x := 1 \div x$ is dual to the operation $\neg x$ and it is called a *dual pseudocomplementation*, just like in the case of double Stone algebras. $\circ x$ is said to be a *dual complement* of x. Thus, a double Heyting algebra $\langle H,\wedge,\vee,\rightarrow,\div,0,1\rangle$ is regular, if $\neg x = \neg y$ and $\circ x = \circ y$ implies $x=y$. Double Heyting algebras are regular as double pseudocomplemented lattices. It was shown that regular double pseudocomplemented lattices are Heyting algebras (Katrinak, 1973). In any Brouwerian algebra we have $0 = x \div x$ and $1 = \circ 0$. Brouwerian algebras can be also defined as lattices with the greatest element that are closed for pseudodifference (McKinsey & Tarski, 1946; Rasiowa & Sikorski, 1970). The following holds (McKinsey & Tarski, 1946):

Lemma 2
Let $\langle H,\wedge,\vee,1\rangle$ be a lattice with the greatest element 1. Then the following conditions are equivalent:

1. $\langle H,\wedge,\vee,\div,0,1\rangle$ is a Brouwerian algebra,
2. $x \div y \leq z \Leftrightarrow x \leq y \vee z$ for all $x, y, z \in H$,
3. $x \div y$ is the least element $b \in H$, such that $x \leq y \vee b$ for all $x, y \in H$.

Operation $\div$ is not exactly a dual operation to the relative pseudocomplement $\rightarrow$ since an element $x \div y$ is not the least element c such that $y \leq x \vee c$. Such operation is usually denoted by $\leftarrow$. In order to get axioms defining this operation, one has to interchange operations $\wedge$ and $\vee$ in the axioms H3, H4, H5. Double Heyting algebras are defined also with usage of $\leftarrow$. However, if algebra $\langle H,\wedge,\vee,\div,0,1\rangle$ is a bounded lattice satisfying additionally axioms H6, H7, H8, and operation $\leftarrow$ is defined as $x \leftarrow y := y \div x$, then $\leftarrow$ is actually determined by dual counterparts of axioms H3, H4, H5. Thus, in the case of double Heyting algebras, the way of defining it is not essential. We follow an approach used in Rauszer (1974). Let us recall that a lattice $\langle L,\wedge,\vee,0,1\rangle$ is relatively pseudocomplemented if there is the greatest element $a \in L$ such that $x \wedge a \leq y$ for all $x, y \in H$. Lattice $\langle L,\wedge,\vee,0,1\rangle$ is said to be closed for pseudodifference if there is the least element $b \in L$, such that $x \leq y \vee b$, for all $x, y \in H$.

Proposition 9
A complete lattice $\langle H,\wedge,\vee,0,1\rangle$ is relatively pseudocomplemented (i.e., it is a Heyting algebra) if and only if it is infinitely meet distributive, that is, for any $x \in H$ and $\{y_i\}_{i \in I} \subseteq H$ the following equation holds:

$$x \wedge \bigvee_{i \in I} y_i = \bigvee_{i \in I}(x \wedge y_i)$$

Proposition 10
A complete lattice $\langle H,\wedge,\vee,0,1\rangle$ is closed for a pseudodifference (i.e., it is a Brouwerian algebra) if and only if it is infinitely join distributive that is for any $x \in H$ and $\{y_i\}_{i \in I} \subseteq H$ the following equation holds:

$$x \vee \bigwedge_{i \in I} y_i = \bigwedge_{i \in I}(x \vee y_i)$$

Now we can observe that algebra of rough sets of any equivalence approximation space is a double Heyting algebra (see Pagliani, 1998). More exactly, the following holds:

Theorem 9
Let $\langle \Re_R(U),\wedge,\vee,\neg,\circ,(\emptyset,\emptyset),(U,U)\rangle$ be the algebra of rough sets of an approximation space (U,R). Then the structure $\langle \Re_R(U),\wedge,\vee,\rightarrow,\div,(\emptyset,\emptyset),(U,U)\rangle$ with operations $\rightarrow$, $\div$ defined for any $(R_*(X), R^*(X)), (R_*(Y), R^*(Y)) \in \Re_R(U)$ as follows:

$$(R_*(X), R^*(X)) \rightarrow (R_*(Y), R^*(Y))$$
$$:= \left(\bigcup_{Z \in I(X,Y)} R_*(Z), \bigcup_{Z \in I(X,Y)} R^*(Z) \right),$$

$$(R_*(X), R^*(X)) \div (R_*(Y), R^*(Y))$$
$$:= \left(\bigcap_{Z \in D(X,Y)} R_*(Z), \bigcap_{Z \in D(X,Y)} R^*(Z) \right),$$

where

$$I(X,Y)$$
$$:= \{Z \subseteq U : (R_*(X), R^*(X))$$
$$\wedge (R_*(Z), R^*(Z)) \leq (R_*(Y), R^*(Y))\},$$

$$D(X,Y)$$
$$:= \{Z \subseteq U : (R_*(X), R^*(X))$$
$$\leq (R_*(Y), R^*(Y)) \vee (R_*(Z), R^*(Z))\},$$

is a complete and atomic regular double Heyting algebra that satisfies the Stone identity ($\neg x \vee \neg\neg x = 1$).

This theorem was first putted by Pagliani (1998). His justification was appeal to the more general, algebraic result of Katriniak (1973). We present another proof that is direct and elementary in the sense that it is not based on any additional theorems. This proof uses a standard technique and presents the form of the relative pseducomplementation and the dual relative pseudocomplementation within the algebra of rough sets.

Proof. Let $\langle \mathfrak{R}_R(U), \wedge, \vee, \neg, \circ, (\varnothing,\varnothing), (U,U) \rangle$ be the algebra of rough sets of an approximation space (U, R) and let operations $\rightarrow, \div$ be defined as previously stated. Thus, the structure $\langle \mathfrak{R}_R(U), \wedge, \vee, (\varnothing,\varnothing), (U,U) \rangle$ is complete and atomic distributive lattice, in which infinite operations have the following form:

$$\bigwedge_{i \in I}(R_*(X_i), R^*(X_i)) = \left(\bigcap_{i \in I} R_*(X_i), \bigcap_{i \in I} R^*(X_i)\right),$$

$$\bigvee_{i \in I}(R_*(X_i), R^*(X_i)) = \left(\bigcup_{i \in I} R_*(X_i), \bigcup_{i \in I} R^*(X_i)\right),$$

where $\{(R_*(X_i), R^*(X_i))\}_{i \in I} \subseteq \mathfrak{R}_R(U)$. Thus, the family of rough sets $\mathfrak{R}_R(U)$ is closed for operations $\rightarrow, \div$. Let us note that the lattice $\langle \mathfrak{R}_R(U), \wedge, \vee, (\varnothing,\varnothing), (U,U) \rangle$ is a complete sublattice of the Boolean algebra $\langle Def(G_R(U)) \times Def(G_R(U)), \wedge, \vee, \neg, (\varnothing,\varnothing), (U,U) \rangle$, which is a Cartesian power of the algebra $\langle Def(G_R(U)), \cap, \cup, ', (\varnothing,\varnothing), (U,U) \rangle$, an algebra of definable sets in (U, R). So, the lattice $\langle \mathfrak{R}_R(U), \wedge, \vee, (\varnothing,\varnothing), (U,U) \rangle$ is infinitely meet and infinitely join distributive. It means that $\vee\{(R_*(Z), R^*(Z)) : (R_*(X), R^*(X)) \wedge (R_*(Z), R^*(Z)) \leq (R_*(Y), R^*(Y))\}$ is the greatest element $A \in \mathfrak{R}_R(U)$ such that $(R_*(X), R^*(X)) \wedge A \leq (R_*(Y), R^*(Y))$. We can see that $(R_*(X), R^*(X)) \rightarrow (R_*(Y), R^*(Y)) =$

$$= \left(\bigcup_{Z \in I(X,Y)} R_*(Z), \bigcup_{Z \in I(X,Y)} R^*(Z)\right)$$

$$= \bigvee_{Z \in I(X,Y)} (R_*(Z), R^*(Z))$$

$$= \vee\{(R_*(Z), R^*(Z)) : (R_*(X), R^*(X)) \wedge (R_*(Z), R^*(Z)) \leq (R_*(Y), R^*(Y))\}.$$

Therefore, $(R_*(X), R^*(X)) \rightarrow (R_*(Y), R^*(Y))$ is the greatest element $A \in \mathfrak{R}_R(U)$ such that $(R_*(X), R^*(X)) \wedge A \leq (R_*(Y), R^*(Y))$. So, by Lemma 1, $\langle \mathfrak{R}_R(U), \vee, \wedge, \rightarrow, (\varnothing,\varnothing), (U,U) \rangle$ is a Heyting algebra.

Since $\langle \mathfrak{R}_R(U), \wedge, \vee, (\varnothing,\varnothing), (U,U) \rangle$ is infinitely join distributive, then the element $\wedge\{(R_*(Z), R^*(Z)) : (R_*(X), R^*(X)) \leq (R_*(Y), R^*(Y)) \vee (R_*(Z), R^*(Z))\}$ is the least element $A \in \mathfrak{R}_R(U)$ such that $(R_*(X), R^*(X)) \leq (R_*(Y), R^*(Y)) \vee A$. Thus, using Lemma 2, one can show, analogically, that the structure $\langle \mathfrak{R}_R(U), \wedge, \vee, \div, (\varnothing,\varnothing), (U,U) \rangle$ is a Brouwerian algebra. Therefore, the structure $\langle \mathfrak{R}_R(U), \wedge, \vee, \rightarrow, \div, (\varnothing,\varnothing), (U,U) \rangle$ is a double Heyting algebra.

Let operations $\neg$ and $\circ$ be defined as in Theorem 6. It is then easy to show the following $(R_*(X), R^*(X)) \rightarrow (\varnothing,\varnothing) = (R^*(X)', R^*(X)')$ and $(U,U) \div (R_*(Y), R^*(Y)) = (R_*(Y)', R_*(Y)')$. So $\neg(R_*(X), R^*(X)) = (R_*(X), R^*(X)) \rightarrow (\varnothing,\varnothing)$ and $\langle \mathfrak{R}_R(U), \wedge, \vee, \rightarrow, \div, (\varnothing,\varnothing), (U,U) \rangle$. Since the algebra of rough sets $\langle \mathfrak{R}_R(U), \vee, \wedge, \neg, \circ, (\varnothing,\varnothing), (U,U) \rangle$ is regular, complete, and atomic, then double Heyting algebra $\langle \mathfrak{R}_R(U), \wedge, \vee, \rightarrow, \div, (\varnothing,\varnothing), (U,U) \rangle$ is also regular, complete, and atomic. Since algebra $\langle \mathfrak{R}_R(U), \wedge, \vee, \neg, \circ, (\varnothing,\varnothing), (U,U) \rangle$ satisfies the weak law of excluded middle, then algebra $\langle \Re_R(U), \wedge, \vee, \rightarrow, \div, (\emptyset,\emptyset), (U,U) \rangle$ satisfies it too, that is, the following holds: $(R_*(X), R^*(X)) \rightarrow (\emptyset,\emptyset) \vee (((R_*(X), R^*(X)) \rightarrow (\emptyset,\emptyset)) \rightarrow (\emptyset,\emptyset)) = (U,U)$. Therefore, we have shown that $\langle \Re_R(U), \wedge, \vee, \rightarrow, \div, (\emptyset,\emptyset), (U,U) \rangle$ is a complete and atomic regular double Heyting algebra that satisfies the Stone identity. q.e.d.

Now we can see that algebras of rough sets based on incomplete exact knowledge are also Heyting algebras. However, they preserve the weak law of excluded middle. Our question whether inexactness of knowledge makes algebras of definable sets or algebras of rough sets Heyting algebras violating the weak law of excluded middle remains open. In the third section, we have considered tolerance classes as knowledge granules, together with approximation operators based on them. Let us define a new family of knowledge granules that are also constructed from tolerance classes.

Definition 18

Let (U, τ) be a tolerance space. Then a set A_τ is defined as follows:

$$A_\tau := \{\bigcap \Gamma : \Gamma \subseteq H_\tau\}.$$

If $a \in U$, then $A_\tau(a) := \{A \in A_\tau : a \in A\}$.

Thus, elements of A_τ are intersections of classes of a tolerance τ. One can think about τ-classes as basic colors and about elements of A_τ that are intersections of τ-classes as derived colors (see Lin, 2006).

Let us now consider the family A_τ of knowledge granules of tolerance space (U, τ), that is: $G_\tau(U) := A_\tau$. So, we assume that definable sets over this family are unions of intersections of classes. The family $Def(G_\tau(U))$ of all definable sets is closed for arbitrary unions. Notice that $\emptyset \in Def(G_\tau(U))$ because $\bigcup \emptyset = \emptyset$. Since the family of classes H_τ is a covering of U, and so the family A_τ is, thus $U \in Def(G_\tau(U))$. Since the family of knowledge granules is closed for arbitrary intersections, a natural question arises: whether the family of definable sets is also closed for arbitrary intersections. Let us recall the notion of Alexandrov topology (Aleksandroff, 1937): a family of subsets $\Theta \subseteq \wp(U)$ is an *Alexandrov topology* if $\emptyset, U \in \Theta$ and a family Θ is closed for arbitrary unions and intersections of sets. So, if the family $U \in Def(G_\tau(U))$ is closed for arbitrary intersections, then it is an Alexandrov topology on a set U. In order to show this, we need one of generalized distributivity laws (Tarski, 1935):

Lemma 3

Let $\{F_j\}_{j \in J}$ be an indexed family of sets. Let M be a set of indices of the form $M = \bigcup_{i \in I} J_i$ where $J_i \subseteq J$ for each $i \in I$ and $K := \{Z \in \wp(M) : Z \cap J_i \neq \emptyset \text{ for all } i \in I\}$. Then we have:

$$\bigcap_{i \in I} \bigcup_{j \in J_i} F_j = \bigcup_{Z \in K} \bigcap_{j \in Z} F_j$$

Now we can prove the following proposition:

Proposition 11

Let (U, τ) be a tolerance space and let $G_\tau(U) := A_\tau$ be a family of granules. Then the family of definable sets $Def(G_\tau(U))$ is an Alexandrov topology on a set U.

Proof. We show that the family $Def(G_\tau(U))$ is closed for arbitrary intersections. Let the family $A\tau$ be indexed by indices from a set $J: A_\tau = \{F_j\}_{j \in J}$. Let $\{X_i\}_{i \in I} \subseteq Def(G_\tau(U))$. Thus, for each $i \in I$ there is a family $J_i \subseteq J$ such that $X_i = \bigcup_{j \in J_i} F_j$. Then by Lemma 3:

$$\bigcap_{i \in I} X_i = \bigcap_{i \in I} \bigcup_{j \in J_i} F_j = \bigcup_{Z \in K} \bigcap_{j \in Z} F_j.$$

Let us note that for any $Z \in K$, the family $\{F_j\}_{j \in Z}$ is a family of knowledge granules, that is, $\{F_j\}_{j \in Z} \subseteq A_\tau$, thus $\bigcap_{j \in Z} F_j \in A_\tau$ by definition of $A\tau$. Then $\bigcup_{Z \in K} \bigcap_{j \in Z} F_j \in Def(G_\tau(U))$, so

$\bigcap_{i\in I} X_i \in Def(G_\tau(U))$. Since the family $\{X_i\}_{i\in I}$ was chosen arbitrarily, therefore the family of definable sets $Def(G_\tau(U))$ is closed for arbitrary intersections and it is an Alexandrov topology on the set U. q.e.d.

Let us define new approximation operators based on the family of granules $G_\tau(U) := A_\tau$.

Definition 19
Let (U,τ) be a tolerance space and let $G_\tau(U) := A_\tau$. For any $X \subseteq U$ we define operators $T_\tau : \wp(U) \longrightarrow \wp(U), T^\tau : \wp(U) \longrightarrow \wp(U)$, and $BN_T : \wp(U) \longrightarrow \wp(U)$:

$T_\tau(X) := \bigcup\{A \in A_\tau : A \subseteq X\}$,
$T^\tau(X) := \Sigma^\tau(X) \setminus T_\tau(X')$,
$BN_T(X) := T^\tau(X) \setminus T_\tau(X)$,

where $\Sigma^\tau(X) := \bigcup\{A \in A_\tau : A \cap X \neq \emptyset\}$. $T_\tau(X)$ and $T^\tau(X)$ are called *T-lower* and *T-upper approximations* of X, respectively, while $BN_T(X)$ is called *T-boundary* of X.

Proposition 12
Let (U,τ) be a tolerance space and let $G_\tau(U) := A_\tau$. Then tolerance approximation operators T_τ and T^τ are topological interior and closure operators of the Alexandrov topological space $(U, Def(G_\tau(U)))$, so particularly $X \in Def(G_\tau(U)) \Leftrightarrow T_\tau(X) = X$.

Proof. By Proposition 11, $Def(G_\tau(U))$ is an Alexandrov topology on U. Notice, that by the definition of T_τ the set $T_\tau(X)$ is the greatest open set contained in X. Therefore, T_τ is a topological interior operator determined by Alexandrov topology $Def(G_\tau(U))$, that is $X \in Def(G_\tau(U)) \Leftrightarrow T_\tau(X) = X$. Notice that for any $x \in U$, we have $x \in T^\tau(X)$, if for every $A \in A_\tau(x)$, there is $A \cap X \neq \emptyset$. Let $X \subseteq U$, then $T^\tau(X)' = T_\tau(X')$, so $T^\tau(X) = T_\tau(X')'$. Since $T_\tau(X') \in Def(G_\tau(U))$, then $T^\tau(X)$ is a complement of an open set, so $T^\tau(X)$ is a closed set. Since $T_\tau(X')$ is the greatest open set disjoint with X, then $T^\tau(X)$ is the least closed set containing X. Since $X \subseteq U$ was chosen in an arbitrary way, T^τ is a topological closure operator of the topology $Def(G_\tau(U))$. q.e.d.

The family of closed sets of Alexandrov topological space $(U, Def(G_\tau(U)))$ will be denoted by $Cl(G_\tau(U))$. Note that $Cl(G_\tau(U))$ is closed for arbitrary unions and intersections (since $Def(G_\tau(U))$ is closed too). Thus $\langle Cl(G_\tau(U)), \cap, \cup, \emptyset, U \rangle$ is a complete distributive lattice.

Theorem 10
Let (U,τ) be a tolerance space and let $G_\tau(U) := A_\tau$ be a family of granules. Then the algebra of definable sets $\langle Def(G_\tau(U)), \cap, \cup, \rightarrow, \div, \emptyset, U \rangle$, where operations $\rightarrow, \div$ are defined for any $X, Y \subseteq U$ as follows:

$X \rightarrow Y := \bigcup\{A \in A_\tau : X \cap A \subseteq Y\}$,
$X \div Y := \bigcap\{A \in A_\tau : X \subseteq Y \cup A\}$,

is a complete atomic double Heyting algebra.

Proof. By Proposition 11 the family of definable sets $Def(G_\tau(U))$ is Aleksandrov topology on U. Thus, algebra $\langle Def(G_\tau(U)), \cap, \cup, \emptyset, U \rangle$ is a complete distributive lattice of sets, infinitely meet distributive and infinitely join distributive. The set $\bigcup\{A \in A_\tau : X \cap A \subseteq Y\}$ is the greatest set in $Def(G_\tau(U))$ such that $X \cap \bigcup\{A \in A_\tau : X \cap A \subseteq Y\} \subset Y$ and the set $\bigcap\{A \in A_\tau : X \subseteq Y \cup A\}$ is the least element $Z \in Def(G_\tau(U))$ such that $X \subseteq Y \cup Z$. From Propositions 9 and 10, we have that $\langle Def(G_\tau(U)), \cap, \cup, \rightarrow, \emptyset, U \rangle$ is a Heyting algebra and $\langle Def(G_\tau(U)), \cap, \cup, \div, \emptyset, U \rangle$ is a Brouwerian algebra. Hence, $\langle Def(G_\tau(U)), \cap, \cup, \rightarrow, \div, \emptyset, U \rangle$ is a complete double Heyting algebra.

Notice that since the family $Def(G_\tau(U))$ is closed for arbitrary intersections, then for every $x \in U$, $\bigcap A_\tau(x) \in A_\tau$ and $\bigcap A_\tau(x) \in Def(G_\tau(U))$. Then the set $\bigcap A_\tau(x)$ is the least open set containing x. Thus, the family $\{\bigcap A_\tau(x) : x \in U\}$ is the least base of topology $Def(G_\tau(U))$; hence, it is the family of atoms of algebra $\langle Def(G_\tau(U)), \cap, \cup, \rightarrow, \div, \emptyset, U \rangle$ as a lattice. Since the family $\{\bigcap A_\tau(x) : x \in U\}$ is a base of topology $Def(G_\tau(U))$, thus every set $Z \in Def(G_\tau(U))$

contains at least one atom. Therefore, we have shown that $\langle Def(G_\tau(U)), \cap, \cup, \rightarrow, \div, \emptyset, U \rangle$ is a complete atomic double Heyting algebra. *q.e.d.*

The next question is whether, for tolerance spaces (U, τ), their double Heyting algebras of definable sets over A_τ, with approximation operators T_τ and T^τ, are regular, and whether they satisfy the law of excluded middle. The negative answer to this question is presented in the following example.

Example 9

Let $U := \{a,b,c,d,e,f\}$. Let $\tau := \Delta_U \cup \{a,b,c\}^2 \cup \{b,c,d\}^2 \cup \{e,f\}^2$. We have $\tau \in Tol(U)$ and $H_\tau = \{\{a,b,c\},\{b,c,d\},\{e,f\}\}$. Therefore, $A_\tau = \{\{a,b,c\},\{b,c,d\},\{e,f\},\{b,c\}\}$ as well as $Def(G_\tau(U)) = A_\tau \cup \{\{a,b,c,d\}, \{a,b,c,e,f\}, \{b,c,d,e,f\}\{b,c,e,f\},\emptyset,U\}$. Let $\neg X := X \div \emptyset$ and $\circ X := U \div X$. For the sets $\{b\} \neq \{c\}$, we have $\neg\{b\} = T_\tau(\{b\}') = \{e,f\} = T_\tau(\{c\}') = \neg\{c\}$ and $\circ\{b\} = U = \circ\{c\}$. Therefore, Heyting algebra $\langle Def(G_\tau(U)), \cup, \cap, \rightarrow, \div, \emptyset, U \rangle$ is not regular.

Let us consider the set $\{a,f\}$. We have $\neg\{a,f\} = T_\tau(\{a,f\}') = T_\tau(\{b,c,d,e\}) = \{b,c,d\}$ and $\neg\neg\{a,f\} = \neg\{b,c,d\} = T_\tau(\{b,c,d\}') = T_\tau(\{a,e,f\}) = \{e,f\}$. Further, we obtain the following: $\{a,f\} \cup \neg\{a,f\} = \{a,b,c,d,f\} \neq U$ and $\neg\{a,f\} \cup \neg\neg\{a,f\} = \{b,c,d,e,f\} \neq U$. Therefore, Heyting algebra $\langle Def(G_\tau(U)), \cup, \cap, \rightarrow, \div, \emptyset, U \rangle$ does not satisfy the law of excluded middle and the weak law of excluded middle.

Let us now describe the structure of tolerance rough sets based on the family $G_\tau(U) := A_\tau$.

Theorem 11

Let (U, τ) be a tolerance space, and the family of rough sets:

$$\Re_\tau(U) := \{(T_\tau(X), T^\tau(X)) : X \subseteq U\},$$

be equipped with the operations $\wedge, \vee$ defined in the following way:

$$(T_\tau(X), T^\tau(X)) \wedge (T_\tau(Y), T^\tau(Y)) := (T_\tau(X) \cap T_\tau(Y), T^\tau(X) \cap T^\tau(Y)),$$

$$(T_\tau(X), T^\tau(X)) \vee (T_\tau(Y), T^\tau(Y)) := (T_\tau(X) \cup T_\tau(Y), T^\tau(X) \cup T^\tau(Y)).$$

Then the structure $\langle \Re_\tau(U), \wedge, \vee, \rightarrow, \div, \emptyset, U \rangle$ is a complete atomic double Heyting algebra in which infinite lattice operations have the following form:

$$\bigwedge_{i \in I}(T_\tau(X_i), T^\tau(X_i)) = \left(\bigcap_{i \in I} T_\tau(X_i), \bigcap_{i \in I} T^\tau(X_i)\right),$$

$$\bigvee_{i \in I}(T_\tau(X_i), T^\tau(X_i)) = \left(\bigcup_{i \in I} T_\tau(X_i), \bigcup_{i \in I} T^\tau(X_i)\right),$$

where $\{(T_\tau(X_i), T^\tau(X_i))\}_{i \in I} \subseteq \Re_\tau(U)$ and operations $\rightarrow, \div$ are defined for any $(T_\tau(X), T^\tau(X))$, $(T_\tau(Y), T^\tau(Y)) \in \Re_R(U)$ as follows:

$$(T_\tau(X), T^\tau(X)) \rightarrow (T_\tau(Y), T^\tau(Y))$$
$$:= \left(\bigcup_{Z \in I(X,Y)} T_\tau(Z), \bigcup_{Z \in I(X,Y)} T^\tau(Z)\right),$$

$$(T_\tau(X), T^\tau(X)) \div (T_\tau(Y), T^\tau(Y))$$
$$:= \left(\bigcap_{Z \in I(X,Y)} T_\tau(Z), \bigcap_{Z \in D(X,Y)} T^\tau(Z)\right),$$

where

$I(X,Y) := \{Z \subseteq U : (T_\tau(X), T^\tau(X)) \wedge (T_\tau(Z), T^\tau(Z)) \leq (T_\tau(Y), T^\tau(Y))\}$

$D(X,Y) := \{Z \subseteq U : (T_\tau(X), T^\tau(X)) \leq (T_\tau(Y), T^\tau(Y)) \vee (T_\tau(Z), T^\tau(Z))\}$

Proof. Let (U, τ) be a tolerance space and let $\Re_\tau(U) := \{(T_\tau(X), T^\tau(X)) : X \subseteq U\}$. Notice that for any family $\{X_i\}_{i \in I} \subseteq \wp(U)$ the following inclusions hold: $\{T_\tau(X_i)\}_{i \in I} \subseteq Def(G_\tau(U))$ and $\{T^\tau(X_i)\}_{i \in I} \subseteq Cl(G_\tau(U))$. Since families $Def(G_\tau(U))$ and $Cl(G_\tau(U))$ are closed for arbitrary unions and intersections, algebras $\langle Def(G_\tau(U)), \cap, \cup, \varnothing, U \rangle$ and $\langle Cl(G_\tau(U)), \cap, \cup, \varnothing, U \rangle$ are complete atomic distributive lattices of sets that are also infinitely meet distributive and infinitely join distributive. So, the lattice DC = $\langle Def(G_\tau(U)) \times Cl(G_\tau(U)), \wedge, \vee, (\varnothing, \varnothing), (U, U) \rangle$ is also a complete atomic and distributive lattice that is infinitely meet distributive and infinitely join distributive. Note that $RS = \langle \Re_\tau(U), \wedge, \vee, (\varnothing, \varnothing), (U, U) \rangle$ is a complete sublattice of DC. Since lattice DC is complete, RS is complete too. So, $\langle \Re_\tau(U), \wedge, \vee, (\varnothing, \varnothing), (U, U) \rangle$ is a complete, atomic distributive lattice that is also infinitely meet distributive and infinitely join distributive, where the infinite operations have the following form:

$$\bigwedge_{i \in I}(T_\tau(X_i), T^\tau(X_i)) = \left(\bigcap_{i \in I} T_\tau(X_i), \bigcap_{i \in I} T^\tau(X_i) \right),$$

$$\bigvee_{i \in I}(T_\tau(X_i), T^\tau(X_i)) = \left(\bigcup_{i \in I} T_\tau(X_i), \bigcup_{i \in I} T^\tau(X_i) \right),$$

where $\{(T_\tau(X_i), T^\tau(X_i))\}_{i \in I} \subseteq \Re_\tau(U)$. Since $\langle \Re_\tau(U), \wedge, \vee, (\varnothing, \varnothing), (U, U) \rangle$ is infinitely join distributive, then $\bigwedge \{(T_\tau(Z), T^\tau(Z)) : (T_\tau(X), T^\tau(X)) \leq (T_\tau(Y), T^\tau(Y)) \vee (T_\tau(Z), T^\tau(Z))\}$ is the least element $B \in \Re_\tau(U)$ such that $(T_\tau(X), T^\tau(X)) \leq (T_\tau(Y), T^\tau(Y)) \vee B$. One can show that $(T_\tau(X), T^\tau(X)) \div (T_\tau(Y), T^\tau(Y)) =$

$$= \left(\bigcap_{Z \in I(X,Y)} T_\tau(Z), \bigcap_{Z \in I(X,Y)} T^\tau(Z) \right),$$

$$= \bigwedge_{Z \in D(X,Y)} (T_\tau(Z), T^\tau(Z)),$$

$$= \bigwedge \{(T_\tau(Z), T^\tau(Z)) : (T_\tau(X), T^\tau(X)) \leq (T_\tau(Y), T^\tau(Y)) \vee (T_\tau(Z), T^\tau(Z))\}.$$

Consequently, $(T_\tau(X), T^\tau(X)) \div (T_\tau(Y), T^\tau(Y))$ is the least element $B \in \Re_\tau(U)$ that satisfies $(T_\tau(X), T^\tau(X)) \leq (T_\tau(Y), T^\tau(Y)) \vee B$. By Lemma 2, the structure $\langle \Re_\tau(U), \wedge, \vee, \div, (\varnothing, \varnothing), (U, U) \rangle$ is a Brouwerian algebra.

Since $\langle \Re_\tau(U), \wedge, \vee, (\varnothing, \varnothing), (U, U) \rangle$ is infinitely meet distributive as a lattice, then the element $\bigvee \{(T_\tau(Z), T^\tau(Z)) : (T_\tau(X), T^\tau(X)) \wedge (T_\tau(Z), T^\tau(Z)) \leq (T_\tau(Y), T^\tau(Y))\}$ is the greatest element $B \in \Re_\tau(U)$ such that $(T_\tau(X), T^\tau(X)) \wedge B \leq (T_\tau(Y), T^\tau(Y))$. So, using Lemma 1, one can show analogically that the structure $\langle \Re_\tau(U), \wedge, \vee, \rightarrow, (\varnothing, \varnothing), (U, U) \rangle$ is a Heyting algebra. Therefore, the structure $\langle \Re_\tau(U), \wedge, \vee, \rightarrow, \div, (\varnothing, \varnothing), (U, U) \rangle$ is a double Heyting algebra and, consequently, it is a complete atomic double Heyting algebra.

The final question is of the same kind as the one after Theorem 10, but now stated for Heyting algebras for rough sets, instead of definable sets. The question of regularity of these algebras we left open, while in the case of the weak law of excluded middle, the answer is negative:

Example 10

Let $U := \{a, b, c, d, e, f\}$. Let $\tau := \Delta_U \cup \{a,b\}^2 \cup \{b,c,d,e\}^2 \cup \{e,f\}^2$. We have $\tau \in Tol(U)$, $H_\tau = \{\{a,b\}, \{b,c,d,e\}, \{e,f\}\}$ and $A_\tau = \{\{a,b\}, \{b\}, \{b,c,d,e\}, \{e,f\}\}$. Further, we have: (See Box 2).

Let us define $\neg(T_\tau(X), T^\tau(X))$ as $(T_\tau(X), T^\tau(X)) \rightarrow (\varnothing, \varnothing)$ for any $X \subseteq U$. Let us consider $B := (T_\tau(\{b\}), T^\tau(\{b\})) = (\{b\}, \{a,b,c\})$. Then one can easily check that:

Box 2.

$T_\tau(\{b\}) = \{b\}$	$T^\tau(\{b\}) = \{a,b,c\}$	$(T_\tau(\{b\}), T^\tau(\{b\})) = (\{b\}, \{a,b,c\})$
$T_\tau(\{e\}) = \{e\}$	$T^\tau(\{e\}) = \{e\}$	$(T_\tau(\{e\}), T^\tau(\{e\})) = (\varnothing, \{e\})$
$T_\tau(\{a,b\}) = \{a,b\}$	$T^\tau(\{a,b\}) = \{a,b,c\}$	$(T_\tau(\{a,b\}), T^\tau(\{a,b\})) = (\{a,b\}, \{a,b,c\})$

$\neg B = (T_\tau(X), T^\tau(X)) \to (\emptyset, \emptyset)$
$= (\emptyset, \{e\}) = (T_\tau(\{e\}), T^\tau(\{e\}))$,

$\neg\neg B = (\emptyset, \{e\}) \to (\emptyset, \emptyset)$
$= (\{a,b\}, \{a,b,c\}) = (T_\tau(\{a,b\}), T^\tau(\{a,b\}))$.

Thus:

$B \vee \neg B = (\{b\}, \{a,b,c\}) \vee (\emptyset, \{e\})$
$= (\{b\} \cup \emptyset, \{a,b,c\} \cup \{e\})$
$= (\{b\}, \{a,b,c,e\}) \neq (U, U)$

$\neg B \vee \neg\neg B = (\emptyset, \{e\}) \vee (\{a,b\}, \{a,b,c\})$
$= (\emptyset \cup \{a,b\}, \{e\} \cup \{a,b,c\})$
$= (\{a,b\}, \{a,b,c,e\}) \neq (U, U)$

Hence, the considered double Heyting algebra of rough sets $\langle \Re_\tau(U), \wedge, \vee, \to, \div, (\emptyset, \emptyset), (U, U) \rangle$ violates the law of excluded middle as well as the weak law of excluded middle.

CONCLUSION

Information can be represented by information systems. These systems can be analyzed by means of different information relations. The choice of a given information relation usually depends on the goal of the analysis. According to Pawlak's approach to knowledge as ability to discern between objects in the universe of discourse, information relations can be regarded as representations of knowledge. Information relations can be analyzed from the point of view of their formal properties. It is in fact the analysis of knowledge represented by a given information relation. Some of the properties of information systems follow from the definitions of relations, while some others may be determined by the contents of information systems. For example, all similarity relations are tolerances by their definitions, while some of them are transitive tolerances or even identities on the specific universes of given information systems. As a summary, some of the properties of knowledge depend on the goal of analysis (which was the reason for a choice of a given information relation), while some other properties depend on the contents of information represented within information systems.

In this chapter, we classified information systems with respect to such features as determinacy and completeness, which depend on the contents of information systems. From an abstract point of view, we can omit the origins of those features, that is, their definitions and families of attributes behind them. Instead, we can take into account their formal properties as relations on universes of given information systems. From this point of view, it is important, for instance, whether a given elation is transitive or not; however, the reasons for its transitivity are not essential. At that more abstract level, knowledge is represented by approximation spaces, as a base for constructing granules of knowledge. In this process, called granulation, the objects of a given approximation space are collected into different sets called granules. It is based on a given relation representing knowledge and granules can be interpreted as the sets consisting of objects being indiscernible or similar with respect to the given criteria.

In the case of transitive tolerances representing strong indiscernibility, knowledge granules are equivalence classes. In the case of tolerances in general, as we have shown, there are various ways of granulating the universe. The essential property of granules from an abstract point of view is exactness of knowledge: whether knowledge granules are mutually disjoint or not. In the case of tolerance relations, this property is equivalent to transitivity of relations, so it can be interpreted as a formal counterpart of exactness of knowledge. This is a basis for the proposed classification of types of knowledge, where the abstract granules are basic building blocks for constructing concepts—definable sets.

On the basis of knowledge granules, the operators enabling to approximate vague concepts are constructed. The problem how to treat vague concepts in the computer science was an inspiration for introducing the rough set theory. We presented algebraic foundations of the rough set theory from that perspective, with selected rough set-related algebraic structures arising in different types of knowledge. We examined them with respect to satisfying or violating the algebraic version of the *tertium non datur* (excluded middle) principle. We also pointed out that the weak law of excluded middle gives the reason for differentiating between two types of vagueness: weak and strong. We were especially interested in informational representation of abstract algebraic structures, to strengthen their correspondence to particular types of knowledge arising from information systems. From this perspective, one of the most interesting future research directions is to answer the question of characterization of algebras of rough sets that do not satisfy the weak law of excluded middle.

ACKNOWLEDGMENT

Research of the first author was supported by the Faculty of Psychology, Warsaw University, research grant no. BST 1069/16. The first author wishes to thank Yiyu Yao and the Staff of Department of Computer Science, University of Regina, for valuable discussions and providing good working conditions and friendship atmosphere during his visit in the academic year 2005/2006. Both authors wish to thank Andrzej Skowron for his valuable remarks.

REFERENCES

Alexandroff, P. (1937). Diskrete räume. *Matematičeskij Sbornik, 2,* 501-518.

Banerjee M., & Chakraborty, M. (1993). Rough algebra. *Bulletin of the Polish Academy of Sciences, Mathematics, 41*(4), 293-297.

Banerjee M., & Chakraborty, M. (1996). Rough sets through algebraic logic. *Fundamenta Informaticae, 28*(3-4), 211-221.

Banerjee M., & Chakraborty, M. (2003). Algebras from rough sets. In S. K. Pal, L. Polkowski, & A. Skowron, (Eds.), *Rough-neural computing: Techniques for computing with words* (pp.157-184). Heidelberg, Germany: Springer Verlag.

Bazan, J., Skowron, A., & Swiniarski, R., (2006). Rough sets and vague concepts approximations: From sample approximation to adaptative learning. *Transactions on Rough Set, V, Journal Subline, Lectures Notes in Computer Science, 4100* (pp. 39-63).

Birkhoff, G. (1940) *Lattice theory* (1st ed.). Providence, RI: Colloquium Publications XXV, American Mathematical Society.

Birkhoff, G. (1967) *Lattice theory* (3rd ed.). Providence, RI: Colloquium Publications XXV, American Mathematical Society.

Bonikowski, Z. (1994). Algebraic structures of rough sets. In *Rough sets, fuzzy sets and knowledge discovery (RSKD'93). Workshops in computing* (pp. 242-247). Berlin, London: Springer-Verlag and British Computer Society.

Burris, S., & Sankappanavar, H. P. (1981). *A course in universal algebra*. Berlin-Heidelberg, Germany: Springer-Verlag.

Cattaneo, G. (1997). Generalized rough sets. Preclusivity fuzzy-intuitionistic (BZ) lattices. *Studia Logica, 58,* 47-77.

Cattaneo, G. (1998). Abstract approximation spaces for rough theories. In L. Polkowski, & A. Skowron, (Eds.). *Rough sets in knowledge discovery* (pp. 59-98). Heidelberg: Physica-Verlag.

Chuchro, M. (1994). On rough sets in topological Boolean algebras. In W. Ziarko (Ed.), *Rough sets, fuzzy sets and knowledge discovery. Proceedings* of the International Workshop on Rough Sets and Knowledge Discovery (RSKD'93) (pp 157 - 160). Berlin, Heidelberg: Springer-Verlag.

Comer, S. (1991). An algebraic approach to the approximation of information. *Fundamenta Informaticae, 14*, 492-502.

Comer, S. (1995). Perfect extensions of regular double Stone algebras. *Algebra Universalis, 34*, 96-109.

Davey, B.A., & Priestley, H.A. (2001). *Introduction to lattices and order.* Cambridge, UK: Cambridge University Press.

Demri, S. (1999). A logic with relative knowledge operators. *Journal of Logic, Language and Information, 8*(2), 167-185.

Demri, S. (2000). The nondeterministic information logic NIL is PSPACE-complete. *Fundamenta Informaticae, 42*(3-4), 211-234.

Demri, S., & Orłowska, E. (1998). Informational representability of models for information logics. In E. Orłowska (Ed.). *Incomplete information: Rough set analysis* (pp. 383-409). Heidelberg, Germany, New York: Springer-Verlag.

Demri, S., & Orłowska, E. (2002). *Incomplete information: Structures, inference, complexity.* Berlin, Heidelberg, Germany: Springer-Verlag.

Dubois, D., Godo, L., Prade, H., & Esteva, F. (2005). An information-based discussion of vagueness. In H. Cohen, C. Lefebvre (Eds.), *Handbook of Categorization in cognitive science* (pp. 892-913). Elsevier.

Düntsch, I. (1997). A logic for rough sets. *Theoretical Computer Science, 179*, 427-436.

Frege, G. (1903). *Grundgesetzen der Arithmetik, 2.* Jena, Germany: Verlag von Herman Pohle.

Düntsch, I., & Gediga, G. (1997). Algebraic aspects of attribute dependencies in information systems. *Fundamenta Inforamticae, 29*, 119-133.

Düntsch, I., & Gediga, G. (2000). *Rough set data analysis: A road to non-invasive knowledge discovery.* Bangor, UK: Methodos Publishers.

Düntsch, I., & Orłowska, E. (2001). Boolean algebras arising from information systems. Proceedings of the Tarski Centenary Conference, Banach Centre, Warsaw, June 2001. *Annals of Pure and Applied Logic, 217*(1-3), 77-89.

Gehrke, M., & Walker, E. (1992). On the structure of rough sets. *Bulletin of the Polish Academy of Sciences, Mathematics, 40*, 235-245.

Gratzer, G. (1971). *Lattice theory: First concepts and distirbutive lattices.* San Francisco: W. H. Freeman and Company.

Gratzer, G. (1978). *General lattice theory.* Basel, Switzerland: Birkhauser.

Greco, A., Matarazzo, B., & Słowiński, R. (2002). Rough approximation by dominance relations. *International Journal of Intelligent Systems, 17*, 153-171.

Greco, S., Pawlak, Z., & Słowiński, R. (2002). Generalized decision algorithms, rough inference rules and flow graphs. In J.J. Alpigini, J.F. Peters, A. Skowron, N. Zhong (Eds.), *Rough sets and current trends in computing, 3rd International Conference (RSCTC'2002). Lectures Notes in Artificial Intelligence, 2475* (pp. 93-104). Berlin, Heildelberg, Germany: Springer-Verlag.

Grzymała-Busse, J. W. (1988). Knowledge acquisition under uncertainty - A rough set approach. *Journal of Intelligent and Robotic Systems, 1*(1), 3-16.

Grzymała-Busse, J.W., & Hu, M. (2001). A comparison of several approaches to missing attribute values in data mining. In W. Ziarko, & Y. Yao (Eds.), *Proceedings of the 2nd International Conference on Rough Sets and Current Trends in Computing (RSCTC'2000). Lectures Notes in Artificial Intelligence, 2005* (pp. 340-347). Heildelber, Germany: Springer-Verlag.

Grzymała-Busse, J. W., & Stefanowski, J. (1999). Two approaches to numerical attribute discretization for rule induction. In *Proceedings: The 5th International Conference of the Decision Sciences Institute (ICDCI'99)* (pp. 1377-1379). Athens, Greece.

Grzymała-Busse, J.W., & Stefanowski, J. (2001). Three discretization methods for rule induction. *International Journal of Intelligent Systems, 16*(1), 29-38.

Grzymała-Busse, J.W., Stefanowski, J., & Ziarko, W. (1996). Rough sets: Facts versus misconceptions. *Informatica, 20*, 455-465.

Hempel, C.G. (1939). Vagueness and logic. *Philosophy of Science, 6*, 163-180.

Inuiguchi, M., Hirano, S., & Tsumoto, S. (Eds.). (2003). *Rough set theory and granular computing.* Berlin, Heidelberg, Germany: Springer Verlag.

Iwiński, T. B. (1987). Algebraic approach to rough sets. *Bulletin of the Polish Academy of Sciences, Mathematics, 35*, 673-683.

Järvinen, J. (1999). *Knowledge representation and rough sets.* Doctoral dissertation. University of Turku: Turku Center for Computer Science.

Järvinen, J. (2006). *Lattice theory of rough sets.* Unpublished manuscript.

Katriniak, T. (1973). The structure of distributive double p-algebras. Regularity and Congruences. *Algebra Universalis, 3*, 238-246.

Keefe, R. (2000). *Theories of vagueness.* Cambridge, UK: Cambridge University Press.

Kondrad, E., Orłowska, E., & Pawlak, Z. (1981). *Knowledge representation systems.* Technical Report 433, Institute of Computer Science, Polish Academy of Sciences.

Koppelberg, S. (1989). General theory of Boolean algebras. In J.D. Monk & R. Bonett (Eds.), *Handbook of Boolean algebras.* Amsterdam: North Holland.

Lin, T.Y. (1998). Granular computing on binary relations. I: Data mining and neighborhood systems. In L. Polkowski & A. Skowron, (Eds.). *Rough sets in knowledge discovery* (pp. 107-121). Heidelberg: Physica-Verlag.

Lin, T.Y. (1998). Granular computing on binary relations. II: Rough sets representations and belief functions. In L. Polkowski & A. Skowron (Eds.), *Rough sets in knowledge discovery* (pp. 121-140). Heidelberg: Physica-Verlag.

Lin, T.Y. (1999). Granular computing: Fuzzy logic and rough sets. In L.A. Zadeh & J. Kacprzyk (Eds.), *Computing with words in information/intelligent systems* (pp.183-200). Berlin, Heidelberg, Germany: Springer-Verlag.

Lin, T.Y. (2006). A roadmap from rough set theory to granular computing. In G. Wang, J.F. Peters, A. Skowron, & Y.Y. Yao (Eds.), *Rough sets and knowledge technology. Proceedings of the First International Conference, RSKT2006, Chonging China. Lectures Notes in Computer Science, 4062* (pp. 33-41). Berlin, Heidelberg, Germany: Springer-Verlag.

Lin, T.Y., & Cercone, N. (Eds.). (1997). *Rough sets and data mining.* Dodrecht: Kluwer Academic Publisher.

Lipski, W. (1976). Informational systems with incomplete information. In *3rd International Symposium on Automata, Languages and Programming* (pp. 120-130). Edinburgh, Scotland.

McKinsey, J.C.C., & Tarski, A. (1946). On closed elements in closure algebras. *The Annals of Mathematics, 47*, 122-162.

Nakata, M., & Sakai, H. (2005). Rough sets handling missing values probabilistically interpreted. In D. Ślęzak G. Wang, M. S. Szczuka, I. Düntsch,

& Y.Yao (Eds.) *Rough sets, fuzzy sets, data mining, and granular computing, 10th International Conference, Regina, Canada (RSFDGrC 2005), Lectures Notes in Artificial Intelligence 3641* (pp. 325-334). Berlin, Heidelberg: Springer-Verlag.

Nguyen, H.S. (1998). Discretization problems for rough set methods. In L. Polkowski, & A. Skowron, (Eds.), *Rough sets and current trends in computing, 1st International Conference (RSCTC'98), Warsaw, Poland. Lectures Notes in Artificial Intelligence, 1424* (pp. 545-552). Berlin, Heidelberg, Germany: Springer-Verlag.

Obtułowicz, A. (1987). Rough sets and Heyting algebra valued sets. *Bulletin of the Polish Academy of Sciences, Mathematics, 35*(9-10), 667-671.

Orłowska, E. (1988). Representation of vague information. *Information Systems, 13*, 167-174.

Orłowska, E. (1989). Logic for reasoning about knowledge. *Zeitschrift fur Mathematische Logik und Grundlagen der Mathematik, 35*, 559-568.

Orłowska, E. (1993a). Reasoning with incomplete information: Rough set based information logics. In V. Algar, S. Bergler, & F.Q. Dong (Eds.), *Incompleteness and Uncertainty in Information Systems Workshop* (pp. 16-33). Berlin, Springer-Verlag.

Orłowska, E. (1993b). Rough semantics for non-classical logics. In W. Ziarko (Ed.), *2nd International Workshop on Rough Sets and Knowledge Discovery*, Banff, Canada (pp. 143-148).

Orłowska, E. (1995). Information algebras. In V. S. Alagar, & M. Nivat (Eds.), *Algebraic methodology and software technology, Lectures Notes in Computer Science, 936* (pp. 50-65).

Orłowska, E. (1997). Studying incompleteness of information: A class of information logics. In K. Kijania-Placek & J. Woleński (Eds.), *The Lvov-Warsaw School and contemporary philosophy* (pp. 303-320). Dordrecht: Kluwer Academic Publishers.

Orłowska, E. (Ed.). (1998a). *Incomplete information: Rough set analysis*. Heidelberg, Germany, New York: Springer-Verlag.

Orłowska, E. (1998b). Introduction: What you always wanted to know about rough sets. In E. Orłowska (Ed.), *Incomplete information: Rough set analysis* (pp. 1-20). Heidelberg, Germany, New York: Springer-Verlag.

Orłowska, E., & Pawlak, Z. (1984). Representation of nondeterministic information. *Theoretical Compter Science, 29*, 27-39.

Pagliani, P. (1996). Rough sets and Nelson algebras. *Fundamenta Inforamticae, 27*(2-3), 205-219.

Pagliani, P. (1998a). Rough set theory and logic-algebraic structures. In E. Orłowska (Ed.). *Incomplete information: Rough set analysis* (pp. 109-192). Heidelberg, Germany, New York: Springer-Verlag.

Pagliani, P. (1998b). Intrinsic co-Heyting boundaries and information incompleteness in rough set analysis. In L. Polkowski, & A. Skowron, (Eds.), *Rough sets and current trends in Computing, 1st International Conference (RSCTC'98)*, Warsaw, Poland. *Lectures Notes in Artificial Intelligence, 1424* (pp. 123-130). Berlin, Heidelberg, Germany: Springer-Verlag.

Pal, S.K., Polkowski, L., & Skowron, A.(Eds.), (2004). *Rough-neural computing: Techniques for computing with words*. Heidelberg, Germany, Springer Verlag.

Pawlak, Z. (1973). *Mathematical foundations of informational retrieval* (Research Report 101). Institute of Computer Science, Polish Academy of Sciences.

Pawlak, Z. (1981). Information systems: Theoretical foundations. *Information Systems, 6*, 205-218.

Pawlak, Z. (1982). Rough sets. *International Journal of Computer and Information Sciences, 11*, 341-356.

Pawlak, Z. (1991). *Rough sets. Theoretical aspects of reasoning about data*. Dodrecht: Kluwer Academic Publisher.

Pawlak, Z. (2003). Elementary rough set granules: toward a rough set processor. In S.K. Pal, L. Polkowski, & A. Skowron (Eds.), *Rough-neural computing: Techniques for computing with words* (pp. 5-13). Heidelberg, Germany, Springer Verlag.

Pawlak, Z. (2004). Some issues on rough sets. *Transactions on Rough Set, I, Journal Subline, Lectures Notes in Computer Science, 3100*, 1-58.

Peters, J.F. (2007). Near sets. Special theory about nearness of objects. *Fundamenta Informaticae, 75*, in press.

Pogonowski, J. (1981). *Tolerance spaces with applications to linguistics*. Poznań: Adam Mickiewicz University Press.

Polkowski, L. (2002). *Rough sets: Mathematical foundations*. Heidelberg, Germany: Physica-Verlag.

Polkowski, L., & Skowron, A. (1996). Rough mereology: A new paradigm for approximate reasoning. *International Journal of Approximate Reasoning, 15*, 333-365.

Polkowski, L., & Skowron, A. (Eds.). (1998a). *Rough Sets and Current Trends in Computing, 1st International Conference (RSCTC'98), Warsaw, Poland. Lectures Notes in Artificial Intelligence, 1424*. Berlin, Heidelberg, Germany: Springer-Verlag.

Polkowski, L., & Skowron, A. (Eds.). (1998b). *Rough sets in knowledge discovery*. Heidelberg: Physica-Verlag.

Polkowski, L., Skowron, A., & Żytkow, J. (1995). Rough foundations for rough sets. In T.Y. Lin, & A.M. Wildberger (Eds.), *Soft computing: Rough sets, fuzzy logic, neural networks, uncertainty management, knowledge discovery* (pp. 55-58). San Diego, CA: Simulation Councils, Inc.

Pomykała, J., & Pomykała, J.A. (1988). The Stone algebra of rough sets. *Bulletin of the Polish Academy of Sciences, Mathematics, 36*, 451-512.

Rasiowa, H., & Sikorski, R. (1970). *The mathematics of metamathematics*. Warsaw, Poland: Polish Scientific Publisher.

Rauszer, C. (1974). Semi-Boolean algebras and their applications to intuitionistic logic with dual operators. *Fundamenta Mathematicae, 83*, 219-249.

Russell, B. (1923). Vagueness. *The Australasian Journal of Psychology and Philosophy, 1*, 84-92.

Sakai, H., & Nakata, M. (2005). Discernibility functions and minimal rules in non-deterministic information systems. In D. Ślęzak G. Wang, M. S. Szczuka, I. Düntsch, & Y.Yao (Eds.), *Rough sets, fuzzy sets, data mining, and granular computing, 10th International Conference*, Regina, Canada *(RSFDGrC 2005). Lectures Notes in Artificial Intelligence, 3641* (pp. 254-264). Berlin, Heidelberg: Springer-Verlag.

Skowron, A. (2001). Toward intelligent systems: Calculi of information granules. *Bulletin of the International Rough Set Society, 5*, 9-30.

Skowron, A. (2003). Approximation spaces in rough neurocomputing. In M. Inuiguchi, S. Hirano, S. Tsumoto (Eds.), *Rough set theory and granular computing* (pp. 13-22). Berlin, Heidelberg, Germany: Springer Verlag.

Skowron, A. (2004). Vague concepts: A rough-set approach. In B. De Beats, R. De Caluwe, G. De Tre, J. Fodor, J. Kacprzyk, & S. Zadrożny (Eds.), *Data and knowledge engineering: Proceedings of*

EUROFUSE2004 (pp. 480-493). Warszawa, Poland: Akademicka Oficyna Wydawnicza EXIT.

Skowron, A. (2005). Rough sets and vague concepts. *Fundamenta Informaticae, 64,* 417-431.

Skowron, A., & Rauszer, C. (1991). The discernibility matrices and functions in information systems. In R. Słowiński (Ed.), *Intelligent decision support, handbook of applications and advances of the rough set theory* (pp. 331-362). Dordrecht, Holland: Kluwer Academic Publisher.

Skowron, A., & Stepaniuk, J. (1996). Tolerance approximation spaces. *Fundamenta Informaticae, 27,* 245-253.

Skowron, A., & Stepaniuk, J. (2003). Informational granules and rough-neural computing. In S. K. Pal, L. Polkowski, A. Skowron (Eds.), *Rough-neural computing: Techniques for computing with words* (pp. 43-84). Heidelberg, Germany, Springer Verlag.

Skowron, A., Stepaniuk, J., & Peters, J.F. (2003). Towards discovery of relevant patterns from parametrized schemes of information granule construction. In M. Inuiguchi, S. Hirano, & S. Tsumoto (Eds.), *Rough set theory and granular computing* (pp. 97-108). Berlin, Heidelberg, Germany: Springer Verlag.

Skowron, A., Stepaniuk, J., Peters, J.F., & Swiniarski, R. (2006). Calculi of approxiamtion spaces. *Fundamenta Informaticae, 72,* 363-378.

Ślęzak, D. (1999). Decomposition and synthesis of decision tables with respect to generalized decision functions. In S.K. Pal, & A. Skowron, (Eds.), *Rough fuzzy hybridization – A new trend in decision making.* Berlin, Heidelberg, Germany: Springer-Verlag.

Ślęzak, D. (2002). Approximate entropy reducts. *Fundamenta Informaticae, 53*(3-4), 365-390.

Ślęzak, D. (2003). Approximate Markov boundaries and Bayesian Networks: Rough set approach. In M. Inuiguchi, S. Hirano, & S. Tsumoto (Eds.), *Rough set theory and granular computing* (pp. 109-121). Berlin, Heidelberg, Germany: Springer Verlag.

Ślęzak, D. (2005). Rough sets and Bayes factor. *Transactions on Rough Set, III, Journal Subline, Lectures Notes in Computer Science, 3400,* 202-229.

Ślęzak, D., & Wasilewski, P. (in preparation). *Granular sets.* Unpublished manuscript.

Słowiński, R. (Ed.). (1992). *Intelligent decision support, Handbook of applications and advances of the rough set theory.* Dordrecht, Holland: Kluwer Academic Publisher.

Słowiński, R., & Stefanowski, J. (1989). Rough classification in incomplete information systems. *Mathematical and Computer Modelling, 12*(10-11), 1347-1357.

Słowiński, R., & Stefanowski, J. (1994). Handling various types of uncertainty in the rough set approach. In W. Ziarko (Ed.), *Rough Sets, Fuzzy Sets and Knowledge Discovery. Proceedings of the International Workshop on Rough Sets and Knowledge Discovery (RSKD'93)* (pp. 366-376). Berlin, Heidelberg: Springer-Verlag.

Słowiński, R., & Stefanowski, J. (1996). Rough set reasoning about uncertain data. *Fundamenta Informaticae, 27,* 229-243.

Słowiński, R., & Vanderpooten, D. (1997). Similarity relation as a basis for rough approximations. In P. P. Wang (Ed.), *Machine intelligence and soft computing, IV* (pp. 17-33). Raleigh, NC: Bookwrights.

Słowiński, R., & Vanderpooten, D. (2000). A generalized definition of rough approximations based on similarity. *IEEE Transactions on Data and Knowledge Engineering, 12*(2), 331-336.

Smith, N.J.J. (2005). Vagueness as closeness. *Australasian Journal of Philosophy, 83,* 157-183.

Stefanowski, J. (1998). Handling continuous attributes in discovery of strong decision rules. In L. Polkowski, & A. Skowron, (Eds.), *Rough Sets and Current Trends in Computing, 1st International Conference (RSCTC'98)*, Warsaw, Poland. *Lectures Notes in Artificial Intelligence, 1424* (pp. 394-401). Berlin, Heidelberg, Germany: Springer-Verlag.

Stefanowski, J., & Tsoukias, A. (2001). Incomplete information tables and rough classification. *Computational Intelligence: An International Journal, 17*(3), 545-566.

Stone, P. (1936). The theory of representation for Boolean algebras. *Transactions of the American Mathematical Society, 40*(1), 37-111.

Tarski, A. (1935). Zur Grundlegung der Booleschen Algebra, I. *Fundamenta Mathematicae, 24*, 177-198.

Vakarelov, D. (1989). Modal logics for knowledge representation systems. In A. R. Meyer, & M. Taitslin (Eds.), *Symposium on Logic Foundations of Computer Science, Pereslav-Zalessky* (pp. 257-277). *Lectures Notes in Computer Science, 363.* Berlin, Heidelberg, Germany: Springer-Verlag.

Vakarelov, D. (1991). A modal logic for similarity relations in Pawlak knowledge representation systems. *Fundamenta Informaticae, 15*, 61-79.

Vakarelov, D. (1998). Information systems, similarity and modal logics. In E. Orłowska (Ed.), *Incomplete information: Rough set analysis* (pp. 492-550). Heidelbelrg, Germany: Physica-Verlag.

Wasilewska, A. (1997). Topological rough algebras. In T. Y. Lin & N. Cercone (Eds.), *Rough sets and data mining* (pp. 411-425). Dodrecht: Kluwer Academic Publisher.

Wasilewski, P. (2004). *On selected similarity relations and their applications into cognitive science* (in Polish). Unpublished doctoral dissertation, Department of Logic, Jagiellonian University, Cracow.

Wasilewski, P. (2005). Concept lattices vs. approximation spaces. In D. Ślęzak G. Wang, M. S. Szczuka, I. Düntsch, & Y.Yao (Eds.), *Rough Sets, Fuzzy Sets, Data Mining, and Granular Computing, 10th International Conference, Regina, Canada (RSFDGrC 2005), Lectures Notes in Artificial Intelligence 3641* (pp. 114-123). Berlin, Heidelberg: Springer-Verlag.

Wasilewska, A., & Banerjee, M. (1995). Rough sets and topological quasi-Boolean algebras. In T. Y. Lin (Ed.), *Workshops on Rough Sets and Data Mining at 23rd Annual Computer Science Conference* (pp. 61-67). Nashville, TN.

Wasilewska, A., & Vigneron, L. (1995). Rough equality algebras. In P. P. Wang, (Ed.), *Proceedings of the International Workshop on Rough Sets and Soft Computing at 2nd Annual Joint Conference on Information Sciences (JCIS'95)* (pp. 26-30). Raleigh, NC.

Yao, Y.Y. (1998a). Constructive and algebraic methods of the theory of rough sets. *Information Sciences, 109*(1-4), 21-47.

Yao, Y.Y. (1998b). Relational interpretations of neighborhood operators and rough set approximation operator. *Information Sciences 111*(1-4), 239-259.

Yao, Y.Y. (2001). Information granulation and rough set approximation. *International Journal of Intelligent Systems, 16*(1), 87-104.

Yao, Y.Y. (2004). A partition model of granular computing. *Transactions on Rough Set, I, Journal Subline, Lectures Notes in Computer Science, 3100*, 232-253.

Ziarko, W. (1993). Variable precision rough set model. *Journal of Computer and Systems Sciences, 46*, 35-59.

Chapter II
Rough Sets and Boolean Reasoning

Hung Son Nguyen
Warsaw University, Poland

ABSTRACT

This chapter presents the Boolean reasoning approach to problem solving and its applications in rough sets. The Boolean reasoning approach has become a powerful tool for designing effective and accurate solutions for many problems in decision making, approximate reasoning, and optimization. In recent years, Boolean reasoning has become a recognized technique for developing many interesting concept approximation methods in rough set theory. This chapter presents a general framework for concept approximation by combining the classical Boolean reasoning method with many modern techniques in machine learning and data mining. This modified approach-called "the approximate Boolean reasoning" methodology-has been proposed as an even more powerful tool for problem solving in rough set theory and its applications in data mining. Through some most representative applications in many KDD problems, including feature selection, feature extraction, data preprocessing, classification of decision rules, and decision trees, association analysis, the author hopes to convince that the proposed approach not only maintains all the merits of its antecedent, but also owns the possibility of balancing between quality of the designed solution and its computational time.

INTRODUCTION

Concept approximation problem is one of the most important issues in machine learning and data mining. Classification, clustering, association analysis, and regression are examples of well-known problems in data mining that can be formulated as concept approximation problems. A great effort of many researchers has been done to design newer, faster, and more efficient methods for solving the concept approximation problem.

Rough set theory has been introduced by Pawlak (1991) as a tool for concept approximation under uncertainty. The idea is to approximate the concept by two descriptive sets called *lower and upper approximations*. The lower and upper approximations must be extracted from available training data. The main philosophy of rough set approach to concept approximation problem is based on minimizing the difference between upper and lower approximations (also called the *boundary region*). This simple, but brilliant idea, leads to many efficient applications of rough sets in machine learning and data mining like feature selection, rule induction, discretization, or classifier construction (Skowron, Pawlak, Komorowski, & Polkowski, 2002).

The problem considered in this chapter is the creation of a general framework for concept approximation. The need for such a general framework arises in machine learning and data mining. This chapter presents a solution to this problem by introducing a general framework for concept approximation that combines rough set theory, Boolean reasoning methodology, and data mining. This general framework for approximate reasoning is called *rough sets and approximate Boolean reasoning* (RSABR). The contribution of this chapter is the presentation of the theoretical foundation of RSABR, as well as its application in solving many data mining problems and knowledge discovery in databases (KDD) such as feature selection, feature extraction, data preprocessing, classification of decision rules and decision trees, and association analysis.

As Boolean algebra has a fundamental role in computer science, the Boolean reasoning approach is also an ideological method in artificial intelligence. In recent years, Boolean reasoning approach shows to be a powerful tool for designing effective and accurate solutions for many problems in rough set theory.

The chapter is organized as follows. Section 2 introduces the basic notions about Boolean algebras, Boolean functions, and the main principle of the Boolean reasoning methodology to problem solving. Section 3 presents some fundamental applications of ABR approach in rough set theory including reduct calculation, decision rule induction, and discretization. Section 4 extends section 3 with some applications of ABR in data mining.

BOOLEAN REASONING APPROACH TO PROBLEM SOLVING

The main subject of this section is related to the notion of *Boolean functions*. We consider two equivalent representations of Boolean functions, namely *the truth table form*, and *the Boolean expressions form*. The latter representation is derived from George Boole's formalism (1854), which eventually became *Boolean algebra* (Boole, 1854). We also discuss some special classes of Boolean expressions that are useful in practical applications.

Boolean Algebra

Boolean algebra was an attempt to use algebraic techniques to deal with expressions in the propositional calculus. Today, these algebras find many applications in electronic design. They were first applied to switching by Claude Shannon in the 20th century (Shannon, 1938, 1940). Boolean algebra is also a convenient notation for representing Boolean functions.

Boolean algebras are algebraic structures that "capture the essence" of the logical operations AND, OR, and NOT, as well as the corresponding set-theoretic operations intersection, union, and complement. As Huntington recognized, there are various equivalent ways of characterizing Boolean algebras (Huntington, 1933). One of the most convenient definitions is the following.

Definition 2.1. (*Boolean algebra*) *A tuple* $B = (B, +, \cdot, 0, 1)$, *where B is a nonempty set, + and · are binary operations, 0 and 1 are distinct elements of B, is called the Boolean algebra if the following axioms hold:* (see Box 1), *for any elements* $a, b, c \in B$.

Box. 1

Commutative laws:	$a + b = b + a$	$a \cdot b = b \cdot a$
Distributive laws:	$(a \cdot b + c) = a \cdot b + a \cdot c$	$a + (b \cdot c) = (a + b) \cdot (a + c)$
Identity elements: 0	$a + 0 = a$	$a \cdot 1 = a$
Complements: $\forall_{a \in B} \exists_{\bar{a} \in B}$	$a + \bar{a} = 1$	$a \cdot \bar{a} = 0$

The operations "+" (*Boolean "addition"*), "·" (*Boolean "multiplication"*), and " ¯ " (*Boolean complementation*) are known as *Boolean operations*. The set B is called *the universe* or *the carrier*. The elements 0 and 1 are called the *zero and unit elements* of B, respectively. Any algebraic equality derived from the axioms of Boolean algebra remains true when the operations + and · are interchanged and the identity elements 0 and 1 are interchanged. Because of the duality principle, for any given theorem we immediately get also its dual.

A Boolean algebra is called *finite* if its universe is a finite set. Although Boolean algebras are quintuples, it is customary to refer to a Boolean algebra by its carrier.

The structure of Boolean algebras is characterized by the well-known Stone representation theorem. It has been shown in Selman (Selman, Kautz, &. McAllester, 1997) that every Boolean algebra is isomorphic to the algebra of clopen (i.e., simultaneously closed and open) subsets of its Stone space. Due to the properties of Stone space for finite algebras, this result means that every finite Boolean algebra is isomorphic to the Boolean algebra of subsets of some finite set S.

Example: The following structures are most popular Boolean algebras:

1. Two-value (or binary) Boolean algebra $B_2 = (\{0,1\}, +, \cdot, 0, 1)$ is the smallest, but the most important, model of general Boolean algebra. It has only two elements, 0 and 1. The binary operations + and · are defined by $x + y = \max\{x, y\}$ and $x \cdot y = \min\{x, y\}$, respectively. In the binary Boolean algebra, 0 is the complement of 1 and 1 is the complement of 0.
2. The power set of any given set S forms a Boolean algebra with the two operations $+ := \cup$ (union) and $\cdot := \cap$ (intersection). The smallest element 0 is the empty set and the largest element 1 is the set S itself.
3. The set of all subsets of S that are either finite or cofinite is a Boolean algebra.

An n-variable mapping $f : B^n \to B$ is called *a Boolean function* if and only if it can be expressed by a Boolean expression, that is, an expression involving constants, variables, Boolean operations, and corresponding parentheses. Formally, Boolean expressions can be defined inductively, starting with constants, variables, and using three elementary operations as building blocks. Thus, for an arbitrary Boolean algebra B, not every mapping $f : B^n \to B$ is a Boolean function.

Boolean Functions in Binary Boolean Algebra

In this chapter we concentrate on the binary Boolean algebra only.

Binary Boolean algebra has some specific properties that do not necessarily satisfy for general Boolean algebras. For example, in the binary Boolean algebra, if $xy=0$ then either $x=0$ or $y=0$.

Moreover, any mapping, $f : \{0,1\}^n \to \{0,1\}$ is called an n-variable *switching function*. One can show that every switching function is a Boolean function, and the number of n-variable switching functions is equal to 2^{2^n}.

It is important to understand that every Boolean function can be represented by numerous Boolean expressions, whereas every Boolean expression represents a unique function. As a matter of fact, for the binary Boolean algebra B_2, the number of n-variables Boolean functions is bounded by 2^{2^n}, whereas the number of n-variable Boolean expressions is infinite. These remarks motivate the distinction that we draw between functions and expressions. We say that two Boolean expressions ϕ and ψ are semantically equivalent if they represent the same Boolean function. When this is the case, we write $\phi = \psi$.

Binary Boolean algebra plays a crucial role for the verification problem, due to the fact that any Boolean function is uniquely determined by its 0, 1 assignments of variables (see the *Löwenheim-Müller Verification Theorem,* Brown,1990; Rudeanu ,1974). Therefore, any identity can be verified by all 0, 1 substitutions of arguments.

Let us mention some more well-known identities that are useful for further consideration (Brown, 1990). Thus, for any elements x, y, z of an arbitrary Boolean algebra we have: (see Box 2).

Binary Boolean algebra also has applications in propositional calculus, interpreting 0 as false, 1 as true, + as the logical OR (disjunction), · as the logical AND (conjunction), and $^-$ as the logical NOT (complementation, negation).

Some Classes of Boolean Expressions, Normal Forms

Let us remember that Boolean expressions are formed by *letters*, that is, constants and variables using Boolean operations like conjunction, disjunction, and complementation.

Expressions in binary Boolean algebras are quite specific, because almost all of them are constant free (except the two constant expressions 0 and 1). Some common subclasses of Boolean expressions are as follow:

- A *term* is either 1 (the unit element), a single literal (a letter or its complement), or a conjunction of literals in which no letter appears more than once.
- A *clause* is either 0 (the zero element), a single literal, or a conjunction of literals in which no letter appears more than once.
- A Boolean expression is said to be in *disjunctive normal form* (DNF) if it is a disjunction of terms.

Box. 2

Associative laws:	$(x + y) + z = x + (y \cdot z)$ and $(x \cdot y) \cdot z = x \cdot (y \cdot z)$
Idempotence:	$x + x = x$ and $x \cdot x = x$
Prop. of 0 and 1:	$x + 1 = 1$ and $x \cdot 0 = 0$
Absorption laws:	$(y \cdot x) + x = x$ and $(y + x) \cdot x = x$
Involution:	$\overline{\overline{(x)}} = x$
DeMorgan laws:	$\overline{(x + y)} = \overline{x} \cdot \overline{y}$ and $\overline{(x \cdot y)} = \overline{x} + \overline{y}$
Consensus laws:	$(x + y) \cdot (\overline{x} + z) \cdot (y + z) = (x + y) \cdot (\overline{x} + z)$
	and $(x \cdot y) + (\overline{x} \cdot z) + (y \cdot z) = (x \cdot y) + (\overline{x} \cdot z)$

- DNF has a dual *conjunctive normal form* (CNF) that is a conjunction of clauses.

A *monomial* is a Boolean function that can be expressed by a term. It is easy to show that there are exactly 3^n possible terms over n variables. There are also 3^n possible clauses. If f can be represented by a term, then (by De Morgan laws) $\bar{f}$ can be represented by a clause, and vice versa. Thus, terms and clauses are dual of each other. From psychological experiments it follows that CNF expressions seem easier for humans to learn than DNF expressions.

Any Boolean function can be represented in both CNF and DNF. For our convenience, let us introduce the notation $x^0 = \bar{x}$ and $x^1 = x$ for any Boolean variable x. For any sequence $\mathbf{b} = (b_1,...,b_n) \in \{0,1\}^n$ and any vector of Boolean variables $\mathbf{X} = (x_1,...,x_n)$, we define the *minterm* and *the maxterm* of $\mathbf{X}$ by:

$$m_\mathbf{b}(\mathbf{X}) = X^\mathbf{b} = x_1^{b_1} x_2^{b_2} ... x_n^{b_n};$$

$$s_\mathbf{b}(\mathbf{X}) = \overline{m_\mathbf{b}(\mathbf{X})} = x_1^{\bar{b_1}} + x_2^{\bar{b_2}} + ... + x_n^{\bar{b_n}}.$$

From the Shannon's expansion theorem (49) one can show that any switching function can be represented by:

Minterm canonical form $f(\mathbf{X}) = \sum_{\mathbf{b} \in f^{-1}(1)} m_\mathbf{b}(\mathbf{X})$ (1)

Maxterm canonical form $f(\mathbf{X}) = \prod_{\mathbf{a} \in f^{-1}(0)} s_\mathbf{a}(\mathbf{X})$ (2)

Implicants and Prime Implicants

Given a function f and a term t, we define the *quotient* of f with respect to t, denoted by f/t, to be the function formed from f by imposing the constraint $t = 1$. For example, let f be a Boolean function given by $f(x_1, x_2, x_3, x_4) = \bar{x_1} x_2 x_4 + x_2 \bar{x_3} x_4 + x_1 \bar{x_2} x_4$.

The quotient of f with respect to $x_1 \bar{x_3}$ is $f/x_1 \bar{x_3} = f(1, x_2, 0, x_4) = x_2 x_4 + \bar{x_2} x_4$.

It is clear that the function f/t can be represented by a formula that does not involve any variable appearing in t. Let us define two basic notions in Boolean function theory called *implicant* and *prime implicant*.

Definition 2.2. A term t is *an implicant* of a function f if $f/t = 1$. An implicant t is *a prime implicant* of f if the term t' formed by taking any literal out of t is no longer an implicant of f (the prime implicant cannot be "divided" by any term and remain an implicant).

Let us observe that each term in a DNF expression of a function is an implicant because it "implies" the function (if the term has value 1, so does the DNF expression).

In a general Boolean algebra, for two Boolean functions h and g we write $h \ll g$ if and only if the identity $h\bar{g} = 0$ is satisfied. This property can be verified by checking whether $h(\mathbf{X})\overline{g(\mathbf{X})} = 0$ for any zero-one vector $\mathbf{X} = (\alpha_1,...,\alpha_n) \in \{0,1\}^n$. A term t is *an implicant* of a function f if and only if $t \ll f$.

Thus, both $x_2 \bar{x_3}$ and $x_1 \bar{x_3}$ are prime implicants of $f = x_2 \bar{x_3} + \bar{x_1} \bar{x_3} + x_2 x_1 \bar{x_3} + x_1 x_2 x_3$, but $x_2 x_1 \bar{x_3}$ is not.

Monotone Boolean Function

A Boolean function $\phi: \{0,1\}^n \to \{0,1\}$ is called "*monotone*" if $\forall_{\mathbf{x},\mathbf{y} \in \{0,1\}^n} (\mathbf{x} \le \mathbf{y}) \Rightarrow (\phi(\mathbf{x}) \le \phi(\mathbf{y}))$.

It has been shown that monotone functions can be represented by a Boolean expression without negations. Thus, a *monotone expression* is an expression without negation.

A monotone formula ϕ in disjunctive normal form is *irredundant* if and only if no term of ϕ covers another term of ϕ. For a monotone formula, the disjunction of all its prime implicants yields an equivalent monotone DNF. On the other hand, every prime implicant must appear in every equivalent DNF for a monotone formula. Hence, the smallest DNF for a monotone formula is unique and equals the disjunction of all its prime implicants. This is not the case for nonmonotone formulas, where the smallest DNF is a subset of the set of all prime implicants.

APPROXIMATE BOOLEAN REASONING METHODOLOGY

Boolean reasoning approach is a general framework for solving many complex decisions or optimization problems. The standard Boolean reasoning approach, called syllogistic reasoning, was described by Brown (1990). SAT planning, the planning method proposed by Henry Kautz and Bart Selman (1992, 1996), is one of the most famous applications of Boolean reasoning methodology.

In this section, we present a general framework of approximate Boolean reasoning (ABR) methodology that is an extension of the standard Boolean reasoning approach.

The greatest idea of Boole's algebraic approach to logic was to reduce the process of reasoning to the process of calculation. In Boolean algebras, a system of logical equations can be transformed to a single equivalent Boolean equation.

Boole and other 19th century logicians based their symbolic reasoning on an equation of 0-normal form, that is, $f(x_1, x_2, ..., x_n) = 0$. Blake (1937) showed that the consequents of this equation are directly derived from the prime implicants of f. Thus, the representation of f as a disjunction of all its prime implicants is called the *Blake Canonical Form* of a Boolean function f and denoted by $BCF(f)$, that is, $BCF(f) = t_1 + t_2 + ... + t_k$, where $\{t_1, ..., t_k\}$ is the collection of all prime implicants of the function f. This observation enables one to develop an interesting Boolean reasoning method called *syllogistic reasoning* that extracts conclusions from a collection of Boolean data.

Quine (1952, 1959, 1961) also appreciated the importance of the concept of prime implicants in his research related to the problem of minimizing the complexity of Boolean formulas.

The basic idea of the Boolean reasoning methodology is to represent constraints and facts of the considered problem in form of a collection of Boolean equations. By this way, the problem can be encoded by a single Boolean equation of form $f(x_1, x_2, ..., x_n) = 0$ (or dually $f = 1$), where f is assumed to be in the clausal form. In the next step, depending on the formulation of the problem, the set of all or some prime implicants of f is generated and applied to solve the problem. Let us recall some representative examples of the Boolean reasoning approach to problem solving:

- The most famous problem related to Boolean functions, so called the satisfiability problem (SAT), is to check whether there exists such an evaluation of variables that a given Boolean function becomes satisfied. In other words, the problem is to solve the equation $f(x_1, ..., x_n) = 1$. SAT is the first problem that has been proved to be NP complete (the Cook's theorem). This important result is used to prove the NP hardness of many other problems by showing the polynomial transformation of SAT to the studied problem.

  From a practical point of view, any SAT-solver (heuristical algorithm for SAT) can be used to design heuristic solutions for problems in the class NP by transformation of the considered problem to a Boolean formula. Therefore, instead of solving a couple of hard problems, the main effort is limited to create efficient heuristics for the SAT problem.

- One of the possible solutions for scheduling problem is based on SAT solver. In this method, the specification of scheduling problem is formulated by a Boolean function, where each variable encodes one possible assignment of tasks, resources, time slots, and so forth. The encoding function is satisfiable if and only if there exists a correct schedule for the given specification (Selman, 1997).

The general scheme of the approximate Boolean reasoning approach is as follows:

1. **Modeling:** Represent the problem Π (or its simplification) by a collection of Boolean equations.
2. **Reduction:** Condense the equations into an equivalent problem encoded by a single Boolean equation of the form $f_\Pi(x_1,...,x_n) = 1$ or a simplified problem encoded by an equation $f'_\Pi = 1$.
3. **Development:** Generate an approximate solution of the formulated problem over f_Π (or f'_Π).
4. **Reasoning:** Apply an approximate reasoning method to solve the original problem.

Analogically to symbolic approaches in other algebras, Step 1 is performed by introducing some variables, and describing the problem in the language of Boolean algebra. After that, obtained description of the problem is converted into Boolean equations. Steps 2 and 3 are independent of the problem to be solved and are more or less automated. In Step 2, three types of problems over Boolean equation can be considered:

1. Search for all solutions (all prime implicants) of f.
2. SAT: check whether any solution of $f = 1$ exists.
3. MINPRIME$_{mon}$: search for the minimal prime implicant of a monotone Boolean function f.

The mentioned problems have the high computational complexity, as all of them are NP hard. In this chapter, we concentrate on the MINPRIME$_{mon}$ problem because it is useful for solving optimization problems. Since most of the problems in data mining can be formulated as optimization problems, we will show in the next sections some applications of the Boolean reasoning approach to

Figure 1. General scheme of approximate Boolean reasoning approach

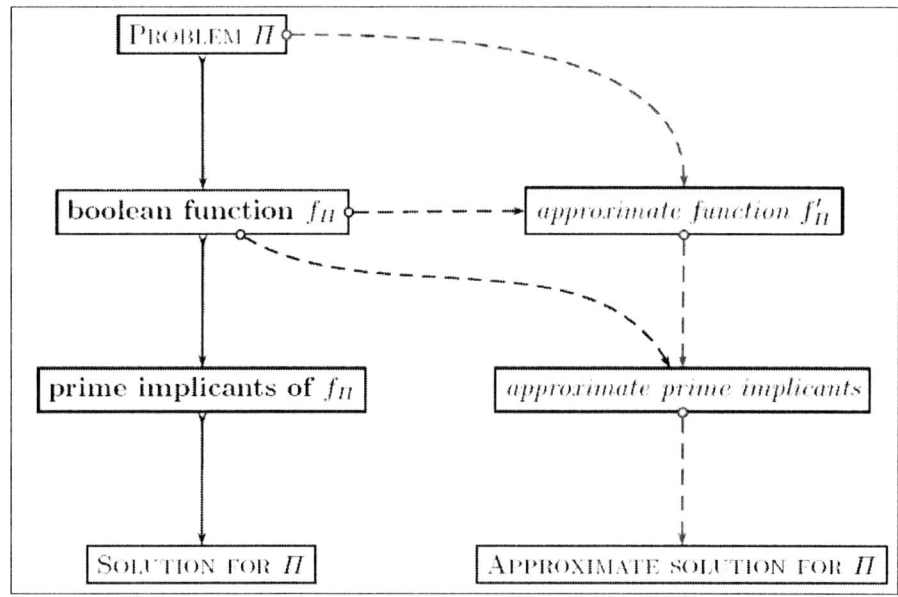

data mining, where the minimal prime implicants play a crucial role.

The investigation in MINPRIME$_{mon}$ is very important not only for application of the Boolean reasoning approach to optimization problems. For example, the mentioned SAT problem can be also transformed to MINPRIME$_{mon}$ as follows:

Let ϕ be a Boolean formula in negation normal form[1] (NNF) with n variables $x_1,...,x_n$. Let $r(\phi)$ denote the formula obtained by replacing all appearances of $\overline{x_i}$ in ϕ by the new variable y_i (for $i = 1,2,...,n$), and let $c(\phi)$ denote the conjunction $\prod_{i=1}^{n}(x_i + y_i)$.

One can show that can ϕ is satisfied if and only if monotone Boolean formula $r(\phi) \cdot c(\phi)$ has a prime implicant consisting of at most n variables. The main idea is to prove that a vector $\mathbf{a} = (a_1,...,a_n) \in \{0,1\}^n$ is a satisfied evaluation for ϕ (i.e., $f_\phi(\mathbf{a}) = 1$) if and only if the term:

$$t_\mathbf{a} = \prod_{a_i=1} x_i \cdot \prod_{a_i=0} y_i$$

is a prime implicant of $r(\phi) \cdot c(\phi)$.

Therefore, every heuristic algorithm A for MINPRIMI$_{mon}$ problem can be used to solve (in approximate way) the SAT problem for an arbitrary formula ϕ.

Since the minimal prime implicant problem is NP hard, it cannot be solved (in general case) by exact methods only. It is necessary to create some heuristics to search for short prime implicants of large and complicated Boolean functions.

The high computational complexity of prime implicant problems means that the standard Boolean reasoning approach is not applicable in many real-world problems, particularly in data mining, where a large amount of data is one of the major challenges.

The natural approach to managing hard problems is to search for an approximate instead of an exact or optimal solution. The first attempt might be related to calculation of prime implicants, as it is the most complex step in the Boolean reasoning schema. In the next section, we describe some well-known approximate algorithms for prime implicant problems.

Each approximate method is characterized by two parameters: the quality of approximation and the computation time. Searching for the proper balance between those parameters is the biggest challenge of modern heuristics. We have proposed a novel method, called the *approximate Boolean reasoning method*, to extend this idea. In the approximate Boolean reasoning approach to problem solving, not only calculation of prime implicants, but every step in the original scheme (Figure 1) is approximately performed to achieve an approximate solution.

Searching for prime implicants of monotone functions.

In case of minimal prime implicant for monotone functions, the input Boolean function is assumed to be given in the CNF form, that is, it is presented as a conjunction of clauses, for example:

$$\psi = (x_1 + x_2 + x_3)(x_2 + x_4)$$
$$(x_1 + x_3 + x_5)(x_1 + x_5)(x_5 + x_6)(x_1 + x_2) \quad (3)$$

Searching for the minimal prime implicant can be treated as the minimal hitting set problem, that is, the problem of searching for minimal set of variables X such that for every clause of the given formula, at least one of its variables must occur in X.

Let us recall that every monotone Boolean function can be expanded by a variable x_i as follows:

$$\phi(x_1,...,x_n) = x_i \cdot \phi/x_i + \phi/\overline{x_i}$$
$$= x_i\phi(x_1,...,x_{i-1},1,x_{i+1},...,x_n)$$
$$+ (x_1,...,x_{i-1},0,x_{i+1},...,x_n)$$

The basic problem in many existing heuristics for minimal prime implicant is to evaluate the chance that the variable x_i belongs to the minimal prime implicant of ϕ. We can do that by defining an evaluation function $Eval(x_i;\phi)$ that takes under consideration two formulas ϕ/x_i and $\phi/\overline{x_i}$, that is:

$$Eval(x_i;\phi) = F(\phi/x_i, \phi/\overline{x_i}) \qquad (4)$$

The algorithm should decide either to select the best variable x_{best} and continue with ϕ/x_{best} or to remove the worst variable x_{worst} and continue with $\phi/\overline{x_{worst}}$.

One can apply an idea of DPLL algorithms to solve the minimal prime implicant for monotone formulas. The algorithm of searching for minimal prime implicant starts from an empty term t, and in each step, it might choose one of the following actions:

- **Unit clause:** If $\phi/t\overline{x_i}$ degenerates for some variable x_i, that is, $\phi/t\overline{x_i} = 0$, then x_i must occur in every prime implicant of ϕ/t. Such a variable is called *the core variable*. The core variable can be quickly recognized by checking whether there exists a unit clause, that is, a clause that consists of one variable only. If x_i is a core variable, then the algorithm should continue with ϕ/tx_i.
- **Final step:** If there exists variable x_i such that ϕ/tx_i degenerates, then x_i is the minimal prime implicant of ϕ/t, the algorithm should return tx_i as a result and stop here;
- **Heuristic decision:** If none of previous rules cannot be performed, the algorithm should use the evaluation function (equation (4)) to decide how to continue the searching process. The decision is related to adding a variable x_i to t and continuing the search with formula ϕ/tx_i or rejecting a variable x_j and continuing the search with formula $\phi/t\overline{x_j}$.

Let us mention some most popular heuristics that have been proposed for minimal prime implicant problem for monotone Boolean functions:

1. **Greedy algorithm:** This simple method (see Garey & Johnson, 1979) is using the number of unsatisfied clauses as a heuristic function. In each step, the greedy method selects the variable that most frequently occurs within clauses of the given function and removes all those clauses that contain the selected variable. For the function in equation (3), x_1 is the most preferable variable by the greedy algorithm. The result of greedy algorithm for this function might be $x_1 x_4 x_6$, while the minimal prime implicant is $x_2 x_5$.

2. **Linear programming:** The minimal prime implicant can also be resolved by converting the given function into a system of linear inequations and applying the integer linear programming (ILP) approach to this system, (see Manquinho, Flores, Silva, &. Oliveira, 1997; Pizzuti, 1996).

    Assume that an input monotone Boolean formula is given in CNF. The idea is to associate with each Boolean variable x_i an integer variable t_i. Each monotone clause $x_{i_1} + ... + x_{i_k}$ is replaced by an equivalent inequality: $t_{i_1} + ... + t_{i_k} \geq 1$ and the whole CNF formula is replaced by a set of inequalities $\mathbf{A} \cdot \mathbf{t} \geq \mathbf{b}$. The problem is to minimize the number of variables with the value one assigned. The resulting ILP model is as follows:

    $\min(t_1 + t_2 + ... + t_n)$
    $s.t. \quad \mathbf{A} \cdot \mathbf{t} \geq \mathbf{b}$

3. **Simulated annealing:** many optimization problems are resolved by a Monte Carlo search method called simulated annealing. In case of minimal prime implicant problem, the search space consists of all subsets of variables, and the cost function for a given subset X of Boolean variables is defined by two factors: (1) the number of clauses that

are uncovered by X, and (2) the size of X, see (Sen. S., 1993).

ROUGH SETS AND BOOLEAN REASONING

Rough-Set Preliminaries

The main idea of rough set theory is based on approximating the unknown concept by a pair of sets called lower and upper approximations. The lower approximation contains those objects that certainly, according to the actual knowledge of the learner, belong to the concept, the upper approximation contains those objects that possibly belong to the concept.

Let $C \subseteq X$ be a concept and let $\mathbf{S} = (U, A, dec)$ be a decision table describing the training set $U \subseteq X$. Any pair $P = (\mathbf{L}, \mathbf{U})$ is called *rough approximation of* C (see Bazan, Nguyen, Skowron, &. Szczuka, 2003) if it satisfies the following conditions:

1. $\mathbf{L} \subseteq \mathbf{U} \subseteq X$;
2. $\mathbf{L}, \mathbf{U}$ are expressible in the language L;
3. $\mathbf{L} \cap U \subseteq C \cap U \subseteq \mathbf{U} \cap U$;
4. $\mathbf{L}$ is maximal and $\mathbf{U}$ is minimal among those L-definable sets satisfying 3.

The sets $\mathbf{L}$ and $\mathbf{U}$ are called the *lower approximation* and the *upper approximation* of the concept C, respectively. The set $\mathbf{BN} = \mathbf{U} - \mathbf{L}$ is called the *boundary region of approximation* of C. For objects $x \in \mathbf{U}$, we say that "probably, x is in . C. The concept C is called *rough* with respect to its approximations $(\mathbf{L}, \mathbf{U})$ if $\mathbf{L} \neq \mathbf{U}$, otherwise C is called *crisp* in X.

The first definition of rough approximation was introduced by Pawlak in his pioneering book on rough set theory (Pawlak, 1991). For any subset of attributes $B \subset A$, the set of objects U is divided into *equivalence classes* by the *indiscernibility relation,* and the upper and lower approximations are defined as unions of corresponding equivalence classes. This definition can be called *the attribute-based rough approximation*. A great effort by many researchers in RS Society has been made to modify and to improve this classical approach. One can find many interesting methods for rough approximation like variable RS model (Ziarko, 1993), tolerance-based rough approximation, approximation space (Skowron & Stepaniuk, 1996), or classifier-based rough approximations (Bazan et. al., 2003).

The condition (4) in this list can be substituted by inclusion, to a degree, to make it possible to induce approximations of higher quality on the concept of the whole universe X. In practical applications, it is hard to fulfill the last condition. Hence, by using some heuristics, we construct suboptimal instead of maximal or minimal sets. This condition is the main inspiration for all applications of rough sets in data mining and decision support systems.

Rough Sets and Boolean Reasoning Approach to Attribute Selection

Feature selection has been an active research area in pattern recognition, statistics, and data mining communities. The main idea of feature selection is to select a subset of the most relevant attributes for classification task, or to eliminate features with little or no predictive information. Feature selection can significantly improve the comprehensibility of the resulting classifier models and often build a model that generalizes better to unseen objects (Liu & Motoda H., 1999). Further, it is often the case that finding the correct subset of predictive features is an important problem in its own right.

In rough set theory, the feature selection problem is defined in terms of reducts (Pawlak, 1991). We will generalize this notion and show an application of the ABR approach to this problem.

In general, reducts are minimal subsets (with respect to the set inclusion relation) of attributes

47

that contain a necessary amount of *information* about a considered problem. The notion of information is as abstractive as the notion of energy in physics, and we will not able to define it exactly. Instead of explicit information, we have to define some *objective properties* for all subsets of attributes. Such properties can be expressed in different ways, for example, by logical formulas or, as in this section, by a *monotone evaluation function*, which is described as follows.

For a given information system $S=(U, A)$, the function $\mu_S : P(A) \longrightarrow \Re^+$ where $P(A)$ is the power set of A, is called *the monotone evaluation function* if the following conditions hold:

1. The value of $\mu_S(B)$ can be computed using information set $INF(B)$ for any $B \subset A$
2. For any $B, C \subset A$, if $B \subset C$, then $\mu_S(B) \leq \mu_S(C)$

Definition 3.1. (μ-*reduct*) *Any set* $B \subseteq A$ *is called the reduct relative to a monotone evaluation function* μ, *or briefly* μ-*reduct, if B is the smallest subset of attributes that* $\mu(B) = \mu(A)$, *that is,* $\mu(B') \leq \mu(B)$ *for any proper subset* $B' \subseteq B$.

This definition is general for many different definitions of reducts. Let us mention some well-known types of reducts used in rough set theory.

Basic Types of Reducts in Rough Set Theory

For any subset of attributes $B \subset A$ of a given information system $S=(U, A)$, we define the *B-indiscernibility relation* (denoted by $IND_S(B)$) by:

$$IND_S(B) = \{(x, y) \in U \times U : inf_B(x) = inf_B(y)\}.$$

Relation $IND_S(B)$ is an equivalence relation. Its equivalence classes can be used to define the lower and upper approximations of concepts in rough set theory (Pawlak, 1982, 1991).

The complement of indiscernibility relation is called *B-discernibility relation* and is denoted by $DISC_S(B)$. Hence:

$$DISC_S(B) = U \times U - IND_S(B)$$
$$= \{(x, y) \in U \times U : inf_B(x) \neq inf_B(y)\}$$
$$= \{(x, y) \in U \times U : \exists_{a \in B} a(x) \neq a(y)\}.$$

It is easy to show that $DISC_S(B)$ is monotone, that is, for any $B, C \subset A$:

$$B \subset C \Rightarrow DISC_S(B) \subset DISC_S(C).$$

Intuitively, any reduct (in rough set theory) is a minimal subset of attributes that preserves the discernibility between information vectors of objects. The following notions of reducts are often used in rough set theory.

Definition 3.2. (*information reducts*) *Any minimal subset B of A such that* $DISC_S(A) = DISC_S(B)$ *is called the information reduct (or reduct, for short) of S. The set of all reducts of a given information system S is denoted by* $RED(S)$.

In the case of decision tables, we are interested in the ability of describing decision classes by using subsets of condition attributes. This ability can be expressed in terms of *generalized decision function* $\partial_B : U \to P(V_{dec})$, where:

$$\partial_B(x) = \{i : \exists_{x' \in U} [(x' IND(B)x) \wedge (d(x') = i)]\}$$

Definition 3.3. (*decision reducts*) *A set of attributes* $B \subseteq A$ *is called a decision-oriented reduct (or a relative reduct) of decision table S if and only if:*

- $\partial_B(x) = \partial_A(x)$ *for all object* $x \in U$
- *Any proper subset of B does not satisfy the previous condition*

that is, B is a minimal subset (with respect to the inclusion relation $\subseteq$) of the attribute set satisfying the property $\forall_{x \in U} \partial_B(x) = \partial_A(x)$.

A set $C \subset A$ of attributes is called *superreduct* if there exists a reduct B such that $B \subset C$. One can show that:

1. Information reducts for a given information system $\mathbf{S} = (U, A)$ are exactly those reducts with respect to *discernibility function*, which is defined for arbitrary subset of attributes $B \subset A$ as a number of pairs of objects discerned by attributes from B, that is:

$$\mathrm{disc}(B) = \frac{1}{2}\mathrm{card}(DISC_\mathbf{S}(B)).$$

2. Relative reducts for decision tables $\mathbf{S} = (U, A \cup \{dec\})$ are exactly those reducts with respect to *relative discernibility function*, which is defined by:

$$\mathrm{disc}_{dec}(B) = \frac{1}{2}\mathrm{card}\big(DISC_\mathbf{S}(B) \cap DISC_\mathbf{S}(\{dec\})\big).$$

The relative discernibility function returns the number of pairs of objects from different classes, which are discerned by attributes from B.

Many other types of reducts, for example, frequency-based reducts (Slezak, 2000) or entropy reducts (Slezak 2002), can be defined by selection of different monotone evaluation functions.

Boolean reasoning approach for reduct problem.

There are two problems related to the notion of "reduct," that have been intensively explored in rough set theory by many researchers (see, e.g., Bazan, 1998; Jensen, Shen, & Tuso, 2005; Kryskiewicz, 1997; Slezak, 2000; Slezak, 2002; Wroblewski, 1996). The first problem is related to searching for reducts with the minimal cardinality called *the shortest reduct problem*. The second problem is related to searching for all reducts. It has been shown that the first problem is NP hard (see Skowron & Rauszer, 1992) and second is at least NP hard. Some heuristics have been proposed for those problems. Here we present the approach based on Boolean reasoning as proposed in Skowron and Rauszer (1992) (see Figure 2).

Given a decision table $\mathbf{S} = (U, A \cup \{dec\})$, where $U = \{u_1, u_2, ..., u_n\}$ and $A = \{a_1, ..., a_k\}$. By *discernibility matrix* of the decision table $\mathbf{S}$ we mean the $(n \times n)$ matrix $\mathbf{M}(\mathbf{S}) = \big[M_{i,j}\big]_{i,j=1}^n$, where $M_{i,j} \subset A$ is the set of attributes discerning u_i and u_j, that is, $M_{i,j} = \{a_m \in A : a_m(u_i) \neq a_m(u_j)\}$. Let us denote by $VAR_\mathbf{S} = \{x_1, ..., x_k\}$ a set of Boolean variables corresponding to attributes $a_1, ..., a_k$. For any subset of attributes $B \subset A$, we denote by $X(B)$ the set of Boolean variables corresponding to attributes from B. We will encode reduct problem as a problem of searching for the corresponding set of variables.

For any two objects $u_i, u_j \in U$ such that $M_{i,j} \neq \emptyset$ the Boolean clause χ_{u_i, u_j}, called *discernibility clause*, is defined as follows:

$$\chi_{u_i, u_j}(x_1, ..., x_k) = \sum_{a_m \in M_{i,j}} x_m \quad (5)$$

The objective is to create a Boolean function of which any prime implicant corresponds to a

Figure 2. The Boolean reasoning scheme for solving reduct problem

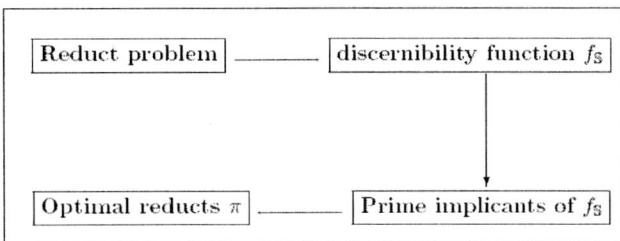

reduct of **S**. In case of the information reduct problem, this function is called the *discernibility function* and denoted by f_s. In the case of the relative reduct problem, it is called the *decision-oriented discernibility function* and denoted by f_S^{dec}. The definitions of f_s and f_S^{dec} are as follows:

$$f_S(x_1,...,x_k) = \prod_{M_{i,j} \neq \emptyset} (\chi_{u_i,u_j}(x_1,...,x_k))$$

$$f_S^{dec}(x_1,...,x_k) = \prod_{dec(u_i) \neq dec(u_j)} (\chi_{u_i,u_j}(x_1,...,x_k))$$

The following properties were proved in Skowron and Rauszer, (1992):

4. A subset of attributes B is a reduct of **S** if and only if the term $T_{X(B)} = \prod_{a_i \in B} x_i$ is a prime implicant of the discernibility function f_s.

5. A subset of attributes B is a relative reduct of decision table **S** if and only if $T_{X(B)}$ is the prime implicant of the relative discernibility function f_S^{dec}.

Example

Let us consider the decision table "weather" presented in Table 1.

This table consists of 4 attributes: a_1, a_2, a_3, a_4, hence, the set of corresponding Boolean variables consists of:

$$VAR_S = \{x_1, x_2, x_3, x_4\}$$

The discernibility matrix can be treated as a board containing $n \times n$ boxes. Noteworthy is the fact that the discernibility matrix is symmetrical with respect to the main diagonal because $M_{i,j} = M_{j,i}$, and that sort-

Box 1.

$$f(x_1, x_2, x_3, x_4) = (x_1)(x_1 + x_4)(x_1 + x_2)(x_1 + x_2 + x_3 + x_4)(x_1 + x_2 + x_4)$$
$$(x_2 + x_3 + x_4)(x_1 + x_2 + x_3)(x_4)(x_2 + x_3)(x_2 + x_4)$$
$$(x_1 + x_3)(x_3 + x_4)(x_1 + x_2 + x_4).$$

Table 1. The exemplary "weather" decision table

date	outlook	temperature	humidity	windy	play
ID	a_1	a_2	a_3	a_4	dec
1	sunny	hot	high	FALSE	no
2	sunny	hot	high	TRUE	no
3	overcast	hot	high	FALSE	yes
4	rainy	mild	high	FALSE	yes
5	rainy	cool	normal	FALSE	yes
6	rainy	cool	normal	TRUE	no
7	overcast	cool	normal	TRUE	yes
8	sunny	mild	high	FALSE	no
9	sunny	cool	normal	FALSE	yes
10	rainy	mild	normal	FALSE	yes
11	sunny	mild	normal	TRUE	yes
12	overcast	mild	high	TRUE	yes
13	overcast	hot	normal	FALSE	yes
14	rainy	mild	high	TRUE	no

Table 2. The compact form of discernibility matrix

M	1	2	6	8
3	a_1	a_1, a_4	a_1, a_2, a_3, a_4	a_1, a_2
4	a_1, a_2	a_1, a_2, a_4	a_2, a_3, a_4	a_1
5	a_1, a_2, a_3	a_1, a_2, a_3, a_4	a_4	a_1, a_2, a_3
7	a_1, a_2, a_3, a_4	a_1, a_2, a_3	a_1	a_1, a_2, a_3, a_4
9	a_2, a_3	a_2, a_3, a_4	a_1, a_4	a_2, a_3
10	a_1, a_2, a_3	a_1, a_2, a_3, a_4	a_2, a_4	a_1, a_3
11	a_2, a_3, a_4	a_2, a_3	a_1, a_2	a_3, a_4
12	a_1, a_2, a_3	a_1, a_2	a_1, a_2, a_3	a_1, a_4

ing all objects according to their decision classes causes a shift of all empty boxes nearby to the main diagonal. In case of the decision table with two decision classes, the discernibility matrix can be rewritten in a more compact form, as shown in Table 2.

The discernibility function is constructed from the discernibility matrix by taking a conjunction of all discernibility clauses. After reducing all repeated clauses, we have: (see Box 1).

One can find relative reducts of the decision table by searching for its prime implicants. The straightforward method calculates all prime implicants by translation to DNF. One can do it as follows:

- Remove those clauses that are absorbed by some other clauses (using absorbtion rule: $p(p+q) \equiv p$): $f = (x_1)(x_4)(x_2 + x_3)$
- Translate f from CNF into DNF $f = x_1 x_4 x_2 + x_1 x_4 x_3$.
- Every monomial corresponds to a reduct. Thus we have 2 reducts: $R_1 = \{a_1, a_2, a_4\}$ and $R_2 = \{a_1, a_3, a_4\}$.

Approximate Algorithms for Reduct Problem

Every heuristic algorithm for the prime implicant problem can be applied to the discernibility function to solve the minimal reduct problem. One such heuristic was proposed in Skowron and Rauszer (1992) and was based on the idea of greedy algorithm (see Sect. 1), where each attribute is evaluated by its discernibility measure, that is, the number of pairs of objects that are discerned by the attribute, or, equivalently, the number of its occurrences in the discernibility matrix. The pseudocode of this algorithm is presented in Algorithm 1.

The reader may have a feeling that the greedy algorithm for reduct problem has quite a high complexity because two main operations:

- $disc(B)$ – number of pairs of objects discerned by attributes from B;
- $isCore(a)$ – check whether a is a core attribute;

are defined by the discernibility matrix, which is a complex data structure containing $O(n^2)$ cells, and each cell can contain up to $O(m)$ attributes, where n is the number of objects and m is the number of attributes of the given decision table. This suggests that the two main operations need at least $O(mn^2)$ computational time.

Fortunately, both operations can be performed more efficiently. It has been shown (Nguyen & Nguyen, 1996) that both operations can be calculated in time $O(mn \log n)$ without the necessity to store the discernibility matrix. In Nguyen (2006),

```
Algorithm 1: Searching for short reduct
begin
    B := ∅;
    //Step 1. Initializing B by core attributes
    for a ∈ A do
        if isCore(a) then
            | B := B ∪ {a};
        end
    end
    //Step 2. Including attributes to B
    repeat
        a_max := arg max disc_dec(B ∪ {a});
                  a∈A-B
        eval(a_max) := disc_dec(B ∪ {a_max}) - disc_dec(B);
        if (eval(a_max) > 0) then
            | B := B ∪ {a};
        end
    until (eval(a_max) == 0) OR (B == A);
    //Step 3. Elimination
    for a ∈ B do
        if (disc_dec(B) = disc_dec(B - {a})) then
            | B := B - {a};
        end
    end
end
```

we present an effective implementation of this heuristic that can be applied to large data sets.

Let us illustrate the idea by using discernibility matrix (Table 2) from the previous section.

- First we have to calculate the number of occurrences of each attribute in the discernibility matrix:

$$eval(a_1) = disc_{dec}(a_1) = 23$$
$$eval(a_2) = disc_{dec}(a_2) = 23$$
$$eval(a_3) = disc_{dec}(a_3) = 18$$
$$eval(a_4) = disc_{dec}(a_4) = 16$$

Thus a_1 and a_2 are the two most preferred attributes.

- Assume that we select a_1. Now we are taking under consideration only those cells of the discernibility matrix that do not contain a_1. There are 9 such cells only, and the number of occurrences are as the following:

$$eval(a_2) = disc_{dec}(a_1, a_2) - disc_{dec}(a_1) = 7$$
$$eval(a_3) = disc_{dec}(a_1, a_3) - disc_{dec}(a_1) = 7$$
$$eval(a_4) = disc_{dec}(a_1, a_4) - disc_{dec}(a_1) = 6$$

- If this time we select a_2, then there remain only 2 cells, and, both contain a_4;
- Therefore, the greedy algorithm returns the set $\{a_1, a_2, a_4\}$ as a reduct of sufficiently small size.

There is another reason for choosing a_1 and a_4, because they are *core attributes*[2]. It has been shown that an attribute is a core attribute if and only if it occurs in the discernibility matrix as a singleton (Skowron & Rauszer, 1992). Therefore, core attributes can be recognized by searching for all single cells of the discernibility matrix.

Decision Rule Induction

Classification is one of the most popular applications of rough set theory in data mining. The idea is to construct from data some rough-set-based classification algorithms, so called "rough classifiers," that are featured by the ability of giving the description of the lower and upper approximations of decision classes. Unlike standard classification algorithms, in order to classify a new unseen object, rough classifiers can return the vector of its rough membership to decision classes.

Decision rules are important components of many rough classifiers. Therefore, searching for a good collection of rules is the fundamental problem in rough sets.

In this section, we investigate the application of the Boolean reasoning method to the rule induction problem.

Decision rules are logical formulas that indicate the relationship between conditional and decision attributes. In this chapter we consider only those decision rules $\mathbf{r}$ whose premise is a Boolean monomial of descriptors, that is:

$$\mathbf{r} \equiv (a_{i_1} = v_1) \wedge ... \wedge (a_{i_m} = v_m) \Rightarrow (dec = k) \quad (6)$$

Every decision rule $\mathbf{r}$ of the form (6) can be characterized by the following features:

length($\mathbf{r}$)	The number of descriptors in the premise of $\mathbf{r}$
[$\mathbf{r}$]	The carrier of $\mathbf{r}$, that is, the set of objects from U satisfying the premise of $\mathbf{r}$
support($\mathbf{r}$)	the number of objects satisfying the premise of $\mathbf{r}$: *support*($\mathbf{r}$) = *card*([$\mathbf{r}$])
conf($\mathbf{r}$)	The confidence of $\mathbf{r}$: $confidence(\mathbf{r}) = \frac{|[\mathbf{r}] \cap DEC_k|}{|[\mathbf{r}]|}$

The decision rule $\mathbf{r}$ is called *consistent* with A if *confidence*($\mathbf{r}$) = 1.

In the rough set approach to concept approximation, decision rules are used to define finer rough approximation compared to attribute-base rough approximation. Each decision rule is supported by some objects and, inversely, the information vector of each object can be reduced to obtain a minimal consistent decision rule.

The Boolean reasoning approach to decision rule construction from a given decision table $\mathbf{S} = (U, A \cup \{dec\})$ is very similar to the minimal reduct problem. The only difference occurs in the encoding step, that is:

- **Encoding:** For any object $u \in U$ in, we define a function $f_u(a_1^*, ..., a_m^*)$, called *discernibility function for u* by:

$$f_u(a_1^*, ..., a_m^*) = \prod_{u_j : dec(u_j) \neq dec(u)} \chi_{u, u_j}(x_1, ..., x_k) \quad (7)$$

where χ_{u, u_j} is the discernibility clause for objects u and u_j (see equation (5)).

- **Heuristics:** all heuristics for minimal prime implicant problem can be applied to Boolean functions in equation (7). Because there are n such functions, where n is a number of objects in the decision table, the well-known heuristics may show to be time consuming.

Example

Let us consider the decision table that is shown in Example 1. Consider the object number 1:

1. The discernibility function is determined as follows:

$$f_1(x_1, x_2, x_3, x_4) = x_1(x_1 + x_2)(x_1 + x_2 + x_3)$$
$$(x_1 + x_2 + x_3 + x_4)$$
$$(x_2 + x_3)(x_1 + x_2 + x_3)(x_2 + x_3 + x_4)$$
$$(x_1 + x_2 + x_4)$$

2. After transformation into DNF we have:

$$f_1(x_1, x_2, x_3, x_4) = x_1(x_2 + x_3)$$
$$= x_1 x_2 + x_1 x_3$$

3. Hence, there are two object-oriented reducts, that is, $\{a_1, a_2\}$ and $\{a_1, a_3\}$. The corresponding decision rules are:

$(a_1 = \text{sunny}) \wedge (a_2 = \text{hot}) \Rightarrow (dec = \text{no})$
$(a_1 = \text{sunny}) \wedge (a_3 = \text{high}) \Rightarrow (dec = \text{no})$

Let us notice that all rules have the same decision class; precisely, the class of the considered object. If we wish to obtain minimal consistent rules for the other decision classes, we should repeat the algorithm for another object.

Discretization of Real-Value Attributes

Discretization of real-value attributes is an important task in data mining, particularly for the classification problem. Empirical results show that the quality of classification methods depends on the discretization algorithm used in the preprocessing step. In general, discretization is a process of searching for partition of attribute domains into intervals and unifying the values over each interval. Hence, the discretization problem can be defined as a problem of searching for a relevant set of cuts (i.e., boundary points of intervals) on attribute domains.

In Nguyen and Nguyen (1998), Boolean reasoning approach to real-value attribute discretization problem was presented. The problem is to search for a minimal set of cuts on real-value attributes that preserves the discernibility between objects.

More precisely, for a given decision table $S = (U, A \cup \{dec\})$, a cut $(a;c)$ on an attribute $a \in A$ is said to *discern* a pair of objects $x, y \in U$ (or objects x and y are *discernible by* $(a;c)$) if $(a(x) - c)(a(y) - c) < 0$. Two objects are discernible by a set of cuts C if they are discernible by at least one cut from C.

A set of cuts C is *consistent with* S (or S-consistent, for short) if and only if for any pair of objects $x, y \in U$ such that $dec(x) \neq dec(y)$, the following condition holds:

IF x, y are discernible by A THEN x, y are discernible by C.

The discretization process made by a consistent set of cuts is called the *compatible discretization*. We are interested in searching for consistent sets of cuts of the size as small as possible.

Formally, the optimal discretization problem is defined as follows:

OptiDisc: optimal discretization problem
input: A decision table S.
output: S-optimal set of cuts.

The corresponding decision problem can be formulated as:

DiscSize: k-cuts discretization problem
input: A decision table S and an integer k.
question: Decide whether there exists a S-irreducible set of cuts P such that $card(P) < k$.

The following fact has been shown in (23).

Theorem 3.4. *The problem DiscSize is polynomially equivalent to the PrimiSize problem.*

As a corollary, we can prove the following Theorem.

Theorem 3.5. *Computational complexity of discretization problems:*

1. The problem DiscSize is NP complete.
2. The problem OptiDisc is NP hard.

This fact means that we cannot expect a polynomial time-searching algorithm for optimal discretization, unless P = NP.

Discretization Method Based on Rough Set and Boolean Reasoning

Given a decision table $S = (U, A \cup \{dec\})$ and a set of candidate cuts C, the discretization problem is encoded as follows:

- **Variables:** Each cut $(a,c) \in C$ is associated with a Boolean variable $x_{(a,c)}$
- **Encoding:** Similarly to the reduct problem, a discernibility function between $u_i, u_j \in U$, where $i, j = 1,...,n$, is defined by:

$$\phi_{i,j} = \sum_{(a,c) \text{ discerns } u_i \text{ and } u_j} x_{(a,c)}$$

and the discretization problem is encoded by the following Boolean function:

$$\phi = \prod_{dec(u_i) \neq dec(u_j)} \phi_{i,j} \qquad (8)$$

- **Greedy heuristics:** Again all mentioned heuristics for prime implicant problem can be applied to optimal discretization problem, but we have to take under our attention their computational complexity. The encoding Boolean function (equation (8)) consists of $O(nm)$ possible variables, and $O(n^2)$ clauses, where n and m are numbers of objects and attributes, respectively. Therefore, the time complexity of searching for best cut is $O(n^2 m)$.

We have shown that the presented MD heuristic can be implemented more efficiently. The idea is based on a special data structure for efficiently storing the partition of objects made by the actual set of cuts. This data structure, called *DTree*, a shortcut of discretization tree, is a modified decision tree structure. It contains the following methods:

- *Init*(**S**): Initializes the data structure for the given decision table.
- *Conflict*(): Returns the number of pairs of undiscerned objects.
- *GetBestCut*(): Returns the best cut point with respect to the discernibility measure.
- *InsertCut*(*a*,*c*): Inserts the cut (*a*;*c*) and updates the data structure.

It has been shown that except *Init*(**S**), the time complexity of all other methods is $O(nk)$, where n is the number of objects and k is the number of attribute, (see Nguyen, 1997; Nguyen, Skowron, & Synak, 1998). The method *Init*(**S**) requires $O(nk \log n)$ computation steps because it prepares each attribute by sorting objects with respect to this attribute.

MD heuristic can be efficiently implemented using *DTree* structure as follows: (see Algorithm 2).

This improved algorithm has been implemented in ROSETTA (Ohrn, Komorowski, Skowron, & Synak, 1998) and RSES (Bazan, Nguyen, Nguyen, Synak, & . Wróblewski., 2000; Bazan & Szczuka., 2001) systems.

APPLICATIONS OF ROUGH SETS AND BOOLEAN REASONING METHODOLOGY IN DATA MINING

In the previous section, we have presented the most typical applications of the Boolean reasoning

---

Algorithm 2: Implementation of **MD-heuristic** using *DTree* structure

Input: Decision table $\mathbb{S} = (U, A, dec)$.
Output: The semi-optimal set of cuts.
begin
    $DTree$ **D** := new $DTree()$;
    D.Init($\mathbb{S}$);
    while (D.*Conflict()* > 0) **do**
        Cut c := D.GetBestCut();
        if *(c.quality == 0)* **then**
            break;
        end
        D.InsertCut(c.attribute, c.cutpoint);
    end
    D.PrintCuts();
end

approach in rough set methods. These methods were discovered by Skowron and Rauszer (1992; Skowron, 1993), who showed that Boolean reasoning approach can offer an elegant and complete solution for both attribute reduction and minimal decision rules construction problems (Skowron et al., 2002). Rough set theory and Boolean reasoning approach have become a fundamental of many accurate concept approximation methods. The fact that most data mining tasks can be formulated as concept approximation problems suggests a straightforward application of rough set in data mining.

This section presents some applications of rough set and approximate Boolean reasoning (ABR) approach to modern problems in data mining.

DECISION TREES

Decision tree is the name of a classification method that is derived by the fact that it can be represented by an oriented tree structure, where each *internal node* is labeled by a *test* on an information vector, each *branch* represents an outcome of the test, and *leaf nodes* represent decision classes or class distributions.

Usually, tests in decision trees are required to have a small number of possible outcomes. In general, the following types of tests are considered in the literature:

1. **Attribute-based tests:** This type consists of tests defined by symbolic attributes, that is, for each attribute $a \in A$ we define a test t_a such that for any object u from the universe, $t_a(u) = a(u)$;
2. **Value-based tests:** for each attribute $a \in A$ and for each value $v \in V_a$ we define a test $t_{a=v}$ such that for any object u:

$$t_{a=v}(u) = \begin{cases} 1 & \text{if } a(u) = v \\ 0 & \text{otherwise;} \end{cases}$$

3. **Cut-based tests:** Tests of this type are defined by cuts on real value attributes. For each attribute $a \in A$ and for each value $c \in R$ we define a test $t_{a>c}$ such that for any object:

$$u: t_{a>c}(u) = \begin{cases} 1 & \text{if } a(u) > c \\ 0 & \text{otherwise;} \end{cases}$$

4. **Value set-based tests:** For each attribute $a \in A$ and for each set of values $S \subset V_a$ we define a test $t_{a \in S}$ such that for any object u from the universe, $t_{a \in S}(u) = \begin{cases} 1 & \text{if } a(u) \in S \\ 0 & \text{otherwise} \end{cases}$. This is a generalization of the previous types.
5. **Hyperplane-based tests:** Tests of this type are defined by linear combinations of continuous attributes. A test $t_{w_1 a_1 + \ldots + w_k a_k > w_0}$ where $a_1, \ldots, a_k$ are continuous attributes and $w_0, w_1, \ldots, w_k$ are real numbers, is defined by $t_{w_1 a_1 + \ldots + w_k a_k > w_0}(u) = 1$ if and only if $w_1 a_1(u) + \ldots + w_k a_k(u) > w_0$.

Decision tree is one of the most favorite types of templates in data mining, because of its simple representation and easy readability. Analogous to other classification algorithms, there are two issues related to the decision-tree approach, that is, how to classify new unseen objects using decision tree and how to construct an optimal decision tree for a given decision table.

In order to classify an unknown example, the information vector of this example is tested against the decision tree. The path is traced from the root to a leaf node that holds the class prediction for this example.

A decision tree is called *consistent* with a given decision table **S** if it properly classifies all objects from **S**. A given decision table may have many consistent decision trees. The main objective is to build a decision tree of high prediction accuracy for a given decision table. This requirement is realized by a philosophical principle called Occam's Razor.

This principle, thought up a long time ago by William Occam while shaving, states that the shortest hypothesis, or solution to a problem, should be the one we should prefer (over longer, more-complicated ones). This is one of the fundamental tenets of the way western science works and has received much debate and controversy. The specialized version of this principle applied to the decision trees can be formulated as follows:

The world is inherently simple. Therefore, the smallest decision tree that is consistent with the samples is the one that is most likely to identify unknown objects correctly.

Unfortunately, the problem of searching for shortest tree for a decision table has shown to be NP hard. It means that no computer algorithm can solve this in a feasible amount of time in the general case. Therefore, only heuristic algorithms have been developed to find a good tree, usually very close to the best.

The basic heuristic for construction of decision tree (for example ID3 or later C4.5, see Quinlan, 1993; Quinlan, 1986), CART (Breiman, Friedman, Olshen, & Stone, 1984) is based on the top-down recursive strategy described as follows:

1. It starts with a tree with one node representing the whole training set of objects.
2. If all objects have the same decision class, the node becomes leaf and is labeled with this class.
3. Otherwise, the algorithm selects the best test t_{Best} from the set of all possible tests.
4. The current node is labeled by the selected test t_{Best} and it is branched accordingly to values of t_{Best}. Also, the set of objects is partitioned and assigned to new created nodes.
5. The algorithm uses the same processes (steps 2, 3, 4) recursively for each new node to form the whole decision tree.
6. The partitioning process stops when either all examples in a current node belong to the same class, or no test function has been selected in Step 3.

Developing decision tree induction methods (see Fayyad & Irani, 1993; Quinlan, 1993), we should define some heuristic measures (or heuristic quality functions) to estimate the quality of tests. In tree induction process, the optimal test t_{Best} with respect to the function F is selected as the result of Step 3.

More precisely, let $T = \{t_1, t_2, ..., t_m\}$ be a given set of all possible tests, heuristic measure is a function:

$$F : T \times P(U) \to R$$

where $P(U)$ is the family of all subsets of U. The value $F(t, X)$, where $t \in T$ and $X \subset U$, should estimate the chance that t_i labels the root of the optimal decision tree for X. Usually, the value $F(t, X)$ depends on how the test t splits the set of objects X.

The most famous measure for construction of decision tree is called the "entropy measure." This concept uses class entropy as a criterion to evaluate the list of best cuts that, together with the attribute domain, induce the desired intervals. The class information entropy of the set of N objects X with the class distribution $\langle N_1, ..., N_d \rangle$, where $N_1 + ... + N_d = N$, is defined by:

$$Ent(X) = -\sum_{j=1}^{d} \frac{N_j}{N} \log \frac{N_j}{N}$$

Hence, the entropy of the partition induced by a cut point c on attribute a is defined by:

$$E(a,c;U) = \frac{|U_L|}{n} Ent(U_L) + \frac{|U_R|}{n} Ent(U_R) \quad (9)$$

where $\{U_L, U_R\}$ is a partition of U defined by c. For a given feature a, the cut c_{min}, which minimizes the entropy function over all possible cuts, is selected. The decision tree and discretization methods based on information entropy are reported in Fayyad and Irani (1993) and Quinlan (1993).

In the following sections, we describe some typical applications of rough sets and approximate Boolean reasoning approach to decision-tree methods.

MD Algorithm for Decision-Tree Induction

In Boolean reasoning approach to discretization, qualities of cuts were evaluated by their discernibility properties. In this section, we present an application of discernibility measure in the induction of decision trees. This method of decision-tree induction is called the maximal-discernibility algorithm, or shortly, MD algorithm.

MD algorithm uses discernibility measure to evaluate the quality of tests. Intuitively, a pair of objects is said to be conflict if they belong to different decision classes. An internal conflict of a set of objects $X \subset U$ is defined by the number of conflict pairs of objects from X. Let $(n_1,...,n_d)$ be a counting table of X, then $conflict(X)$ can be computed by:

$$conflict(X) = \sum_{i<j} n_i n_j.$$

If a test t determines a partition of a set of objects X into $X_1, X_2, ..., X_{n_t}$, then discernibility measure for t is defined by:

$$Disc(t, X) = conflict(X) - \sum_{i=1}^{n_t} conflict(X_i). \quad (10)$$

Thus, the more pairs of objects are separated by the test t the larger is the chance that t labels the root of the optimal decision tree for X.

MD algorithm is using two kinds of tests, depending on attribute types. In case of symbolic attributes $a_j \in A$, test functions defined by sets of values, that is:

$$t_{a_j \in V}(u) = 1 \iff \left[a_j(u) \in V \right]$$

where $V \subset V_{a_j}$, are considered. For numeric attributes $a_i \in A$, only test functions defined by cuts:

Figure 3. Geometrical interpretation of the discernibility measure

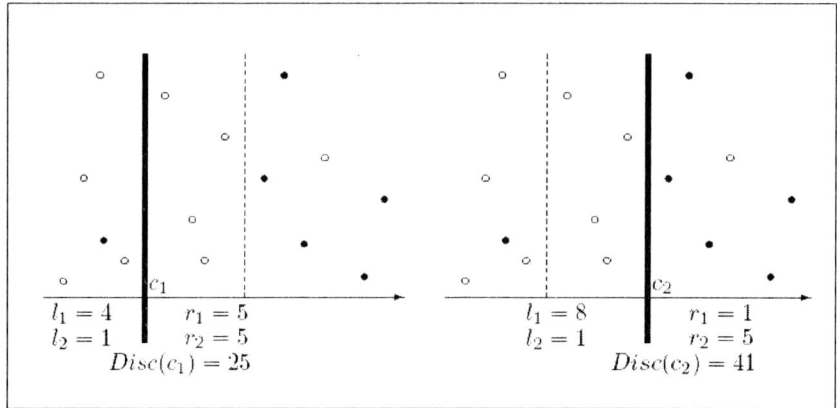

$t_{a_i > c}(u)$

$= True \iff [a_i(u) \leq c] \iff [a_i(u) \in (-\infty; c)]$

where c is a cut in V_{a_i}, are considered.

The high accuracy of decision trees constructed by using discernibility measure and their comparison with entropy-based methods has been reported in Nguyen (1997).

Soft Cuts and Soft Decision Trees

So far, we have presented decision-tree methods working with cuts that are treated as sharp partition, since real values are partitioned into disjoint intervals. One can observe that some objects that are very similar can be treated as very different because they are in different sides of a cut. In this section, we introduce some notions of soft cuts that discern two objects if they are located far enough from cuts.

The formal definition of soft cuts is following:

Definition 4.1. *A soft cut is any triple $p = \langle a, l, r \rangle$, where $a \in A$ is an attribute, $l, r \in \Re$ are called the left and right bounds of p ($l \leq r$); the value $\varepsilon = \frac{r-l}{2}$ is called the uncertain radius of p. We say that a soft cut p discerns pair of objects x_1, x_2 if $a(x_1) < l$ and $a(x_2) > r$.*

The intuitive meaning of $p = \langle a, l, r \rangle$ is such that there is a real cut somewhere between l and r. So we are not sure where one can place the real cut in the interval $[l, r]$. Hence, for any value $v \in [l, r]$ we are not able to check if v is either on the left side or on the right side of the real cut. Then we say that the interval $[l, r]$ is an uncertain interval of the soft cut p. Any normal cut can be treated as soft cut of radius equal to 0.

Any set of soft cuts splits the real axis into intervals of two categories: the intervals corresponding to new nominal values and the intervals of uncertain values called boundary regions. The problem of searching for a minimal set of soft cuts with a given uncertain radius can be solved in a similar way to the case of sharp cuts. We propose some heuristic for this problem in the last section of the chapter. The problem becomes more complicated if we want to obtain as small as possible set of soft cuts with the radius as large as possible. We will discuss this problem in the next chapter.

The test functions defined by traditional cuts can be replaced by soft cuts. We have proposed two strategies being modifications of standard classification method for decision tree with soft cuts (fuzzy separated cuts). They are called fuzzy decision tree and rough decision tree.

In fuzzy decision tree method instead of checking the condition $a(u) > c$ we have to check how strong is the hypothesis that u is on the left or right side of the cut (a, c). This condition can be expressed by $\mu_L(u)$ and $\mu_R(u)$, where μ_L and μ_R are membership functions of left and right

Figure 4. The soft cut

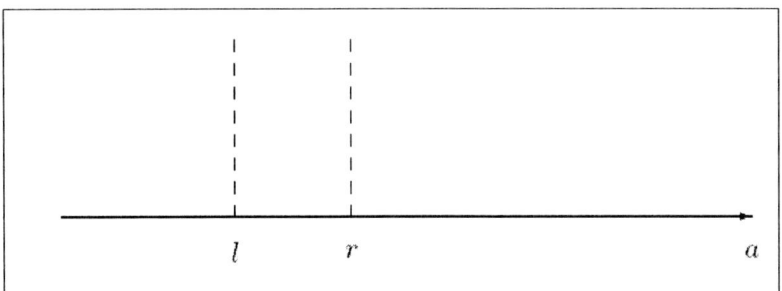

intervals (respectively). The values of those membership functions can be treated as a probability distribution of u in the node labeled by soft cut $(a, c-\varepsilon, c+\varepsilon)$. Then one can compute the probability of the event that object u is reaching a leaf. The decision for u is equal to decision labeling the leaf with largest probability.

In the case of rough decision tree, when we are not able to decide to turn left or right (the value $a(u)$ is too close to c) we do not distribute the probability to children of considered node. We have to compare their answers, taking into account the numbers of supported by them objects. The answer with the most number of supported objects is a decision of given object.

Some efficient methods for decision-tree construction from large data sets, which are assumed to be stored in database servers and to be accessible by SQL queries, were presented in Nguyen (1999, 2001) and Nguyen H.S. and Nguyen S. H (2005). The proposed decision-tree construction methods minimize the number of simple queries necessary to search for the best cuts by using "divide and conquer" search strategy. To make it possible, some novel "approximate measures" defined on intervals of attribute values were developed. Proposed measures are necessary to evaluate a chance that a given interval contains the best cut.

ASSOCIATION ANALYSIS

Let $S = (U, A)$ be an information table. By descriptors (or simple descriptors) we mean the terms of the form $(a=v)$, where $a \in A$ is an attribute and $v \in V_a$ is a value in the domain of a (see (32)). By template we mean the conjunction of descriptors:

$$\mathbf{T} = D_1 \wedge D_2 \wedge ... \wedge D_m,$$

where $D_1, ... D_m$ are either simple or generalized descriptors. We denote by $length(\mathbf{T})$ the number of descriptors being in $\mathbf{T}$.

For the given template with length m:

$$\mathbf{T} = (a_{i_1} = v_1) \wedge ... \wedge (a_{i_m} = v_m)$$

the object $u \in U$ is said to satisfy the template $\mathbf{T}$ if and only if $\forall_j a_{i_j}(u) = v_j$. In this way, the template $\mathbf{T}$ describes the set of objects having the common property: "values of attributes $a_{i_1}, ..., a_{i_m}$ are equal to , respectively." In this sense, one can use templates to describe the regularity in data, that is, patterns, in data mining or granules, in soft computing.

Templates, except for length, are also characterized by their support. The support of a template $\mathbf{T}$ is defined by:

$$support(\mathbf{T}) = |\{u \in U : u \text{ satisfies } \mathbf{T}\}|.$$

From descriptive point of view, we prefer long templates with large support.

The templates that are supported by a predefined number (say *min_support*) of objects are called the frequent templates. This notion corresponds exactly to the notion of frequent itemsets for transaction databases (Agrawal, Imielinski, & Swami, 1993). Many efficient algorithms for frequent itemset generation have been proposed in Agrawal et al., and others (1993; Agrawal, Mannila, Srikant, Toivonen, & Verkamo 1996; Agrawal & Srikant, 1994; Han, Pei, & Yin, 2000; Zaki, 1998). The problem of frequent template generation using rough-set method has been also investigated in Nguyen (2000) and Nugyen et. al. (1998). In the previous section, we considered a special kind of template called decision template or decision rule. Almost all objects satisfying a decision template should belong to one decision class.

Let us assume that the template $\mathbf{T}$, which is supported by at least s objects, has been found (using one of existing algorithms for frequent templates). We assume that $\mathbf{T}$ consists of m descriptors, that is $\mathbf{T} = D_1 \wedge D_2 \wedge ... \wedge D_m$ where D_i (for $i = 1, ..., m$) is a descriptor of the form $(a_i = v_i)$

for some $a_i \in A$ and $v_i \in V_{a_i}$. We denote the set of all descriptors occurring in the template **T** by $DESC(\mathbf{T})$, that is:

$$DESC(\mathbf{T}) = \{D_1, D_2, ..., D_m\}.$$

Any set of descriptors $\mathbf{P} \subseteq DESC(\mathbf{T})$ defines an association rule:

$$R_\mathbf{P} =_{def} \left(\bigwedge_{D_i \in \mathbf{P}} D_i \Rightarrow \bigwedge_{D_j \notin \mathbf{P}} D_j \right).$$

The confidence factor of the association rule $R_\mathbf{P}$ can be redefined as:

$$confidence(R_\mathbf{P}) =_{def} \frac{support(\mathbf{T})}{support(\bigwedge_{D_i \in \mathbf{P}} D_i)},$$

that is, the ratio of the number of objects satisfying **T** to the number of objects satisfying all descriptors from **P**. The length of the association rule $R_\mathbf{P}$ is the number of descriptors from **P**.

In practice, we would like to find as many association rules with satisfactory confidence as possible (i.e., $confidence(R_\mathbf{P}) \geq c$ for a given $c \in (0;1)$). The following property holds for the confidence of association rules:

$$\mathbf{P}_1 \subseteq \mathbf{P}_2 \Rightarrow confidence(R_{\mathbf{P}_1}) \leq confidence(R_{\mathbf{P}_2}).$$

(11)

This property says that if the association rule $R_\mathbf{P}$ generated from the descriptor set **P** has satisfactory confidence then the association rule generated from any superset of **P** also has satisfactory confidence.

For a given confidence threshold $c \in (0;1]$ and a given set of descriptors $\mathbf{P} \subseteq DESC(\mathbf{T})$, the association rule $R_\mathbf{P}$ is called c-representative if:

1. $confidence(R_\mathbf{P}) \geq c$;
2. For any proper subset $\mathbf{P}' \subset \mathbf{P}$ we have $confidence(R_{\mathbf{P}'}) < c$.

From equation (11) one can see that instead of searching for all association rules, it is enough to find all c-representative rules. Moreover, every c-representative association rule covers a family of association rules. The shorter the association rule R is, the bigger is the set of association rules covered by R. First of all, we show the following theorem:

Theorem 2. *For a fixed real number $c \in (0;1]$ and a template **T**, the optimal c–association rules problem, that is, searching for the shortest c-representative association rule from **T** in a given table A, is NP hard.*

Searching for Optimal Association Rules by Rough Set Methods

To solve the presented problem, we show that the problem of searching for optimal association rules

Figure 5. The Boolean reasoning scheme for association rule generation

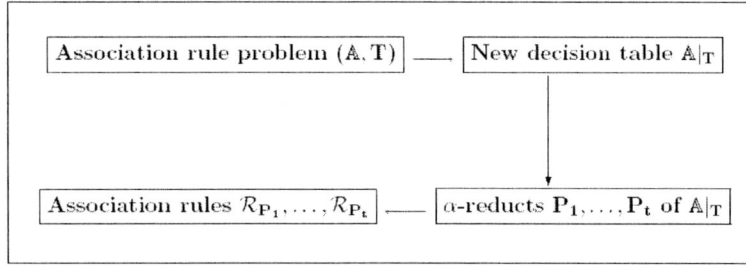

from a given template is equivalent to the problem of searching for local α-reducts for a decision table, which is a well-known problem in rough set theory. We propose the Boolean reasoning approach for association rule generation.

We construct a new decision table $A|_T = (U, A|_T \cup d)$ from the original information table A and the template T as follows:

- $A|_T = \{a_{D_1}, a_{D_2}, ..., a_{D_m}\}$ is a set of attributes corresponding to the descriptors of the template T:

$$a_{D_i}(u) = \begin{cases} 1 & \text{if the object } u \text{ satisfies } D_i, \\ 0 & \text{otherwise;} \end{cases} \quad (12)$$

- The decision attribute d determines whether a given object satisfies the template T, that is:

$$d(u) = \begin{cases} 1 & \text{if the object } u \text{ satisfies } T, \\ 0 & \text{otherwise.} \end{cases} \quad (13)$$

The following theorems describe the relationship between association rules problem and reduct searching problem.

Theorem 3. *For a given information table $A = (U, A)$ and a template T,*

1. The set of descriptors P is a reduct in $A|_T$ if and only if the rule:

$$\bigwedge_{D_i \in P} D_i \Rightarrow \bigwedge_{D_j \notin P} D_j$$

is 100%-representative association rule from T.

2. The set of descriptors P is a reduct in $A|_T$ if and only if the rule:

$$\bigwedge_{D_i \in P} D_i \Rightarrow \bigwedge_{D_j \notin P} D_j$$

is a c-representative association rule obtained from T if and only if P is a α-reduct of $A|_T$, where $\alpha = 1 - \frac{\frac{1}{c}-1}{\frac{n}{s}-1}$, n is the total number of objects from U and $s = support(T)$. In particular, the problem of searching for optimal association rules can be solved using methods for α-reduct finding.

The readers can find the proof of these facts in Nguyen and Zelzak (1999) and Nguyen (2006). Similarly to the traditional reduct problems, searching for minimal α-reducts is a well-known problem in the rough-set theory. One can show that the problem of searching for shortest α-reducts is NP hard (Nguyen & Zelzak, 1999) and the problem of searching for all the α-reducts is at least NP hard. However, there exist many approximate algorithms solving the following problems:

- Searching for shortest reduct (see Skowron & Rauszer, 1992)

Box 3.

$$f(D_1, D_2, D_3, D_4, D_5) = (D_2 \vee D_4 \vee D_5) \wedge (D_1 \vee D_3 \vee D_4) \wedge (D_2 \vee D_3 \vee D_4)$$
$$\wedge (D_1 \vee D_2 \vee D_3 \vee D_4) \wedge (D_1 \vee D_3 \vee D_5)$$
$$\wedge (D_2 \vee D_3 \vee D_5) \wedge (D_3 \vee D_4 \vee D_5) \wedge (D_1 \vee D_5)$$

Box 4.

$$f(D_1, D_2, D_3, D_4, D_5) = (D_3 \wedge D_5) \vee (D_4 \wedge D_5) \vee (D_1 \wedge D_2 \wedge D_3) \vee$$
$$(D_1 \wedge D_2 \wedge D_4) \vee (D_1 \wedge D_2 \wedge D_5) \vee (D_1 \wedge D_3 \wedge D_4)$$

Table 3. The example of information table A and template $\mathbf{T}$ supported by 10 objects and the new decision table $A|_\mathbf{T}$ constructed from A and template $\mathbf{T}$

A	a_1	a_2	a_3	a_4	a_5	a_6	a_7	a_8	a_9
T	**0**	*	**1**	**2**	*	**0**	*	**1**	*
u_1	0	*	1	1	*	2	*	2	*
u_2	0	*	2	1	*	0	*	1	*
u_3	0	*	2	1	*	0	*	1	*
u_4	0	*	2	1	*	0	*	1	*
u_5	1	*	2	2	*	1	*	1	*
u_6	0	*	1	2	*	1	*	1	*
u_7	1	*	1	2	*	1	*	1	*
u_8	0	*	2	1	*	0	*	1	*
u_9	0	*	2	1	*	0	*	1	*
u_{10}	0	*	2	1	*	0	*	1	*
u_{11}	1	*	2	2	*	0	*	2	*
u_{12}	0	*	3	2	*	0	*	2	*
u_{13}	0	*	2	1	*	0	*	1	*
u_{14}	0	*	2	2	*	2	*	2	*
u_{15}	0	*	2	1	*	0	*	1	*
u_{16}	0	*	2	1	*	0	*	1	*
u_{17}	0	*	2	1	*	0	*	1	*
u_{18}	1	*	2	1	*	0	*	2	*

$A|_\mathbf{T}$	D_1	D_2	D_3	D_4	D_5	d
u_1	1	0	1	0	0	0
u_2	1	1	1	1	1	1
u_3	1	1	1	1	1	1
u_4	1	1	1	1	1	1
u_5	0	1	0	0	1	0
u_6	1	0	0	0	1	0
u_7	0	0	0	0	1	0
u_8	1	1	1	1	1	1
u_9	1	1	1	1	1	1
u_{10}	1	1	1	1	1	1
u_{11}	0	1	0	1	0	0
u_{12}	1	0	0	1	0	0
u_{13}	1	1	1	1	1	1
u_{14}	1	1	0	0	0	0
u_{15}	1	1	1	1	1	1
u_{16}	1	1	1	1	1	1
u_{17}	1	1	1	1	1	1
u_{18}	0	1	1	1	0	0

- Searching for a number of short reducts (see, e.g., Wroblewski, 1998)
- Searching for all reducts (see, e.g., (Bazan, 1998))

The algorithms for the first two problems are quite efficient from computational complexity point of view. Moreover, in practical applications, the reducts generated by them are quite closed to the optimal one. Every heuristic algorithm for reduct calculation can be modified for the use of the corresponding α-reduct problem

Example

The following example illustrates the main idea of our method. Let us consider the information table A (Table 3) with 18 objects and 9 attributes.

Assume that the Template

$\mathbf{T} = (a_1 = 0) \wedge (a_3 = 2) \wedge (a_4 = 1) \wedge (a_6 = 0) \wedge (a_8 = 1)$

has been extracted from the information table A. One can see that $support(\mathbf{T}) = 10$ and $length(\mathbf{T}) = 5$. The new decision table $A|_\mathbf{T}$ is presented in Table 3.

The discernibility function for decision table $A|_\mathbf{T}$ is as follows: (see Box 3).

After the condition presented in Table 4 is simplified, we obtain six reducts for the decision table $A|_\mathbf{T}$. (See Box 4).

Table 4. The simplified version of the discernibility matrix $M(A|_T)$; representative association rules with (100%)-confidence and representative association rules with at least (90%)-confidence

$M(A\|_T)$	$u_2, u_3, u_4, u_8, u_9,$ $u_{10}, u_{13}, u_{15}, u_{16}, u_{17}$
u_1	$D_2 \vee D_4 \vee D_5$
u_5	$D_1 \vee D_3 \vee D_4$
u_6	$D_2 \vee D_3 \vee D_4$
u_7	$D_1 \vee D_2 \vee D_3 \vee D_4$
u_{11}	$D_1 \vee D_3 \vee D_5$
u_{12}	$D_2 \vee D_3 \vee D_5$
u_{14}	$D_3 \vee D_4 \vee D_5$
u_{18}	$D_1 \vee D_5$

$\Rightarrow$

%-representative rules	
$D_3 \wedge D_5$	$D_1 \wedge D_2 \wedge D_4$
$D_4 \wedge D_5$	$D_1 \wedge D_2 \wedge D_3$
$D_1 \wedge D_2 \wedge D_3$	$D_4 \wedge D_5$
$D_1 \wedge D_2 \wedge D_4$	$D_3 \wedge D_5$
$D_1 \wedge D_2 \wedge D_5$	$D_3 \wedge D_4$
$D_1 \wedge D_3 \wedge D_4$	$D_2 \wedge D_5$

%-representative rules	
$D_1 \wedge D_2$	$D_3 \wedge D_4 \wedge D_5$
$D_1 \wedge D_3$	$D_3 \wedge D_4 \wedge D_5$
$D_1 \wedge D_4$	$D_2 \wedge D_3 \wedge D_5$
$D_1 \wedge D_5$	$D_2 \wedge D_3 \wedge D_4$
$D_2 \wedge D_3$	$D_1 \wedge D_4 \wedge D_5$
$D_2 \wedge D_5$	$D_1 \wedge D_3 \wedge D_4$
$D_3 \wedge D_4$	$D_1 \wedge D_2 \wedge D_5$

Figure 6. The illustration of 100% and 90% representative association rules

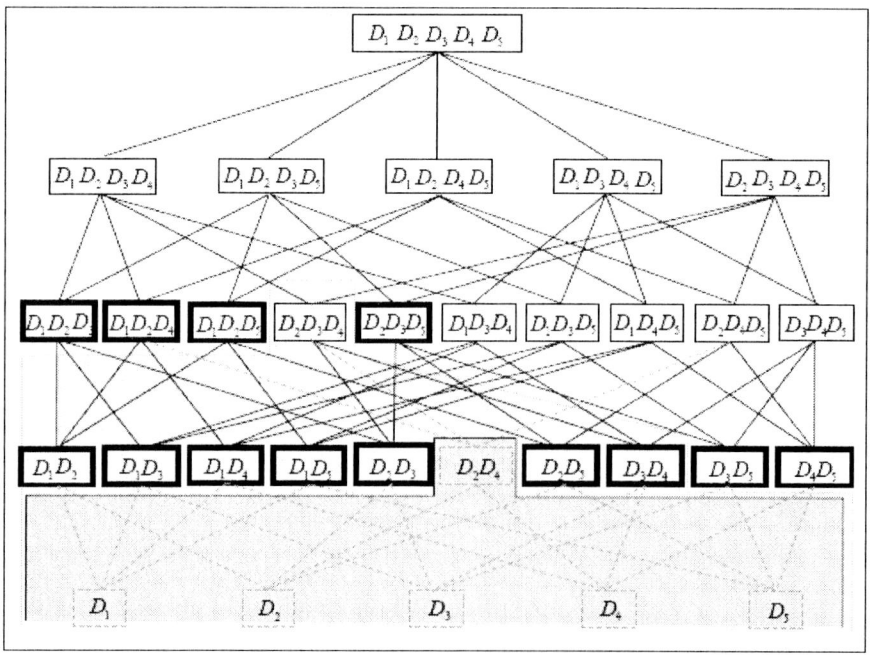

Thus, we have found from template **T** six association rules with (100%)-confidence (see Table 4).

For $c=90\%$, we would like to find α-reducts for the decision table $A|_T$, where:

$$\alpha = 1 - \frac{\frac{1}{c}-1}{\frac{n}{s}-1} = 0.86.$$

Hence, we would like to search for a set of descriptors that covers at least:

$$\lceil (n-s)(\alpha) \rceil = \lceil 8 \wedge 0.86 \rceil = 7,$$

elements of discernibility matrix $M(A|_T)$. One can see that the following sets of descriptors:

$$\{D_1, D_2\}, \{D_1, D_3\}, \{D_1, D_4\}, \{D_1, D_5\},$$
$$\{D_2, D_3\}, \{D_2, D_5\}, \{D_3, D_4\}$$

have nonempty intersection with exactly 7 members of the discernibility matrix $M(A|_T)$. Table 4 presents all association rules achieved from those sets.

In Figure 6, we present the set of all 100%–association rules (light gray region) and 90%–association rules (dark gray region). The corresponding representative association rules are represented in bold frames.

CONCLUSION

The approximate Boolean reasoning methodology was presented as an extension of the original Boolean reasoning scheme. Theoretical foundations of the proposed method, as well as many applications of rough set theory including attribute selection (calculation of reducts), decision rule induction, discretization, and feature extraction have been developed. We also presented some efficient data-mining algorithms based on the approximate Boolean reasoning approach.

We would like to emphasize the fact that approximate Boolean reasoning approach is not only a concrete method for problem solving, but it is a general methodology for development of concept approximation heuristics and data-mining solutions. The secret is embedded in the first step, where the investigated problem is encoded by a Boolean function. The encoding function creates the basis for designing different approximate solutions for the same problem. In many applications, there is no need to construct the encoding function, but the satisfactory knowledge about it facilitates to develop appropriate approximate algorithms that fulfill some predefined requirements about the quality and complexity.

Because of the volume limitation, many other applications of the approximate Boolean reasoning approach and rough set theory were omitted. The reader can find more details about approximate Boolean reasoning methodology in Nguyen (2006) as well as its recent applications in text mining (Manquinho et. al., 1997), mining data sets with continue decision attribute (Nguyen, Łuksza, Mąkosa, & Komorowski, 2005; Nguyen, 1996) and in construction of hierarchical classifiers from data and ontology (Nguyen, Bazan, Skowron, & Nguyen, 2004).

REFERENCES

Agrawal R., Imielinski, T., & Swami A.N. (1993). Mining association rules between sets of items in large databases. In P. Buneman and S. Jajodia (Eds.), *ACM SIGMOD International Conference on Management of Data* (pp. 207-216). Washington, DC.

Agrawal, R. , Mannila H., Srikant, R., Toivonen, H., & Verkamo, A. I. (1996). Fast discovery of association rules. In *Advances in Knowledge Discovery and Data Mining* (pp. 307-328). Menlo Park, CA: AAAI Press/The MIT Press.

Agrawal, R, & Srikant, R. (1994). Fast algorithms for mining association rules. In J.B. Bocca, M. Jarke, & C. Zaniolo (Eds.), *Twentieth Inter-*

national *Conference on Very Large Data Bases VLDB* (pp. 487-499). Morgan Kaufmann.

Bazan. J. (1998). A comparison of dynamic and non-dynamic rough set methods for extracting laws from decision tables. In L. Polkowski & A. Skowron (Eds.), *Rough sets in knowledge discovery 1: Methodology and applications. Studies in fuzziness and soft computing,* 18 (pp. 321-365). Heidelberg, Germany: Springer.

Bazan, J., Nguyen, H.S., Nguyen, S. H., Synak, P., & . Wróblewski. J. (2000). Rough set algorithms in classification problems. In L. Polkowski, T.Y. Lin, &. S. Tsumoto, (Eds.), *Studies in fuzziness and soft computing* (pp. 49-88). Heidelberg, Germany: Springer.

Bazan, J., Nguyen, H.S., Skowron, A , &. Szczuka. M., (2003). A view on rough set concept approximation. In G. Wang, Q. Liu, Y. Yao, & A. Skowron (Eds.), *Proceedings of the Ninth International Conference on Rough Sets, Fuzzy Sets, Data Mining and Granular Computing (RSFDGrC2003)*,Chongqing, China. *Lecture notes in artificial intelligence, 2639* (pp. 181-188). Heidelberg: Springer.

Bazan J., & Szczuka. M. (2001). RSES and RSESlib: A collection of tools for rough set computations. In W. Ziarko & Y. Yao (Eds.), *Proceedings of the 2nd International Conference on Rough Sets and Current Trends in Computing (RSCTC2000)*, Banff, Canada, *Lecture notes in artificial intelligence, 2005* (pp. 106-113). Heidelberg, Germany: Springer.

Blake, A. (1973). *Canonical expressions in Boolean algebra.* PhD thesis, University of Chicago.

Boole. G. (1854). *The law of thought.* MacMillan (also Dover Publications, New-York).

Breiman, L., Friedman, J., Olshen, R., & Stone, C. (1984). *Classification and regression trees.* Monterey, CA: Wadsworth and Brooks.

Brown, F. (1990). *Boolean reasoning.* Dordrecht, Germany: Kluwer Academic Publishers.

Fayyad, U.M., & Irani, K.B. (1993). Multi-interval discretization of continuous-valued attributes for classification learning. *IJCAI,* 1022-1029.

Garey, M.R., & Johnson, D.S. (1979). *Computers and intractability: A guide to the theory of NP completeness.* New York: W.H. Freeman & Co.

Han, J., Pei, J., & Yin, Y. (2000). Mining frequent patterns without candidate generation. In W. Chen, J. Naughton, & P.A. Bernstein (Eds.), *ACM SIGMOD International Conference on Management of Data* (pp. 1-12). ACM Press.

Huntington, E.V. (1933). Boolean algebra. A correction. *Transactions of AMS, 35,* 557-558.

Jensen, R., Shen, Q., & Tuso, A. (2005). Finding rough set reducts with SAT. In D. Ślęzak, G. Wang, M. Szczuka, I. Düntsch, & Y. Yao (Eds.), *Proceedings of the 10th International Conference on Rough Sets, Fuzzy Sets, Data Mining, and Granular Computing (RSFDGrC2005),* Regina, Canada, *Lecture notes in artificial intelligence, 3641* (pp. 194-203). Heidelberg, Germany: Springer.

Kautz, H.A., & Selman B. (1992). Planning as satisfiability. In *Proceedings of the Tenth European Conference on Artificial Intelligence (ECAI'92)* (pp. 359-363).

Kautz, H.A., & Selman. B. (1996). Pushing the envelope: Planning, propositional logic, and stochastic search. In *Proceedings of the Twelfth National Conference on Artificial Intelligence (AAAI'96)* (pp. 1194-1201).

Kryszkiewicz, M. (1997). Maintenance of reducts in the variable precision rough set model. In T.Y. Lin & N. Cercone (Eds.), *Rough sets and data mining – Analysis of imperfect data* (pp. 355-372). Boston: Kluwer Academic Publishers.

Liu, H., & Motoda, H. (Eds.). (1999). *Feature selection for knowledge discovery and data mining.* Kluwer Academic Publishers.

Manquinho, V.M., Flores, P.F., Silva, J.P.M.., &. Oliveira A.L. (1997). Prime implicant computation using satisfiability algorithms. In the *International Conference on Tools with Artificial Intelligence (ICTAI '97)* (pp. 232-239).

Ngo, C.L., & Nguyen H.S. (2005). A method of web search result clustering based on rough sets. In *Proceedings of 2005 IEEE/WIC/ACM International Conference on Web Intelligence* (pp. 673-679). IEEE Computer Society Press.

Nguyen, H.S. (1997). *Discretization of real value attributes, Boolean reasoning approach*. PhD thesis, Warsaw University, Warsaw, Poland.

Nguyen, H.S. (1999). Efficient SQL-querying method for data mining in large data bases. In *Proceedings of Sixteenth International Joint Conference on Artificial Intelligence, IJCAI-99* (pp. 806-811). Stockholm, Sweden: Morgan Kaufmann.

Nguyen, H.S. (2001). On efficient handling of continuous attributes in large data bases. *Fundamenta Informaticae, 48*(1), 61-81.

Nguyen, H.S. (2006a). Approximate Boolean reasoning: Foundations and applications in data mining. Transactions on rough sets V. *Lecture notes on computer science, 4100*, 334-506.

Nguyen, H.S. (2006b). Knowledge discovery by relation approximation: A rough set approach. In *Proceedings of the International Conference on Rough Sets and Knowledge Technology (RSKT)*, Chongqing, China, July 24-26, 2006. *LNAI, 4062* (pp. 103-106). Heidelberg, Germany: Springer.

Nguyen, H.S., Łuksza, M., Mąkosa, E., & Komorowski, J. (2005). An approach to mining data with continuous decision values. In M.A. Kłopotek, S.T. Wierzchoń, & K. Trojanowski (Eds.), *Proceedings of the International IIS: IIPWM'05 Conference*, Gdansk, Poland. *Advances in Soft Computing* (pp. 653-662). Heidelberg, Germany: Springer.

Nguyen, H.S., & Nguyen S.H. (1998). Discretization methods for data mining. In L. Polkowski & A. Skowron (Eds.), *Rough sets in knowledge discovery* (pp. 451-482). Heidelberg; New York: Springer.

Nguyen, H.S., &. Nguyen S.H. (2005). Fast split selection method and its application in decision tree construction from large databases. *International Journal of Hybrid Intelligent Systems, 2*(2), 149-160.

Nguyen, H.S., & Ślęzak D. (1999). Approximate reducts and association rules—Correspondence and complexity results. In A. Skowron, S. Ohsuga, & N. Zhong (Eds.), *New directions in rough sets, data mining and granular-soft computing, Proceedings of RSFDGrC'99*, Yamaguchi, Japan, *LNAI, 1711* (pp. 137-145). Heidelberg, Germany: Springer.

Nguyen, S.H. (2000). Regularity analysis and its applications in Data Mining. In L. Polkowski, T.Y. Lin, & S. Tsumoto (Eds.), *Rough set methods and applications: New developments in knowledge discovery in information systems. Studies in fuzziness and soft computing, 56* (pp. 289-378). Heidelberg, Germany: Springer.

Nguyen, S.H., Bazan, J., Skowron, A., & Nguyen, H.S. (2004). Layered learning for concept synthesis. Transactions on rough sets I. *Lecture Notes on Computer Science, 3100*, 187-208. Heidelberg, Germany: Springer.

Nguyen, S.H., & Nguyen H.S. (1996). Some efficient algorithms for rough set methods. In *Sixth International Conference on Information Processing and Management of Uncertainty on Knowledge Based Systems IPMU'1996*, volume III (pp. 1451-1456). Granada, Spain.

Nguyen, S.H., Skowron, A., & Synak P. (1998). Discovery of data patterns with applications to decomposition and classification problems. In L. Polkowski & A. Skowron (Eds.), *Rough sets in knowledge discovery 2: Applications, case studies and software systems. Studies in fuzziness and soft computing, 19* (pp. 55-97). Heidelberg, Germany: Springer.

Ohrn, A., Komorowski, J., Skowron, A., & Synak P. (1998). The ROSETTA software system. In L. Polkowski & A. Skowron (Eds.), *Rough sets in knowledge discovery 2: Applications, case studies and software systems. Studies in fuzziness and soft computing, 19* (pp. 572-576). Heidelberg, Germany: Springer.

Pawlak, Z. (1982). Rough sets. *International Journal of Computer and Information Sciences, 11*, 341-356.

Pawlak, Z. (1991). Rough sets: Theoretical aspects of reasoning about data. *System theory, knowledge engineering and problem solving, 9*. Dordrecht, The Netherlands: Kluwer Academic Publishers.

Pizzuti C. (1996). Computing prime implicants by integer programming. In *Eighth International Conference on Tools with Artificial Intelligence (ICTAI '96)* (pp. 332-336).

Polkowski, L., Lin, T.Y., &. Tsumoto, S. (Eds.). (2000). Rough set methods and applications: New developments in knowledge discovery in information systems. *Studies in Fuzziness and Soft Computing, 56*. Heidelberg, Germany: Springer.

Quine, W.V.O. (1952). The problem of simplifying truth functions. *American Mathematical Monthly, 59*, 521-531.

Quine, W.V.O. (1959). On cores and prime implicants of truth functions. *American Mathematical, 66*, 755-760.

Quine, W.V.O. (1961). *Mathematical logic*. Cambridge, MA: Harvard University Press.

Quinlan, J. (1993). *C4.5—Programs for machine learning*. Morgan Kaufmann.

Quinlan, R. (1986). Induction of decision trees. *Machine Learning, 1*, 81-106.

Rudeanu. S. (1974). *Boolean functions and equations*. Amsterdam: North-Holland/American Elsevier.

Selman, B., Kautz, H.A., &. McAllester D.A. (1997). Ten challenges in propositional reasoning and search. In *Proceedings of Fifteenth International Joint Conference on Artificial Intelligence* (pp. 50-54).

Sen, S. (1993) . Minimal cost set covering using probabilistic methods. In *SAC '93: Proceedings of the 1993 ACM/SIGAPP Symposium on Applied Computing* (pp. 157-164). New York: ACM Press.

Shannon, C.E. (1938). A symbolic analysis of relay and switching circuits. *Transactions of AIEE, 57*, 713-723.

Shannon, C.E. (1940). *A symbolic analysis of relay and switching circuits*. MIT, Dept. of Electrical Engineering.

Skowron, A. (1993). Boolean reasoning for decision rules generation. In J. Komorowski & Z.W. Raś (Eds.), *Seventh International Symposium for Methodologies for Intelligent Systems ISMIS, Lecture Notes in Artificial Intelligence, 689* (pp. 295-305). Trondheim, Norway: Springer.

Skowron, A., Pawlak, Z., Komorowski, J., & Polkowski L. (2002). A rough set perspective on data and knowledge. In W. Kloesgen & J. Żytkow (Eds.), *Handbook of KDD* (pp. 134-149). Oxford: Oxford University Press.

Skowron, A., & Rauszer C. (1992). The discernibility matrices and functions in information systems. In R. Słowiński (Ed.), *Intelligent decision support – Handbook of applications and advances of the rough sets theory* (pp. 331-362). Dordrecht, The Netherlands: Kluwer Academic Publishers.

Skowron, A., & Stepaniuk J. (1996). Tolerance approximation spaces. *Fundamenta Informaticae, 27*(2-3), 245-253.

Ślęzak, D. (2000). Various approaches to reasoning with frequency-based decision reducts: A survey. In L. Polkowski, T. Y. Lin, &. S. Tsumoto, S. (Eds.), *Studies in Fuzziness and Soft Computing, 56* (pp. 235-285). Heidelberg, Germany: Springer.

Ślęzak, D. (2002). Approximate entropy reducts. *Fundamenta Informaticae, 53*, 365-387.

Stone, M.H. (1963). The theory of representations for Boolean algebras. *Transactions of AMS, 40*, 37–111.

Wróblewski, J. (1996). Theoretical foundations of order-based genetic algorithms. *Fundamenta Informaticae, 28*(3-4), 423-430.

Wróblewski,. J. (1998). Genetic algorithms in decomposition and classification problem. In L. Polkowski & A. Skowron (Eds.), *Rough sets in knowledge discovery 2: Applications, case studies and software systems. Studies in Fuzziness and Soft Computing, 19* (pp. 471-487). Heidelberg, Germany: Springer.

Zaki, M. (1998). Efficient enumeration of frequent sequences. In *Seventh International Conference on Information and Knowledge Management* (pp. 68-75). Washington, DC.

Ziarko, W. (1993). Variable precision rough set model. *Journal of Computer and System Sciences, 46*, 39-59.

Chapter III
Rough Set–Based Feature Selection:
A Review

Richard Jensen
The University of Wales, UK

Qiang Shen
The University of Wales, UK

ABSTRACT

Feature selection aims to determine a minimal feature subset from a problem domain while retaining a suitably high accuracy in representing the original features. Rough set theory (RST) has been used as such a tool with much success. RST enables the discovery of data dependencies and the reduction of the number of attributes contained in a dataset using the data alone, requiring no additional information. This chapter describes the fundamental ideas behind RST-based approaches, and reviews related feature selection methods that build on these ideas. Extensions to the traditional rough set approach are discussed, including recent selection methods based on tolerance rough sets, variable precision rough sets, and fuzzy rough sets. Alternative search mechanisms are also highly important in rough set feature selection. The chapter includes the latest developments in this area, including RST strategies based on hill climbing, genetic algorithms, and ant colony optimization.

INTRODUCTION

The main aim of feature selection (FS) is to determine a minimal feature subset from a problem domain while retaining a suitably high accuracy in representing the original features. In many real world problems, FS is a must due to the abundance of noisy, irrelevant, or misleading features. For instance, by removing these factors, learning from data techniques can benefit greatly. A detailed review of feature selection techniques devised for classification tasks can be found in Dash and Liu (1997).

The usefulness of a feature or feature subset is determined by both its *relevancy* and *redundancy*. A feature is said to be relevant if it is predictive of the decision feature(s), otherwise it is irrelevant. A feature is considered to be redundant if it is highly correlated with other features. Hence, the search for a good feature subset involves finding those features that are highly correlated with the decision feature(s), but are uncorrelated with each other.

A taxonomy of feature selection approaches can be seen in Figure 1. Given a feature set size n, the task of FS can be seen as a search for an 'optimal' feature subset through the competing 2^n candidate subsets. The definition of what an optimal subset is may vary depending on the problem to be solved. Although an exhaustive method may be used for this purpose in theory, this is quite impractical for most datasets. Usually FS algorithms involve heuristic or random search strategies in an attempt to avoid this prohibitive complexity. However, the degree of optimality of the final feature subset is often reduced. The overall procedure for any feature selection method is given in Figure 2 (adapted from Dash & Liu, 1997).

The generation procedure implements a search method (Langley 1994; Siedlecki & Sklansky, 1988) that generates subsets of features for evaluation. It may start with no features, all features, a selected feature set, or some random feature subset. Those methods that start with an initial subset usually select these features heuristically

Figure 1. Aspects of feature selection

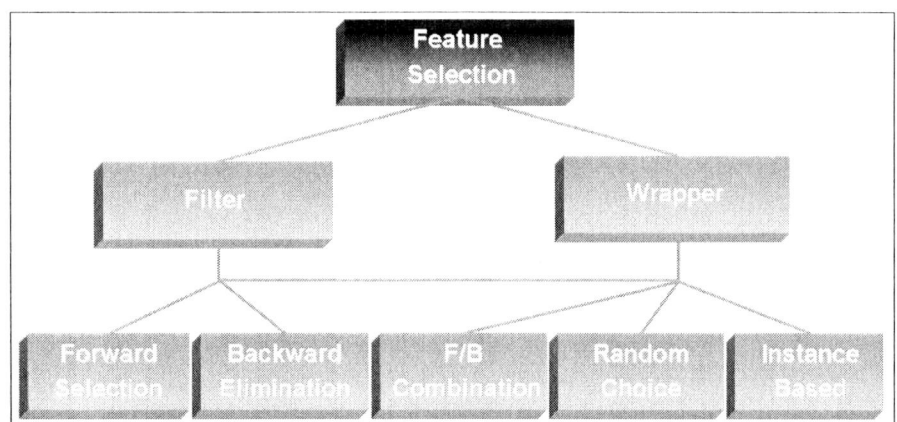

Figure 2. Feature selection

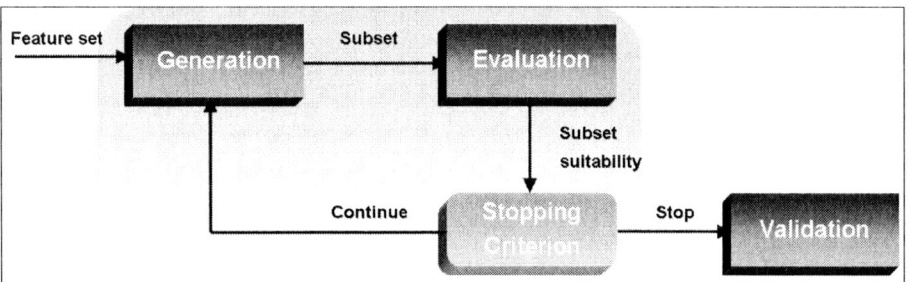

beforehand. Features are added (*forward selection*) or removed (*backward elimination*) iteratively in the first two cases (Dash & Liu, 1997). In the last case, features are either iteratively added, or removed or produced randomly thereafter. An alternative selection strategy is to select instances and examine differences in their features. The evaluation function calculates the suitability of a feature subset produced by the generation procedure and compares this with the previous best candidate, replacing it if found to be better.

A stopping criterion is tested every iteration to determine whether the FS process should continue or not. For example, such a criterion may be to halt the FS process when a certain number of features have been selected if based on the generation process. A typical stopping criterion centred on the evaluation procedure is to halt the process when an optimal subset is reached. Once the stopping criterion has been satisfied, the loop terminates. For use, the resulting subset of features may be validated.

Determining subset optimality is a challenging problem. There is always a trade-off in nonexhaustive techniques between subset minimality and subset suitability; the task is to decide which of these must suffer in order to benefit the other. For some domains (particularly where it is costly or impractical to monitor many features), it is much more desirable to have a smaller, less accurate feature subset. In other areas it may be the case that the modelling accuracy (e.g., the classification rate) using the selected features must be extremely high, at the expense of a nonminimal set of features.

Feature selection algorithms may be classified into two categories based on their evaluation procedure (see Figure 3). If an algorithm performs FS independently of any learning algorithm (i.e., it is a completely separate preprocessor), then it is a *filter* approach. In effect, irrelevant attributes are filtered out before induction. Filters tend to be applicable to most domains, as they are not tied to any particular induction algorithm.

If the evaluation procedure is tied to the task (e.g., classification) of the learning algorithm, the FS algorithm employs the *wrapper* approach. This method searches through the feature subset space using the estimated accuracy from an induction algorithm as a measure of subset suitability. Although wrappers may produce better results, they are expensive to run, and can break down with very large numbers of features. This is due to the use of learning algorithms in the evaluation of subsets, some of which can encounter problems when dealing with large datasets.

This chapter reviews generic filter-based methods to feature selection based on rough set

Figure 3 Filter and wrapper methods

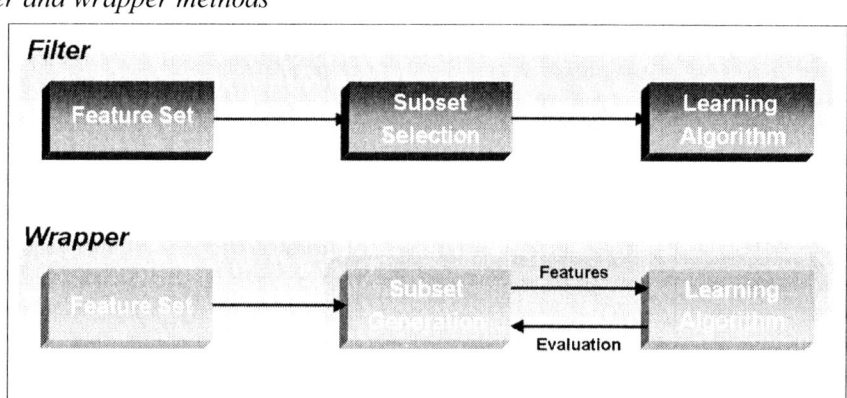

theory. Most developments in the area concentrate on using rough sets within the evaluation function. This can be achieved by employing the rough set dependency degree (including extensions of this measure) or through the use of discernibility functions. The theoretical background and recent approaches for this are found in the next section. Following this, details are given concerning RST-based approaches to handling real-valued and noisy data. Although the techniques reviewed here are primarily for feature selection, their use is not restricted solely to this field. Alternative search mechanisms are presented, in the following section, that attempt to tackle the problem of locating globally optimal subsets. Finally, the chapter is concluded, including a discussion of future research issues.

ROUGH-BASED FEATURE SELECTION

Rough set theory (RST) can be used as a tool to discover data dependencies and to reduce the number of attributes contained in a dataset using the data alone, requiring no additional information (Pawlak, 1991; Polkowski, 2002). Over the past 10 years, RST has become a topic of great interest to researchers, and has been applied to many domains. Given a dataset with discretized attribute values, it is possible to find a subset (termed a *reduct*) of the original attributes using RST that are the most informative; all other attributes can be removed from the dataset with minimal information loss. From the dimensionality reduction perspective, informative features are those that are most predictive of the class attribute.

There are two main approaches to finding rough set reducts: those that consider the degree of dependency and those that are concerned with the discernibility matrix. This section describes the fundamental ideas behind both approaches. To illustrate the operation of these, an example dataset (Table 1) will be used.

Dependency Function-Based Approaches

RSAR

Central to rough set attribute reduction (RSAR) (Chouchoulas & Shen, 2001; Jensen & Shen, 2004b) is the concept of indiscernibility. Let $I = (U, A)$ be an information system, where U is a nonempty set of finite objects (the universe) and A is a nonempty finite set of attributes such that $a: U \rightarrow V_a$ for every $a \in A$. V_a is the set of values that attribute a may take. With any $P \subseteq A$ there is an associated equivalence relation $IND(P)$:

$$IND(P) = \{(x,y) \in U^2 \mid \forall a \in P, a(x) = a(y)\}$$
(1)

The partition of U generated by $IND(P)$ is denoted $U/IND(P)$ (or U/P). If $(x, y) \in IND(P)$, then x and y are indiscernible by attributes from P. The equivalence classes of the P-indiscernibility relation are denoted $[x]_P$. For the illustrative example, if $P = \{b,c\}$, then objects 1, 6, and 7 are indiscernible; as are objects 0 and 4. $IND(P)$ creates the following partition of U:

$$U / IND(P) = U / IND(\{b\}) \otimes U / IND(\{c\})$$

Table 1. An example dataset

$x \in U$	a	b	c	d	e
0	1	0	2	2	0
1	0	1	1	1	2
2	2	0	0	1	1
3	1	1	0	2	2
4	1	0	2	0	1
5	2	2	0	1	1
6	2	1	1	1	2
7	0	1	1	0	1

= {{0,2,4},{1,3,6,7},{5}} ⊗ {{2,3,5},{1,6,7},{0,4}}
= {{2},{0,4},{3},{1,6,7},{5}}

Let $X \subseteq U$. X can be approximated using only the information contained within P by constructing the P-*lower* and P-*upper* approximations of X:

$$\underline{P}X = \{x \mid [x]_P \subseteq X\} \quad (2)$$

$$\overline{P}X = \{x \mid [x]_P \cap X \neq \emptyset\} \quad (3)$$

Let P and Q be equivalence relations over U, then the positive region can be defined as:

$$POS_P(Q) = \bigcup_{X \in U/Q} \underline{P}X \quad (4)$$

The positive region contains all objects of U that can be classified to classes of U/Q using the information in attributes P. For example, let $P = \{b,c\}$ and $Q = \{e\}$, then:

$$POS_P(Q) = \bigcup\{\emptyset,\{2,5\},\{3\}\} = \{2,3,5\}$$

Using this definition of the positive region, the rough set degree of dependency of a set of attributes Q on a set of attributes P is defined in the following way:

For $P, Q \subset A$, it is said that Q depends on P in a degree k ($0 \leq k \leq 1$), denoted $P \rightarrow_k Q$, if:

$$k = \gamma_P(Q) = \frac{|POS_P(Q)|}{|U|} \quad (5)$$

In the example, the degree of dependency of attribute $\{e\}$ from the attributes $\{b,c\}$ is:

$$\gamma_{\{b,c\}}(\{e\}) = \frac{|POS_{\{b,c\}}(\{e\})|}{|U|}$$

$$= \frac{|\{2,3,5\}|}{|\{0,1,2,3,4,5,6,7\}|} = \frac{3}{8}$$

The reduction of attributes is achieved by comparing equivalence relations generated by sets of attributes. Attributes are removed so that the reduced set provides the same predictive capability of the decision feature as the original. A reduct R is defined as a subset of minimal cardinality of the conditional attribute set C such that $\gamma_R(D) = \gamma_C(D)$.

The QUICKREDUCT algorithm, given in Figure 4, attempts to calculate a reduct without exhaustively generating all possible subsets. It starts off with an empty set and adds, in turn, one at a time, those attributes that result in the greatest increase in the rough set dependency metric, until this produces its maximum possible value for the dataset.

According to the algorithm, the dependency of each attribute is calculated, and the best candidate chosen. Attribute d generates the highest dependency degree, so that attribute is chosen and the sets $\{a,d\}$, $\{b,d\}$, and $\{c,d\}$ are evaluated. This process continues until the dependency of the reduct equals the consistency of the dataset (1 if the dataset is consistent). In the example, the algorithm terminates after evaluating the subset $\{b,d\}$. The generated reduct shows the way of reducing the dimensionality of the original dataset by eliminating those conditional attributes that do not appear in the set.

Figure 4. The QUICKREDUCT algorithm

QUICKREDUCT(C,D)
C, the set of all conditional features;
D, the set of decision features.

(1) $R \leftarrow \{\}$
(2) **do**
(3) $T \leftarrow R$
(4) $\forall x \in (C - R)$
(5)   **if** $\gamma_{R \cup \{x\}}(D) > \gamma_T(D)$
(6) $T \leftarrow R \cup \{x\}$
(7) $R \leftarrow T$
(8) **until** $\gamma_R(D) = \gamma_C(D)$
(9) **return** R

This, however, is not guaranteed to find a minimal subset. Using the dependency function to discriminate between candidates may lead the search down a nonminimal path. It is impossible to predict which combinations of attributes will lead to an optimal reduct based on changes in dependency with the addition or deletion of single attributes. It does result in a close-to-minimal subset, though, which is still useful in greatly reducing dataset dimensionality.

VPRS

Variable precision rough sets (VPRS) (Ziarko, 1993) extends rough set theory by the relaxation of the subset operator. It was proposed to analyse and identify data patterns that represent statistical trends rather than functional. The main idea of VPRS is to allow objects to be classified with an error smaller than a certain predefined level. This introduced threshold relaxes the rough set notion of requiring no information outside the dataset itself. Let $X, Y \subseteq U$, the relative classification error is defined by:

$$c(X,Y) = 1 - \frac{|X \cap Y|}{|X|} \qquad (6)$$

Observe that $c(X,Y) = 0$ if and only if $X \subseteq Y$. A degree of inclusion can be achieved by allowing a certain level of error, β, in classification:

$$X \subseteq_\beta Y \text{ iff } c(X,Y) \leq \beta, \quad 0 \leq \beta < 0.5$$

Using $\subseteq_\beta$ instead of $\subseteq$, the β-upper and β-lower approximations of a set X can be defined as:

$$\underline{P_\beta} X = \cup \{[x]_P \in U/P \mid [x]_P \subseteq_\beta X\} \qquad (7)$$

$$\overline{P_\beta} X = \cup \{[x]_P \in U/P \mid c([x]_P, X) < 1-\beta\} \qquad (8)$$

Note that $\underline{P_\beta} X = \underline{P} X$ for $\beta = 0$. The positive, negative, and boundary regions in the original rough set theory can now be extended to:

$$POS_{P,\beta}(Q) = \bigcup_{X \in U/Q} \underline{P_\beta} X$$

$$NEG_{P,\beta}(Q) = U - \bigcup_{X \in U/Q} \overline{P_\beta} X$$

$$BND_{P,\beta}(Q) = \bigcup_{X \in U/Q} \overline{P_\beta} X - \bigcup_{X \in U/Q} \underline{P_\beta} X$$

where P is also an equivalence relation on U. This can then be used to calculate dependencies and thus determine β-reducts. The dependency function becomes:

$$\gamma_{P,\beta}(Q) = \frac{|POS_{P,\beta}(Q)|}{|U|} \qquad (9)$$

Returning to the example dataset in Table 1, the β-positive region can be calculated for $P = \{b,c\}$, $Q = \{e\}$, and $\beta = 0.4$. Setting β to this value means that a set is considered to be a subset of another if at least 60% of their elements are shared. The partitions of the universe of objects for P and Q are:

$U/P = \{\{2\},\{0,4\},\{3\},\{1,6,7\},\{5\}\}$
$U/Q = \{\{0\},\{1,3,6\},\{2,4,5,7\}\}$

For each set $A \in U/P$ and $B \in U/Q$ the value of $c(A,B)$ must be less than β if the equivalence class A is to be included in the β-positive region. Considering $A = \{2\}$ gives:

$c(\{2\},\{0\}) = 1 > \beta$
$c(\{2\},\{1,3,6\}) = 1 > \beta$
$c(\{2\},\{2,4,5,7\}) = 0 < \beta$

So object 2 is added to the β-positive region, as it is a β-subset of $\{2,4,5,7\}$ (and is in fact a traditional subset of the equivalence class). Taking $A = \{1,6,7\}$, a more interesting case is encountered:

$c(\{1,6,7\},\{0\}) = 1 > \beta$
$c(\{1,6,7\},\{1,3,6\}) = 0.3333 < \beta$
$c(\{1,6,7\},\{2,4,5,7\}) = 0.6667 > \beta$

Here the objects 1, 6, and 7 are included in the β-positive region as the set {1,6,7} is a β-subset of {1,3,6}. Calculating the subsets in this way leads to the following β-positive region:

$$POS_{P,\beta}(X) = \{1,2,3,5,6,7\}$$

Compare this with the positive region generated previously: {2,3,5}. Objects 1, 6, and 7 are now included due to the relaxation of the subset operator. If the original dataset contained noise, it could have been the case that these objects did indeed belong to the positive region. Using traditional rough set theory, this would not have been possible due to the inflexibility of the subset operator.

It can be seen that the QUICK REDUCT algorithm outlined previously can be adapted to incorporate the reduction method built upon the VPRS theory. By supplying a suitable β value to the algorithm, the β-lower approximation, β-positive region, and β-dependency can replace the traditional calculations. This will result in a more approximate final reduct, which may be a better generalization when encountering unseen data. Additionally, setting β to 0 forces such a method to behave exactly like RSAR.

Extended classification of reducts in the VPRS approach may be found in Beynon (2000, 2001) and Kryszkiewicz (1994). As yet, there have been no comparative experimental studies between rough set methods and the VPRS method. However, the variable precision approach requires the additional parameter β, which has to be specified from the start. By repeated experimentation, this parameter can be suitably approximated. However, problems arise when searching for true reducts as VPRS incorporates an element of inaccuracy in determining the number of classifiable objects.

Dynamic Reducts

Reducts generated from an information system are sensitive to changes in the system. This can be seen by removing a randomly chosen set of objects from the original object set. Those reducts frequently occurring in random subtables can be considered to be stable; it is these reducts that are encompassed by *dynamic reducts* (Bazan, Skowron, & Synak, 1994).

Let $A = (U, C \cup d)$ be a decision table; then any system $B = (U', C \cup d)$ ($U' \subseteq U$) is called a subtable of A. If F is a family of subtables of A, then

$$DR(A,F) = Red(A,d) \cap \{\bigcap_{B \in F} Red(B,d)\} \quad (10)$$

defines the set of F-dynamic reducts of A. From this definition, it follows that a relative reduct of A is dynamic if it is also a reduct of all subtables in F. In most cases this is too restrictive, so a more general notion of dynamic reducts is required.

By introducing a threshold, $0 \leq \varepsilon \leq 1$, the concept of (F, ε)-dynamic reducts can be defined:

$$DR_\varepsilon(A,F) = \{C \in Red(A,d) : s_F(C) \geq \varepsilon\} \quad (11)$$

where

$$s_F(C) = \frac{|\{B \in F : C \in Red(B,d)\}|}{|F|} \quad (12)$$

is the F-stability coefficient of C. This lessens the previous restriction that a dynamic reduct must appear in *every* generated subtable. Now, a reduct is considered to be dynamic if it appears in a certain proportion of subtables, determined by the value ε. For example, by setting ε to 0.5, a reduct is considered to be dynamic if it appears in at least half of the subtables. Note that if $F = \{A\}$ then $DR(A,F) = Red(A,d)$. Dynamic reducts may then be calculated according to the algorithm given in Figure 5.

Firstly, all reducts are calculated for the given information system, A. Then, the new subsystems A_i are generated by randomly deleting one or more rows from A. All reducts are found for each

Figure 5. Dynamic reduct algorithm

```
DynamicRed(A, ε, its)
  A, the original decision table;
  ε, the dynamic reduct threshold;
  its, the number of iterations.

(1)  R ← {}
(2)  T ← calculateAllReducts(A)
(3)  for j = 1...its
(4)    A_j ← deleteRandomRows(A)
(5)    R ← R ∪ calculateAllReducts(A_j)
(6)  ∀C ∈ T
(7)    if s_F(C, R) ≥ ε
(8)      output C
```

subsystem, and the dynamic reducts are computed using $s_F(C,R)$, which denotes the significance factor of reduct C within all reducts found, R.

Returning to the example decision table (call this A), the first step is to calculate all its reducts. This produces the set of all reducts $A = \{\{b,d\},\{c,d\},\{a,b,d\},\{a,c,d\},\{b,c,d\}\}$. The reduct $\{a,b,c,d\}$ is not included, as this will always be a reduct of any generated subtable (it is the full set of conditional attributes). The next step randomly deletes a number of rows from the original table A. From this, all reducts are again calculated; for one subtable this might be $R = \{\{b,d\},\{b,c,d\},\{a,b,d\}\}$ In this case, the subset $\{c,d\}$ is not a reduct (though it was for the original dataset). If the number of iterations is set to just one, and if ε is set to a value less than 0.5 (implying that a reduct should appear in half of the total number of discovered reducts), then the reduct $\{c,d\}$ is deemed not to be a dynamic reduct.

Intuitively, this is based on the hope that by finding stable reducts, they will be more representative of the real world, that is, it is more likely that they will be reducts for unseen data. A comparison of dynamic and nondynamic approaches can be found in Bazan (1998), where various methods were tested on extracting laws from decision tables. In the experiments, the dynamic method and the conventional RS method both performed well. In fact, it appears that the RS method has, on average, a lower error rate of classification than the dynamic RS method.

A disadvantage of this dynamic approach is that several subjective choices have to be made before the dynamic reducts can be found (for instance the choice of the value of ε; these values are not contained in the data). Also, the huge complexity of finding all reducts within subtables forces the use of heuristic techniques such as genetic algorithms to perform the search. For large datasets, this step may well be too costly.

Han, Hu, and Lin

In Han et al. (2004), a feature selection method based on an alternative dependency measure is presented. The technique was originally proposed to avoid the calculation of discernibility functions or positive regions, which can be computationally expensive without optimizations.

The authors replace the traditional rough set degree of dependency with an alternative measure, the relative dependency, defined as follows for an attribute subset R:

$$\kappa_R(D) = \frac{|U / IND(R)|}{|U / IND(R \cup D)|} \qquad (13)$$

The authors then show that R is a reduct if and only if $\kappa_R(D) = \kappa_C(D)$ and $\forall X \subset R, \ \kappa_X(D) \neq \kappa_C(D)$.

Two algorithms are constructed for feature selection based on this measure. The first (Figure 6) performs backward elimination of features, where attributes are removed from the set of considered attributes if the relative dependency equals 1 upon their removal. Attributes are considered one at a time, starting with the first, evaluating their relative dependency. The second algorithm initially ranks the individual attributes beforehand, using an entropy measure before the backward elimination is performed.

Figure 6. Backward elimination based on relative dependency

```
RelativeReduct(C,D)
  C, the conditional attributes;
  D, the decision attributes;

(1)  R ← C
(2)  ∀a ∈ C
(3)     if (κ_{R-{a}}(D) == 1)
(4)        R ← R - {a}
(6)  return R
```

Returning to the example dataset, the backward elimination algorithm initializes R to the set of conditional attributes, $\{a,b,c,d\}$. Next, the elimination of attribute a is considered:

$$\kappa_{\{b,c,d\}}(D) = \frac{|U/IND(\{b,c,d\})|}{|U/IND(\{b,c,d,e\})|}$$

$$= \frac{|\{\{0\},\{1,6\},\{2\},\{3\},\{4\},\{5\},\{7\}\}|}{|\{\{0\},\{1,6\},\{2\},\{3\},\{4\},\{5\},\{7\}\}|} = 1$$

As the relative dependency is equal to 1, attribute a can be removed from the current reduct candidate $R \leftarrow \{b,c,d\}$. The algorithm then considers the elimination of attribute b from R:

$$\kappa_{\{c,d\}}(D) = \frac{|U/IND(\{c,d\})|}{|U/IND(\{c,d,e\})|}$$

$$= \frac{|\{\{0\},\{1,6\},\{2,5\},\{3\},\{4\},\{7\}\}|}{|\{\{0\},\{1,6\},\{2,5\},\{3\},\{4\},\{7\}\}|} = 1$$

Again, the relative dependency of $\{c,d\}$ evaluates to 1, so attribute b is removed from R, ($R = \{c,d\}$). The next step evaluates the removal of c from the reduct candidate:

$$\kappa_{\{d\}}(D) = \frac{|U/IND(\{d\})|}{|U/IND(\{d,e\})|}$$

$$= \frac{|\{\{0,3\},\{1,2,5,6\},\{4,7\}\}|}{|\{\{0\},\{3\},\{1,6\},\{2,5\},\{4,7\}\}|} = \frac{3}{5}$$

As this does not equal 1, attribute d is not removed from R. The algorithm then evaluates the elimination of attribute d from R ($R = \{c,d\}$):

$$\kappa_{\{c\}}(D) = \frac{|U/IND(\{c\})|}{|U/IND(\{c,e\})|}$$

$$= \frac{|\{\{0,4\},\{1,6,7\},\{2,3,5\}\}|}{|\{\{0\},\{4\},\{1,6\},\{7\},\{2,5\},\{3\}\}|} = \frac{3}{6}$$

Again, the relative dependency does not evaluate to 1; hence, attribute d is retained in the reduct candidate. As there are no further attributes to consider, the algorithm terminates and outputs the reduct $\{c,d\}$.

Zhong, Dong, and Ohsuga

In Zhong et al. (2001), a heuristic filter-based approach is presented based on rough set theory. The algorithm proposed, as reformalised in Figure 7, begins with the core of the dataset (those attributes that cannot be removed without

Figure 7. Heuristic filter-based algorithm

```
select(C,D,O,ε)
  C, the set of all conditional features;
  D, the set of decision features.
  O, the set of objects (instances)
  ε, the reduct threshold

(1)  R ← calculateCore()
(2)  while γ_R(D) < ε
(3)     O ← O - POS_R(D)  //optimization
(4)     ∀x ∈ (C - R)
(5)        v_a = | POS_{R∪{a}}(D) |
(6)        m_a ← | largestEquivClass( POS_{R∪{a}}(D) ) |
(7)     Choose a with largest v_a × m_a
(8)     R ← R ∪ {a}
(9)  return R
```

introducing inconsistencies) and incrementally adds attributes based on a heuristic measure. Additionally, a threshold value is required as a stopping criterion to determine when a reduct candidate is "near enough" to being a reduct. On each iteration, those objects that are consistent with the current reduct candidate are removed (an optimization that can be used with RSAR). As the process starts with the core of the dataset, this has to be calculated beforehand. Using the discernibility matrix for this purpose can be quite impractical for datasets of large dimensionality. However, there are other methods that can calculate the core in an efficient manner (Pawlak 1991). For example, this can be done by calculating the degree of dependency of the full feature set and the corresponding dependencies of the feature set minus each attribute. Those features that result in a dependency decrease are core attributes. There are also alternative methods available that allow the calculation of necessary information about the discernibility matrix without the need to perform operations directly on it (Nguyen & Nguyen, 1996).

EBR

A further technique for rough set feature selection is entropy-based reduction (EBR), developed from work carried out in Jensen and Shen (2004a).

Figure 8. Entropy-based reduction

```
EBR(C,D)
    C, the set of all conditional features;
    D, the set of decision features.

(1) R ← {}
(2) do
(3)     T ← R
(4)     ∀x ∈ (C − R)
(5)         if H(D|R∪{x})<H(D|T)
(6)             T ← R ∪ {x}
(7)     R ← T
(8) until H(D | R) = H(D | C)
(9) return R
```

This approach is based on the entropy heuristic employed by machine learning techniques such as C4.5 (Quinlan, 1993). A similar approach has been adopted in Dash and Liu (1997), where an entropy measure is used for ranking features. EBR is concerned with examining a dataset and determining those attributes that provide the most gain in information. The entropy of attribute A (which can take values $a_1,...,a_m$) with respect to the conclusion C (of possible values $c_1,...,c_n$) is defined as:

$$H(C \mid A) = -\sum_{j=1}^{m} p(a_j) \sum_{i=1}^{n} p(c_i \mid a_j) \log_2 p(c_i \mid a_j)$$

(14)

This can be extended to dealing with *subsets* of attributes instead of individual attributes only. Using this entropy measure, the algorithm used in rough set-based attribute reduction (Chouchoulas & Shen, 2001) can be modified to that shown in Figure 8. This algorithm requires no thresholds in order to function: the search for the best feature subset is stopped when the resulting subset entropy is equal to that of the entire feature set. For consistent data, the final entropy of the subset will be zero. It is interesting to note that any subset with an entropy of 0 will also have a corresponding rough set dependency of 1. Hence, this technique can be used for finding rough set reducts if the data is consistent.

Returning to the example dataset, EBR first evaluates the entropy of each individual attribute:

Subset	Entropy
{a}	1.1887219
{b}	0.75
{c}	0.9387219
{d}	0.75

The subsets with lowest entropy here are {b} and {d}. The algorithm selects attribute b due to it being evaluated first, and adds it to the current feature subset. The next step is to calculate the entropy of all subsets containing *b* and one other attribute:

Subset	Entropy
{a,b}	0.5
{b,c}	0.59436095
{b,d}	0.0

Here, the subset {*b,d*} is chosen, as this results in the lowest entropy. Additionally, the stopping criterion has been met, as this value equals the entropy for the entire feature set ($H(D|\{b,d\}) = 0 = H(D|C)$). The algorithm terminates and returns this feature subset; the dataset can now be reduced to these features only. As the resulting entropy is zero, the returned subset is a rough set reduct.

Other Algorithms

Among the first rough set-based approaches is the *Preset* algorithm (Modrzejewski, 1993), which is another feature selector that uses rough set theory to rank, heuristically, the features, assuming a noise-free binary domain. Since Preset does not try to explore all combinations of the features, it is certain that it will fail on problems whose attributes are highly correlated. There have also been investigations into the use of different reduct quality measures (see Polkowski, Lin, & Tsumoto, 2000 for details).

In Zhang and Yao (2004), a new rough set-based feature selection heuristic, parameterized average support heuristic (PASH), is proposed. Unlike the existing methods, PASH is based on a special parameterized lower approximation that is defined to include all predictive instances. Predictive instances are instances that may produce predictive rules that hold true with a high probability, but are not necessarily always true. The traditional model could exclude predictive instances that may produce such rules. The main advantage of PASH is that it considers the overall quality of the potential rules, thus, producing a set of rules with balanced support distribution over all decision classes. However, it requires a parameter to be defined by the user that adjusts the level of approximation. One of the main benefits of rough set theory is that it does not require such additional information, and hence eliminates the need for user interaction or repeated experimentation.

DISCERNIBILITY MATRIX-BASED APPROACHES

Many applications of rough sets to feature selection make use of discernibility matrices for finding reducts. A discernibility matrix (Skowron & Rauszer, 1992) of a decision table $D = (U, C \cup d)$ is a symmetric $|U| \times |U|$ matrix with entries defined:

$$c_{ij} = \{a \in C \mid a(x_i) \neq a(x_j)\} \quad i,j = 1,...,|U|$$

(15)

Each c_{ij} contains those attributes that differ between objects *i* and *j*. For finding reducts, the decision-relative discernibility matrix is of more interest. This only considers those object discernibilities that occur when the corresponding decision attributes differ. Returning to the example dataset, the decision-relative discernibility matrix found in Table 2 is produced. For example, it can be seen from the table that objects 0 and 1 differ in each attribute. Although some attributes in objects 1 and 3 differ, their corresponding decisions are the same, so no entry appears in the decision-relative matrix. Grouping all entries containing single attributes forms the core of the dataset (those attributes appearing in *every* reduct). Here, the core of the dataset is {*d*}.

From this, the discernibility function can be defined. This is a concise notation of how each object within the dataset may be distinguished from the others. A discernibility function f_D is a Boolean function of *m* Boolean variables $a_1^*,...,$

Table 2. The decision-relative discernibility matrix

$x \in U$	0	1	2	3	4	5	6	7
0								
1	a,b,c,d							
2	a,c,d	a,b,c						
3	b,c		a,b,d					
4	d	a,b,c,d		b,c,d				
5	a,b,c,d	a,b,c		a,b,d				
6	a,b,c,d	b,c		a,b,c,d	b,c			
7	a,b,c,d	d	a,c,d				a,d	

a_m^* (corresponding to the attributes $a_1,..., a_m$) defined as:

$$f_D(a_1^*,...,a_m^*) = \wedge \{\vee c_{ij}^* \mid 1 \leq j \leq i \leq |U|, c_{ij} \neq \emptyset\} \quad (16)$$

where $c_{ij}^* = \{a^* \mid a \in c_{ij}\}$. By finding the set of all prime implicants of the discernibility function, all the minimal reducts of a system may be determined. From Table 2, the decision-relative discernibility function is (with duplicates removed):

$f_D(a,b,c,d)$
$= \{a \vee b \vee c \vee d\} \wedge \{a \vee c \vee d\} \wedge \{b \vee c\} \wedge \{d\}$

$\wedge \{a \vee b \vee c\} \wedge \{a \vee b \vee d\} \wedge \{b \vee c \vee d\} \wedge \{a \vee d\}$

Further simplification can be performed by removing those sets (clauses) that are supersets of others:

$f_D(a,b,c,d) = \{b \vee c\} \wedge \{d\}$

The reducts of the dataset may be obtained by converting this expression from conjunctive normal form to disjunctive normal form (without negations). Hence, the minimal reducts are $\{b,d\}$ and $\{c,d\}$. Although this is guaranteed to discover all minimal subsets, it is a costly operation, rendering the method impractical for even medium-sized datasets.

Johnson Reducer

This is a simple greedy heuristic algorithm that is often applied to discernibility functions to find a single reduct (Øhrn, 1999). Reducts found by this process have no guarantee of minimality, but are generally of a size close to the minimal.

The algorithm begins by setting the current reduct candidate, R, to the empty set. Then, each conditional attribute appearing in the discernibility function is evaluated according to the heuristic measure. For the standard Johnson algorithm, this is typically a count of the number of appearances an attribute makes within clauses; attributes that appear more frequently are considered to be more

Figure 9. Johnson algorithm

```
Johnson(C,f_D)
  C, the set of conditional attributes
  f_D, the discernibility function.

(1)  R ← ∅ ; bestc=0;
(2)  while (f_D not empty)
(3)    for each a ∈ C that appears in f_D
(4)      c = heuristic(a)
(5)      if (c > bestc)
(6)        bestc=c; bestAttr ← a
(7)    R ← R ∪ a
(8)    f_D ← removeClauses(f_D, a)
(9)  return R
```

significant. The attribute with the highest heuristic value is added to the reduct candidate, and all clauses in the discernibility function containing this attribute are removed. As soon as all clauses have been removed, the algorithm terminates and returns the reduct R. R is assured to be a reduct, as all clauses contained within the discernibility function have been addressed.

Variations of the algorithm involve alternative heuristic functions in an attempt to guide search down better paths (Nguyen & Skowron, 1997b; Wang & Wang, 2001). However, no perfect heuristic exists, and hence there is still no guarantee of subset optimality.

Compressibility Algorithm

In Starzyk et al. (Starzyk, Nelson, & Sturtz, 2000), the authors present a method for the generation of all reducts in an information system by manipulating the clauses in discernibility functions. In addition to the standard simplification laws (such as the removal of supersets), the concept of strong compressibility is introduced and applied in conjunction with an expansion algorithm.

The strong compressibility simplification applies where clause attributes are either simultaneously present or absent in all clauses. In this situation, the attributes may be replaced by a single representative attribute. As an example, consider the formula:

$$f_D = \{a \vee b \vee c \vee f\} \wedge \{b \vee d\}$$
$$\wedge \{a \vee d \vee e \vee f\} \wedge \{d \vee c\}$$

The attributes a and f can be replaced by a single attribute as they are both present in the first and third clauses, and absent in the second and fourth. Replacing $\{a \vee f\}$ with g results in:

$$f_D = \{g \vee b \vee c\} \wedge \{b \vee d\} \wedge \{g \vee d \vee e\} \wedge \{d \vee c\}$$

If a reduct resulting from this discernibility function contains the new attribute g, then this attribute may be replaced by either a or f. Here, $\{g,d\}$ is a reduct, and so $\{a,d\}$ and $\{f,d\}$ are reducts of the original set of clauses. Hence, fewer attributes are considered in the reduct-determining process with no loss of information. The complexity of this step is $O(a*c + a^2)$, where a is the number of attributes and c is the number of clauses. The overall algorithm for determining reducts can be found in Figure 10. It uses concepts from Boolean algebra (such as the absorption and expansion laws) with strong compressibility for simplifying the discernibility function.

Returning to the example in Figure 10 and following the algorithm from Step 4, the most commonly occurring attribute can be seen to be d. Hence, the expansion law is applied with respect to this attribute to obtain:

$$f_D = f_1 \vee f_2$$
$$= (\{d\} \wedge \{g \vee b \vee c\}) \vee (\{g \vee b \vee c\} \wedge \{b\}$$
$$\wedge \{g \vee e\} \wedge \{c\})$$

$$= (\{d\} \wedge \{g \vee b \vee c\}) \vee (\{b\} \wedge \{g \vee e\} \wedge \{c\})$$

As all components are in simple form, Step 6 is carried out, where all strongly equivalent classes are replaced by their equivalent attributes:

$$f_D = f_1 \vee f_2$$

$$= (\{d\} \wedge \{a \vee f \vee b \vee c\})$$
$$\vee (\{b\} \wedge \{a \vee f \vee e\} \wedge \{c\})$$

Figure 10. Compressibility algorithm

```
Compressibility(f_D)
  f_D, the discernibility function.

(1)  while (f_D not in simple form)
(2)      applyAbsorptionLaws(f_D)  //remove supersets
(3)      replaceStronglyCompressibleAttributes(f_D)
(4)      a ← mostFrequentAttribute(f_D)
(5)      applyExpansionLaw(a, f_D)
(6)      substituteStronglyCompressibleClasses(f_D)
(7)  Reds ← calculateReducts(f_D)
(8)  return minimalElements(Reds)
```

The corresponding reducts for the function components are as follows (Step 7):

$Red(f_1) = (\{a,d\},\{d,f\},\{b,d\},\{c,d\})$

$Red(f_2) = (\{b,a,c\},\{b,f,c\},\{b,e,c\})$

The reducts for the system are generated by taking the union of the components and determining the minimal elements:

$Red(f_D) = (\{a,d\},\{d,f\},\{b,d\},\{c,d\},$
$\{b,a,c\},\{b,f,c\},\{b,e,c\})$

RSAR-SAT

The problem of finding the smallest feature subsets using rough set theory can be formulated as a propositional satisfiability (SAT) problem. Rough sets allow the generation from datasets of clauses of features in conjunctive normal form. If after assigning truth values to all features appearing in the clauses the formula is satisfied, then those features set to true constitute a valid subset for the data. The task is to find the smallest number of such features so that the CNF formula is satisfied. In other words, the problem here concerns finding a minimal assignment to the arguments of $f(x_1,...,x_n)$ that makes the function equal to 1. There will be at least one solution to the problem (i.e., all x_is set to 1) for consistent datasets. Preliminary work has been carried out in this area (Bakar, Sulaiman, Othman, & Selamat, 2002), though this does not adopt a DPLL-style approach to finding solutions.

The DPLL algorithm for finding minimal subsets can be found in Figure 11, where a search is conducted in a depth-first manner. The key operation in this procedure is the unit propagation Step, unitPropagate(F), in lines (6) and (7). Clauses in the formula that contain a single literal will only be satisfied if that literal is assigned the value 1 (for positive literals). These are called unit clauses. Unit propagation examines the current formula for unit clauses and automatically assigns the appropriate value to the literal they contain. The elimination of a literal can create new unit clauses, and thus, unit propagation eliminates variables by repeated passes until there is no unit clause in the formula. The order of the unit clauses within the formula makes no difference to the results or the efficiency of the process.

Branching occurs at lines (9) to (12) via the function selectLiteral(F). Here, the next literal is chosen heuristically from the current formula, assigned the value 1, and the search continues. If this branch eventually results in unsatisfiability, the procedure will assign the value 0 to this literal instead and continue the search. The importance of choosing good branching literals is well known; different branching heuristics may produce drastically different-sized search trees for the same basic algorithm, thus, significantly affecting the efficiency of the solver. The heuristic currently used within RSAR-SAT is to select the variable that appears in the most clauses in the current set of clauses. Many other heuristics exist for this purpose (Zhang & Malik, 2002), but are not considered here.

A degree of pruning can take place in the search by remembering the size of the currently considered subset and the smallest optimal subset encountered so far. If the number of variables currently assigned 1 equals the number of those

Figure 11. The definition of the DPLL algorithm

```
DPLL(F)
F, the formula containing the current set of clauses.

(1)  if (F contains an empty clause)
(2)      return unsatisfiable
(3)  if (F is empty)
(4)      output current assignment
(5)      return satisfiable
(6)  if (F contains a unit clause {l})
(7)      F' ← unitPropagate(F)
(8)      return DPLL(F')
(9)  x ← selectLiteral(F)
(10) if (DPLL( F ∪ {x}) is satisfiable)
(11)     return satisfiable
(12) else return DPLL( F ∪ {-x})
```

in the presently optimal subset, and the satisfiability of F is still not known, then any further search down this branch will not result in a smaller optimal subset.

Although stochastic methods have been applied to SAT problems (Hoos & Stützle, 1999) these are not applicable here as they provide no guarantee of solution minimality. The DPLL-based algorithm will always find the minimal optimal subset. However, this will come at the expense of time taken to find it. The initial experimentation (Jensen, Shen, & Tuson, 2005) has shown that the method performs well in comparison to RSAR, which often fails to find the smallest subsets.

HANDLING CONTINUOUS VALUES

The reliance on discrete data for the successful operation of RST can be seen as a significant drawback of the approach. Indeed, this requirement of RST implies an objectivity in the data that is simply not present (Koczkodaj, Orlowski, & Marek, 1998). For example, in a medical dataset, values such as *Yes* or *No* cannot be considered objective for a *Headache* attribute, as it may not be straightforward to decide whether a person has a headache or not to a high degree of accuracy. Again, consider an attribute *Blood Pressure*. In the real world, this is a real-valued measurement but for the purposes of RST, must be discretised into a small set of labels such as *Normal*, *High*, and so forth. Subjective judgments are required for establishing boundaries for objective measurements.

In the rough set literature, there are two main ways of handling real-valued attributes: through fuzzy rough sets and tolerance rough sets. Both approaches replace the traditional equivalence classes of crisp rough set theory with alternatives that are better suited to dealing with this type of data. In the fuzzy rough case, fuzzy equivalence classes are employed within a fuzzy extension of rough set theory, resulting in a hybrid approach. In the tolerance case, indiscernibility relations are replaced with similarity relations that permit a limited degree of variability in attribute values. Approximations are constructed based on these tolerance classes in a manner similar to that of traditional rough set theory.

To illustrate the operation of the techniques involved, an example dataset is given in Table 3. The table contains three real-valued conditional attributes and a crisp-valued decision attribute.

Fuzzy Rough Sets

A way of handling this problem is through the use of *fuzzy-rough* sets. Subjective judgments are not entirely removed, as fuzzy set membership functions still need to be defined. However, the method offers a high degree of flexibility when dealing with real-valued data, enabling the vagueness and imprecision present to be modelled effectively.

Fuzzy Equivalence Classes

In the same way that crisp equivalence classes are central to rough sets, *fuzzy* equivalence classes are central to the fuzzy rough set approach (Dubois & Prade, 1992; Thiele, 1998; Yao, 1998). For typical applications, this means that the decision values and the conditional values may all be fuzzy. The concept of crisp equivalence classes can be extended by the inclusion of a fuzzy similarity

Table 3. Example dataset: Crisp decisions

$x \in U$	a	b	c	q
1	-0.4	-0.3	-0.5	no
2	-0.4	0.2	-0.1	yes
3	-0.3	-0.4	-0.3	no
4	0.3	-0.3	0	yes
5	0.2	-0.3	0	yes
6	0.2	0	0	no

relation S on the universe, which determines the extent to which two elements are similar in S. For example, if $\mu_S(x,y) = 0.9$, then objects x and y are considered to be quite similar.

The usual properties of reflexivity ($\mu_S(x,x) = 1$), symmetry ($\mu_S(x,y) = \mu_S(y,x)$), and transitivity ($\mu_S(x,z) \geq \mu_S(x,y) \wedge \mu_S(y,z)$) hold.

Using the fuzzy similarity relation, the fuzzy equivalence class $[x]_S$ for objects close to x can be defined:

$$\mu_{[x]_S}(y) = \mu_S(x,y) \qquad (17)$$

The following axioms should hold for a fuzzy equivalence class F (Höhle, 1988):

$\exists \mu_F(x) = 1$ (μ_F is normalised)
$\mu_F(x) \wedge \mu_S(x,y) \leq \mu_F(y)$
$\mu_F(x) \wedge \mu_F(y) \leq \mu_S(x,y)$

The first axiom corresponds to the requirement that an equivalence class is nonempty. The second axiom states that elements in y's neighbourhood are in the equivalence class of y. The final axiom states that any two elements in F are related via S. Obviously, this definition degenerates to the normal definition of equivalence classes when S is nonfuzzy.

The family of normal fuzzy sets produced by a fuzzy partitioning of the universe of discourse can play the role of fuzzy equivalence classes (Dubois & Prade, 1992). Consider the crisp partitioning of a universe of discourse, U, by the attributes in Q: $U/Q = \{\{1,3,6\},\{2,4,5\}\}$. This contains two equivalence classes ($\{1,3,6\}$ and $\{2,4,5\}$) that can be thought of as degenerated fuzzy sets, with those elements belonging to the class possessing a membership of one, zero otherwise. For the first class, for instance, the objects 2, 4, and 5 have a membership of zero. Extending this to the case of fuzzy equivalence classes is straightforward: objects can be allowed to assume membership values, with respect to any given class, in the interval [0,1]. U/Q is not restricted to crisp partitions only; fuzzy partitions are equally acceptable.

Fuzzy Rough Feature Selection

Fuzzy rough feature selection (FRFS) (Jensen & Shen, 2004a; Jensen & Shen, 2004b; Shen & Jensen, 2004) provides a means by which discrete or real-valued noisy data (or a mixture of both) can be effectively reduced without the need for user-supplied information. Additionally, this technique can be applied to data with continuous or nominal decision attributes and as such, can be applied to regression as well as classification datasets. The only additional information required is in the form of fuzzy partitions for each feature that can be automatically derived from the data.

From the literature, the fuzzy P-lower and P-upper approximations are defined as (Dubois & Prade, 1992):

$$\mu_{\underline{P}X}(F_i) = \inf_x \max\{1 - \mu_{F_i}(x), \mu_X(x)\} \quad \forall i \qquad (18)$$

$$\mu_{\overline{P}X}(F_i) = \sup_x \min\{\mu_{F_i}(x), \mu_X(x)\} \quad \forall i \qquad (19)$$

where F_i denotes a fuzzy equivalence class belonging to U/P. Note that although the universe of discourse in feature selection is finite, this is not the case in general, hence, the use of *sup* and *inf*. These definitions diverge a little from the crisp upper and lower approximations, as the memberships of individual objects to the approximations are not explicitly available. As a result of this, the fuzzy lower and upper approximations are herein redefined as:

$$\mu_{\underline{P}X}(x) = \sup_{F \in U/P} \min(\mu_F(x), \inf_{y \in U} \max\{1 - \mu_F(y), \mu_X(y)\}) \qquad (20)$$

$$\mu_{\overline{P}X}(x) = \sup_{F \in U/P} \min(\mu_F(x), \inf_{y \in U} \max\{1 - \mu_F(y), \mu_X(y)\}) \qquad (21)$$

In implementation, not all $y \in U$ need to be considered; only those where $\mu_F(y)$ is nonzero, that is, where object y is a fuzzy member of (fuzzy) equivalence class F. The tuple $< \underline{P}X, \overline{P}X >$ is called a *fuzzy-rough* set.

The crisp positive region in traditional rough set theory is defined as the union of the lower approximations. By the extension principle (Zadeh, 1975), the membership of an object $x \in U$, belonging to the fuzzy positive region, can be defined by:

$$\mu_{POS_P(Q)}(x) = \sup_{X \in U/Q} \mu_{\underline{P}X}(x) \qquad (22)$$

Object x will not belong to the positive region only if the equivalence class it belongs to is not a constituent of the positive region. This is equivalent to the crisp version, where objects belong to the positive region only if their underlying equivalence class does so. Similarly, the negative and boundary regions can be defined. For this particular feature selection method, the upper approximation is not used, though this may be useful for other methods.

Using the definition of the fuzzy positive region, the new dependency function can be defined as follows:

$$\gamma'_P(Q) = \frac{|\mu_{POS_P(Q)}(x)|}{|U|} = \frac{\sum_{x \in U} \mu_{POS_P(Q)}(x)}{|U|} \qquad (23)$$

As with crisp rough sets, the dependency of Q on P is the proportion of objects that are discernible out of the entire dataset. In the present approach, this corresponds to determining the fuzzy cardinality of $\mu_{POS_P(Q)}(x)$ divided by the total number of objects in the universe. The definition of dependency degree covers the crisp case as its specific instance.

If the fuzzy-rough reduction process is to be useful, it must be able to deal with multiple features, finding the dependency between various subsets of the original feature set. For example, it may be necessary to be able to determine the degree of dependency of the decision feature(s) with respect to $P = \{a,b\}$. In the crisp case, U/P contains sets of objects grouped together that are indiscernible according to both features a and b. In the fuzzy case, objects may belong to many equivalence classes, so the cartesian product of $U/IND(\{a\})$ and $U/IND(\{b\})$ must be considered in determining U/P.

Each set in U/P denotes an equivalence class. For example, if $P = \{a,b\}$, $U/IND(\{a\}) = \{N_a, Z_a\}$ and $U/IND(\{b\}) = \{N_b, Z_b\}$, then:

$$U/P = \{N_a \cap N_b, N_a \cap Z_b, Z_a \cap N_b, Z_a \cap Z_b\}$$

The extent to which an object belongs to such an equivalence class is therefore calculated by using the conjunction of constituent fuzzy equivalence classes, say F_i, $i=1,2,...,n$:

$$\mu_{F_1 \cap F_2 \cap ... \cap F_n}(x) = \min(\mu_{F_1}(x), \mu_{F_2}(x), ..., \mu_{F_n}(x)) \qquad (24)$$

Fuzzy-Rough QuickReduct

A problem may arise when this approach is compared to the crisp approach. In conventional RSAR, a reduct is defined as a subset R of the features that have the same information content as the full feature set A. In terms of the dependency function, this means that the values $\gamma_R(Q)$ and $\gamma_A(Q)$ are identical and equal to 1 if the dataset is consistent. However, in the fuzzy-rough approach this is not necessarily the case, as the uncertainty encountered when objects belong to many fuzzy equivalence classes results in a reduced total dependency.

A possible way of combatting this would be to determine the degree of dependency of a set of decision features, D, upon the full feature set, and use this as the denominator rather than $|U|$ (for normalization), allowing γ' to reach 1. With these issues in mind, a new QuickReduct algorithm has been developed, as given in Figure 12. It employs the new dependency function γ' to choose which features to add to the current reduct candidate in

Figure 12 The fuzzy-rough QUICKREDUCT algorithm

```
FRQUICKREDUCT(C,D)
C, the set of all conditional features;
D, the set of decision features.

(1)   R ← {}; γ'_best = 0
(2)   do
(3)       T ← R
(4)       γ'_prev = γ'_best
(5)       ∀x ∈ (C − R)
(6)           if γ'_{R∪{x}}(D) > γ'_T(D)
(7)               T ← R ∪ {x}
(8)               γ'_best = γ'_T(D)
(9)       R ← T
(10)  until γ'_best == γ'_prev
(11) return R
```

the same way as the original QUICKREDUCT process (see Figure 4). The algorithm terminates when the addition of any remaining feature does not increase the dependency (such a criterion could be used with the original QUICKREDUCT algorithm).

As the new degree of dependency measure is nonmonotonic, it is possible that the QUICKREDUCT style search terminates having reached only a local optimum. The global optimum may lie elsewhere in the search space. This motivates the adoption of alternative search mechanisms. However, the algorithm, as presented in Figure 12, is still highly useful in locating good subsets quickly.

Note that an intuitive understanding of the algorithm implies that, for a dimensionality of n, $(n^2 + n)/2$ evaluations of the dependency function may be performed for the worst-case dataset. However, as FRFS is used for dimensionality reduction prior to any involvement of the system that will employ those features belonging to the resultant reduct, this operation has no negative impact upon the run-time efficiency of the system.

It is also possible to reverse the search; that is, start with the full set of features and incrementally remove the least informative features. This process continues until no more features can be removed without reducing the total number of discernible objects in the dataset. Again, this tends not to be applied to larger datasets, as the cost of evaluating these larger feature subsets is too great.

Application to the Example Dataset

Using Table 3 and the fuzzy sets defined in Figure 13 (for all conditional attributes), and setting $A=\{a\}$, $B=\{b\}$, $C=\{c\}$, and $Q=\{q\}$, the following equivalence classes are obtained:

$U/A = \{N_a, Z_a\}$
$U/B = \{N_b, Z_b\}$
$U/C = \{N_c, Z_c\}$
$U/Q = \{\{1,3,6\}, \{2,4,5\}\}$

Figure 13. Fuzzifications for conditional features

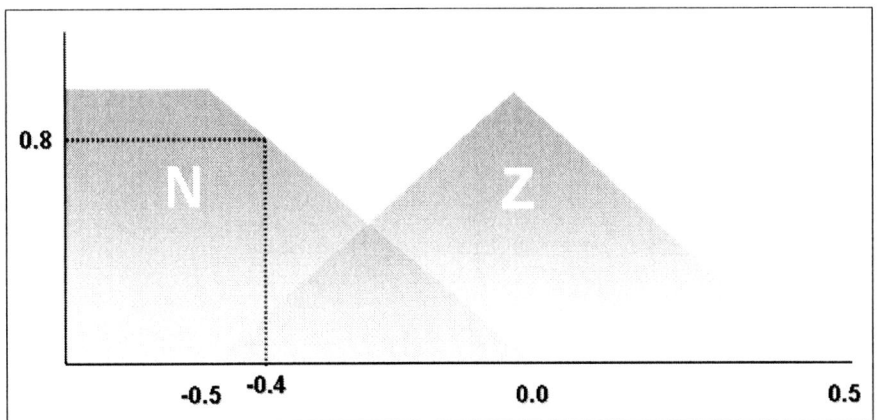

The first step is to calculate the lower approximations of the sets A, B, and C. To clarify the calculations involved, Table 4 contains the membership degrees of objects to fuzzy equivalence classes.

For simplicity, only A will be considered here; that is, using A to approximate Q. For the first decision equivalence class $X = \{1,3,6\}$, $\mu_{\underline{A}\{1,3,6\}}(x)$ needs to be calculated:

$$\mu_{\underline{A}\{1,3,6\}}(x) = \sup_{F \in U/A} \min(\mu_F(x), \inf_{y \in U} \max\{1 - \mu_F(y), \mu_{\{1,3,6\}}(y)\})$$

Considering the first fuzzy equivalence class of A, N_a:

$$\min(\mu_{N_a}(x), \inf_{y \in U} \max\{1 - \mu_{N_a}(y), \mu_{\{1,3,6\}}(y)\})$$

For object 2, this can be calculated as follows. From Table 4 it can be seen that the membership of object 2 to the fuzzy equivalence class N_a, $\mu_{N_a}(2)$, is 0.8. The remainder of the calculation involves finding the smallest of the following values:

$$\max(1 - \mu_{N_a}(1), \mu_{\{1,3,6\}}(1)) = \max(0.2, 1.0) = 1.0$$
$$\max(1 - \mu_{N_a}(2), \mu_{\{1,3,6\}}(2)) = \max(0.2, 0.0) = 0.2$$
$$\max(1 - \mu_{N_a}(3), \mu_{\{1,3,6\}}(3)) = \max(0.4, 1.0) = 1.0$$
$$\max(1 - \mu_{N_a}(4), \mu_{\{1,3,6\}}(4)) = \max(1.0, 0.0) = 1.0$$
$$\max(1 - \mu_{N_a}(5), \mu_{\{1,3,6\}}(5)) = \max(1.0, 0.0) = 1.0$$
$$\max(1 - \mu_{N_a}(6), \mu_{\{1,3,6\}}(6)) = \max(1.0, 1.0) = 1.0$$

From the calculations above, the smallest value is 0.2, hence:

$$\min(\mu_{N_a}(x), \inf_{y \in U} \max\{1 - \mu_{N_a}(y), \mu_{\{1,3,6\}}(y)\})$$
$$= \min(0.8, \inf\{1, 0.2, 1, 1, 1, 1\})$$
$$= 0.2$$

Similarly for Z_a

$$\min(\mu_{Z_a}(x), \inf_{y \in U} \max\{1 - \mu_{Z_a}(y), \mu_{\{1,3,6\}}(y)\})$$
$$= \min(0.2, \inf\{1, 0.8, 1, 0.6, 0.4, 1\})$$
$$= 0.2$$

Thus,

$$\mu_{\underline{A}\{1,3,6\}}(2) = 0.2$$

Calculating the A-lower approximation of $X = \{1,3,6\}$ for every object gives:

$\mu_{\underline{A}\{1,3,6\}}(1) = 0.2$, $\quad \mu_{\underline{A}\{1,3,6\}}(2) = 0.2$
$\mu_{\underline{A}\{1,3,6\}}(3) = 0.4$, $\quad \mu_{\underline{A}\{1,3,6\}}(4) = 0.4$
$\mu_{\underline{A}\{1,3,6\}}(5) = 0.4$, $\quad \mu_{\underline{A}\{1,3,6\}}(6) = 0.4$

The corresponding values for $X = \{2,4,5\}$ can also be determined:

$\mu_{\underline{A}\{2,4,5\}}(1) = 0.2$, $\quad \mu_{\underline{A}\{2,4,5\}}(2) = 0.2$
$\mu_{\underline{A}\{2,4,5\}}(3) = 0.4$, $\quad \mu_{\underline{A}\{2,4,5\}}(4) = 0.4$
$\mu_{\underline{A}\{2,4,5\}}(5) = 0.4$, $\quad \mu_{\underline{A}\{2,4,5\}}(6) = 0.4$

Table 4. Membership values of objects to corresponding fuzzy sets

$x \in U$	a		b		c		q	
	N_a	Z_a	N_b	Z_b	N_c	Z_c	{1,3,6}	{2,4,5}
1	0.8	0.2	0.6	0.4	1.0	0.0	1.0	0.0
2	0.8	0.2	0.0	0.6	0.2	0.8	0.0	1.0
3	0.6	0.4	0.8	0.2	0.6	0.4	1.0	0.0
4	0.0	0.4	0.6	0.4	0.0	1.0	0.0	1.0
5	0.0	0.6	0.6	0.4	0.0	1.0	0.0	1.0
6	0.0	0.6	0.0	1.0	0.0	1.0	1.0	0.0

It is a coincidence here that $\mu_{\underline{A}\{1,3,6\}}(x)$ = $\mu_{\underline{A}\{2,4,5\}}(x)$ for this example. Using these values, the fuzzy positive region for each object can be calculated via using:

$$\mu_{POS_A(Q)}(x) = \sup_{X \in U/Q} \mu_{\underline{A}X}(x)$$

This results in:

$\mu_{POS_A(Q)}(1) = 0.2, \quad \mu_{POS_A(Q)}(2) = 0.2$

$\mu_{POS_A(Q)}(3) = 0.4, \quad \mu_{POS_A(Q)}(4) = 0.4$

$\mu_{POS_A(Q)}(5) = 0.4, \quad \mu_{POS_A(Q)}(6) = 0.4$

The next step is to determine the degree of dependency of Q on A:

$$\gamma'_A(Q) = \frac{\sum_{x \in U} \mu_{POS_A(Q)}(x)}{|U|} = \frac{2}{6}$$

Calculating for B and C gives:

$$\gamma'_B(Q) = \frac{2.4}{6}, \quad \gamma'_C(Q) = \frac{1.6}{6}$$

From this it can be seen that attribute b will cause the greatest increase in dependency degree. This attribute is chosen and added to the potential reduct. The process iterates and the two dependency degrees calculated are:

$$\gamma'_{\{a,b\}}(Q) = \frac{3.4}{6}, \quad \gamma'_{\{b,c\}}(Q) = \frac{3.2}{6}$$

Adding attribute a to the reduct candidate causes the larger increase of dependency, so the new candidate becomes $\{a,b\}$. Lastly, attribute c is added to the potential reduct:

$$\gamma'_{\{a,b,c\}}(Q) = \frac{3.4}{6}$$

As this causes no increase in dependency, the algorithm stops and outputs the reduct $\{a,b\}$. The dataset can now be reduced to only those attributes appearing in the reduct. When crisp RSAR is performed on this dataset (after using the same fuzzy sets to discretize the real-valued attributes), the reduct generated is $\{a,b,c\}$, that is, the full conditional attribute set (Jensen & Shen, 2004b). Unlike crisp RSAR, the true minimal reduct was found using the information on degrees of membership. It is clear from this example alone that the information lost by using crisp RSAR can be important when trying to discover the smallest reduct from a dataset.

Fuzzy Entropy-Guided FRFS

From previous experimentation with crisp rough sets and entropy (Jensen & Shen, 2004b), it was observed that entropy-based methods often found smaller reducts than those based on the dependency function. This provided the motivation for a new fuzzy-rough technique using fuzzy entropy to guide search (Kosko, 1986; Mac Parthláin, Jensen, & Shen, 2006), in order to locate optimal fuzzy-rough subsets.

Fuzzy Entropy

Again, let $I = (U, A)$ be a decision system, where U is a nonempty set of finite objects. $A = \{C \cup D\}$ is a nonempty finite set of attributes, where C is the set of input features and D is the set of classes. An attribute $a \in A$ has corresponding fuzzy subsets $F_1, F_2, ..., F_n$. The fuzzy entropy for a fuzzy subset F_i can be defined as being:

$$H(D|F_i) = \sum_{Q \in U/D} -p(Q|F_i) \log_2 p(Q|F_i) \quad (25)$$

where $p(Q | F_i)$ is the relative frequency of the fuzzy subset F_i of attribute a with respect to the decision Q, and is defined:

$$p(Q|F_i) = \frac{|Q \cap F_i|}{|F_i|} \quad (26)$$

The cardinality of a fuzzy set is denoted by $|.|$. Based on these definitions, the fuzzy entropy for an attribute subset R is defined as follows:

$$E(D|R) = \sum_{F_i \in U/R} \frac{|F_i|}{\sum_{Y_i \in U/R} |Y_i|} . H(D|F_i)$$

This fuzzy entropy can be used to gauge the utility of attribute subsets in a similar way to that of the fuzzy-rough measure. However, the fuzzy entropy measure decreases with increasing subset utility, whereas the fuzzy-rough dependency measure increases. With these definitions, a new feature selection mechanism can be constructed that uses fuzzy entropy to guide the search for the best fuzzy-rough feature subset.

Fuzzy Entropy-Based QuickReduct

Figure 14 shows a fuzzy-rough entropy-based QuickReduct algorithm based on the previously described fuzzy-rough algorithm in Figure 12. FREQuickReduct is similar to the fuzzy-rough algorithm, but uses the entropy value of a data subset to guide the feature selection process. If the fuzzy entropy value of the current reduct candidate is smaller than the previous, then this reduct is retained and used in the next iteration of the loop. It is important to point out that the reduct is evaluated by examining its entropy value; termination only occurs when the addition of any remaining features results in a decrease in the dependency function value (γ'_{prev}). The fuzzy-entropy value therefore is not used as a termination criterion.

The algorithm begins with an empty subset R and with γ'_{prev} initialised to zero. The do-until loop works by examining the entropy value of a subset, and incrementally adding one conditional feature at a time until the dependency function value begins to fall to a value that is lower or equal to that of the last subset. For each iteration, a conditional feature that has not already been evaluated will be temporarily added to the subset R. The entropy of the subset currently being examined (5) is then evaluated and compared with the entropy of T, (the previous subset). If the entropy value of the current subset is lower (6), then the attribute added in (5) is retained as part of the new reduct T (7). The loop continues to evaluate in the above manner by adding conditional features, until the dependency value of the current reduct candidate ($\gamma'_R(D)$) falls to a value lower than or equal to that of the previously evaluated reduct candidate.

Application to the Example Dataset

Employing the same setup as before, but with $D = \{e\}$, the following partitions are obtained:

$U/A = \{N_a, Z_a\}$
$U/B = \{N_b, Z_b\}$
$U/C = \{N_c, Z_c\}$
$U/D = \{\{1,3,6\}, \{2,4,5\}\} = \{Q_1, Q_2\}$

The algorithm begins with an empty subset, and considers the addition of individual features. The attribute that results in the greatest decrease in fuzzy entropy will ultimately be added to the reduct candidate. For attribute a, the fuzzy entropy is calculated as follows ($A = \{a\}$):

$$E(A) = \frac{|N_a|}{|N_a + Z_a|} H(N_a) + \frac{|Z_a|}{|N_a + Z_a|} H(Z_a)$$

For the first part of the summation, the value $H(N_a)$ must be determined. This is achieved in the following way:

Figure 14. The fuzzy-rough entropy-based QuickReduct algorithm

```
FREQuickReduct(C,D)
  C, the set of all conditional features;
  D, the set of decision features.

(1)  T ← {}; γ'_prev = 0
(2)  do
(3)      R ← T
(4)      γ'_prev = γ'_T(D)
(5)      ∀x ∈ (C − R)
(6)          if E(D| R ∪ {x}) < E(D|T)
(7)              T ← R ∪ {x}
(8)  until γ'_T(D) ≤ γ'_prev
(9)  return R
```

$$H(N_a) = \sum_{Q \in U/D} -p(Q|N_a)\log_2 p(Q|N_{ai})$$
$$= -p(Q_1|N_a)\log_2 p(Q_1|N_a)$$
$$- p(Q_2|N_a)\log_2 p(Q_2|N_a)$$

The required probabilities are $p(Q_1|N_a) = 0.6363637$, $p(Q_2|N_a) = 0.3636363$. Hence, $H(N_a) = 0.94566023$. In a similar way, $H(Z_a)$ can be calculated, giving a value of 1.0. To determine the fuzzy entropy for a, the values $\frac{|N_a|}{|N_a+Z_a|}$ and $\frac{|Z_a|}{|N_a+Z_a|}$ must also be determined. This is achieved through the standard fuzzy cardinality, resulting in a fuzzy entropy value of:

$E(A) = (0.47826084 \times H(N_a)) + (0.5217391 \times H(Z_a))$
$= (0.47826084 \times 0.94566023) + (0.5217391 \times 1.0)$
$= 0.9740114$

Repeating this process for the remaining attributes gives:

$E(B) = 0.99629750$
$E(C) = 0.99999994$

From this it can be seen that attribute a will cause the greatest decrease in fuzzy entropy. This attribute is chosen and added to the potential reduct, $R \leftarrow R \cup \{a\}$. This subset is then evaluated using the fuzzy-rough dependency measure, resulting in $\gamma_R(D) = 0.3333333$. The previous dependency value is 0 (the algorithm started with the empty set); hence, the search continues. The process iterates and the two fuzzy entropy values calculated are:

$E(\{a, b\}) = 0.7878490$
$E(\{a, c\}) = 0.9506136$

Adding attribute b to the reduct candidate causes the larger decrease of fuzzy entropy, so the new candidate becomes $\{a,b\}$. The resulting dependency value for this, $\gamma_{\{a,b\}}(D)$ is 0.56666666.

This is, again, larger than the previous dependency value, and so search continues. Lastly, attribute c is added to the potential reduct:

$E(\{a, b, c\}) = 0.7412282$
$(\gamma_{\{a,b,c\}}(D) = 0.56666666)$

As this causes no increase in dependency, the algorithm stops and outputs the reduct $\{a,b\}$. The dataset can now be reduced to only those attributes appearing in the reduct.

Tolerance Rough Sets

Another way of attempting to handle the problem of real-valued data is to introduce a measure of similarity of feature values, and define the lower and upper approximations based on these similarity measures. Such lower and upper approximations define tolerance rough sets (Skowron & Stepaniuk, 1996). By relaxing the transitivity constraint of equivalence classes, a further degree of flexibility (with regard to indiscernibility) is introduced. In traditional rough sets, objects are grouped into equivalence classes if their attribute values are equal. This requirement might be too strict for real-world data, where values might differ only as a result of noise.

Similarity Measures

For the tolerance-based approach, suitable similarity relations must be defined for each attribute, although the same definition can be used for all features if applicable. A standard measure for this purpose, given in Stepaniuk (1998), is:

$$SIM_a(x,y) = 1 - \frac{|a(x) - a(y)|}{|a_{max} - a_{min}|} \quad (27)$$

where a is the attribute under consideration, and a_{max} and a_{min} denote the maximum and minimum values, respectively, for this attribute. When considering more than one attribute, the defined similarities must be combined to provide a measure of the overall similarity of objects. For a subset of features, P, this can be achieved in many ways; two commonly adopted approaches are:

$$(x,y) \in SIM_{P,\tau} \text{ iff } \prod_{a \in P} SIM_a(x,y) \geq \tau \quad (28)$$

$$(x,y) \in SIM_{P,\tau} \text{ iff } \frac{\sum_{a \in P} SIM_a(x,y)}{|P|} \geq \tau \quad (29)$$

where $\tau \in [0,1]$ is a global similarity threshold: τ determines the required level of similarity for inclusion within tolerance classes. It can be seen that this framework allows for the specific case of traditional rough sets by defining a suitable similarity measure (e.g., equality of feature values) and threshold ($\tau = 1$). Further similarity relations are investigated in Nguyen and Skowron (1997a).

Tolerance classes generated by the similarity relation for an object x are defined as:

$$SIM_{P,\tau}(x) = \{y \in U | (x,y) \in SIM_{P,\tau}\} \quad (30)$$

Approximations and Dependency

Lower and upper approximations are then defined in a similar way to traditional rough set theory:

$$\underline{P_\tau}X = \{x \,|\, SIM_{P,\tau}(x) \subseteq X\} \quad (31)$$

$$\overline{P_\tau}X = \{x \,|\, SIM_{P,\tau}(x) \cap X \neq \emptyset\} \quad (32)$$

Positive region and dependency functions then become:

$$POS_{P,\tau}(Q) = \bigcup_{X \in U/Q} \underline{P_\tau}X \quad (33)$$

$$\gamma_{P,\tau}(Q) = \frac{|POS_{P,\tau}(Q)|}{|U|} \quad (34)$$

From these definitions, attribute reduction methods can be constructed that use the tolerance-based degree of dependency, $\gamma_{P,\tau}(Q)$, to gauge the significance of feature subsets. For example, the fuzzy-rough QUICKREDUCT algorithm (Figure 12) can be adapted to perform feature selection based on the tolerance rough set-based measure. The resulting algorithm can be found in Figure 15.

Application to the Example Dataset

To illustrate the operation of the tolerance QUICKREDUCT algorithm, it is applied to the example data given in Table 3. This choice of threshold permits attribute values to differ to a limited extent, allowing close values to be considered as identical. For the decision feature, τ is set to 1 (i.e., objects must have identical values to appear in the same tolerance class), as the decision value is nominal. Setting $A = \{a\}$, $B = \{b\}$, $C = \{c\}$ and $Q = \{q\}$, the following tolerance classes are obtained:

Figure 15. Tolerance QuickReduct algorithm

```
TOLQUICKREDUCT(C,D,τ)
  C, the set of all conditional features;
  D, the set of decision features;
  τ, similarity threshold.

(1)  R ← {}; γ^τ_best = 0
(2)  do
(3)     T ← R
(4)     γ^τ_prev = γ^τ_best
(5)     ∀x ∈ (C − R)
(6)        if γ_{R∪{x},τ}(D) > γ_{T,τ}(D)
(7)           T ← R ∪ {x}
(8)           γ^τ_best = γ_{T,τ}(D)
(9)     R ← T
(10) until γ^τ_best == γ^τ_prev
(11) return R
```

$U/SIM_{A,\tau} = \{\{1,2,3\},\{4,5,6\}\}$
$U/SIM_{B,\tau} = \{\{1,3,4,5\},\{2\},\{6\}\}$
$U/SIM_{C,\tau} = \{\{1\},\{2,4,5,6\},\{3\}\}$
$U/SIM_{Q,\tau} = \{\{1,3,6\}, \{2,4,5\}\}$
$U/SIM_{\{a,b\},\tau} = \{\{1,3\}, \{2\}, \{4,5\}, \{4,5,6\}, \{5,6\}\}$
$U/SIM_{\{b,c\},\tau} = \{\{1,3\}, \{2,6\}, \{4,5,6\}, \{2,4,5\}\}$
$U/SIM_{\{a,b,c\},\tau} = \{\{1,3\}, \{2\},\{4,5,6\}\}$

It can be seen here that some objects belong to more than one tolerance class. This is due to the additional flexibility of employing similarity measures rather than strict equivalence.

Based on these partitions, the degree of dependency can be calculated for attribute subsets, providing an evaluation of their significance. The tolerance QUICKREDUCT algorithm considers the addition of attributes to the currently stored best subset (initially the empty set), and selects the feature that results in the highest increase of this value. Considering attribute b, the lower approximations of the decision classes are calculated as follows:

$\underline{B_\tau}\{1,3,6\} = \{x \mid SIM_{B,\tau}(x) \subseteq \{1,3,6\}\} = \{6\}$
$\underline{B_\tau}\{2,4,5\} = \{x \mid SIM_{B,\tau}(x) \subseteq \{1,3,6\}\} = \{2\}$

Hence, the positive region can be constructed:

$POS_{B,\tau}(Q) = \bigcup_{X \in U/Q} \underline{B_\tau} X$
$= \underline{B_\tau}\{1,3,6\} \cup \underline{B_\tau}\{2,4,5\}$
$= \{2,6\}$

And the resulting degree of dependency is:

$\gamma_{B,\tau}(Q) = \frac{|POS_{B,\tau}(Q)|}{|U|}$
$= \frac{|\{2,6\}|}{|\{1,2,3,4,5,6\}|} = \frac{2}{6}$

For the other conditional features in the dataset, the corresponding dependency degrees are:

$\gamma_{A,\tau}(Q) = \frac{|\{\varnothing\}|}{|\{1,2,3,4,5,6\}|} = \frac{0}{6}$

$\gamma_{C,\tau}(Q) = \frac{|\{1,3\}|}{|\{1,2,3,4,5,6\}|} = \frac{2}{6}$

Following the tolerance QUICKREDUCT algorithm, attribute b is added to the reduct candidate ($R = \{b\}$) and the search continues. The algorithm makes an arbitrary choice here between attributes b and c, as they produce equally high degrees of dependency (although they generate different positive regions). As attribute b was considered before attribute c, it is selected. The algorithm continues by evaluating subsets containing this attribute in combination with the remaining individual attributes from the dataset.

$\gamma_{\{a,b\},\tau}(Q) = \frac{|\{1,2,3,4,5\}|}{|\{1,2,3,4,5,6\}|} = \frac{5}{6}$

$\gamma_{\{b,c\},\tau}(Q) = \frac{|\{1,3\}|}{|\{1,2,3,4,5,6\}|} = \frac{2}{6}$

The subset $\{a,b\}$ is chosen, as this results in a higher dependency degree than $\{b\}$. The algorithm then evaluates the combination of this subset with the remaining attributes (in this example only one attribute, c, remains):

$\gamma_{\{a,b,c\},\tau}(Q) = \frac{|\{1,2,3\}|}{|\{1,2,3,4,5,6\}|} = \frac{3}{6}$

As this value is less than that for subset $\{a,b\}$, the algorithm terminates and outputs the reduct $\{a,b\}$. This is the same subset as that found by the fuzzy-rough method. However, for tolerance-based FS techniques, a suitable similarity measure must be defined for *all* attributes that are considered, and an appropriate value for τ must be determined.

ALTERNATIVE SEARCH MECHANISMS

GA-Based Approaches

Genetic algorithms (GAs) (Holland, 1975) are generally quite effective for rapid search of large, nonlinear, and poorly understood spaces. Unlike classical feature selection strategies, where one solution is optimized, a population of solutions can be modified at the same time (Kudo & Skalansky, 2000; Siedlecki & Sklansky, 1989). This can result in several optimal (or close-to-optimal) feature subsets as output.

A feature subset is typically represented by a binary string with length equal to the number of features present in the dataset. A zero or one in the jth position in the chromosome denotes the absence or presence of the jth feature in this particular subset. The general process for feature selection using GAs can be seen in Figure 16.

An initial population of chromosomes is created; the size of the population and how they are created are important issues. From this pool of feature subsets, the typical genetic operators (crossover and mutation) are applied. Again, the choice of which types of crossover and mutation used must be carefully considered, as well as their probabilities of application. This generates a new feature subset pool that may be evaluated in two different ways. If a filter approach is adopted, the fitness of individuals is calculated using a suitable criterion function. This function evaluates the goodness of a feature subset; a larger value indicates a better subset. Such a criterion function could be Shannon's entropy measure (Quinlan, 1993) or the dependency function from rough set theory (Pawlak, 1991).

For the wrapper approach, chromosomes are evaluated by inducing a classifier based on the feature subset, and obtaining the classification accuracy (or an estimate of it) on the data (Smith & Bull, 2003). To guide the search toward minimal feature subsets, the subset size is also incorporated into the fitness function of both filter and wrapper methods. Indeed, other factors may be included that are of interest, such as the cost of measurement for each feature, and so forth. GAs may also learn rules directly, and in the process, perform feature selection (Cordón, del Jesus, & Herrera, 1999; Jin, 2000; Xiong & Litz, 2002).

A suitable stopping criterion must be chosen. This is typically achieved by limiting the number of generations that take place, or by setting some threshold that must be exceeded by the fitness function. If the stopping criterion is not satisfied, then individuals are selected from the current subset pool and the process repeats.

Figure 16. Feature selection with genetic algorithms

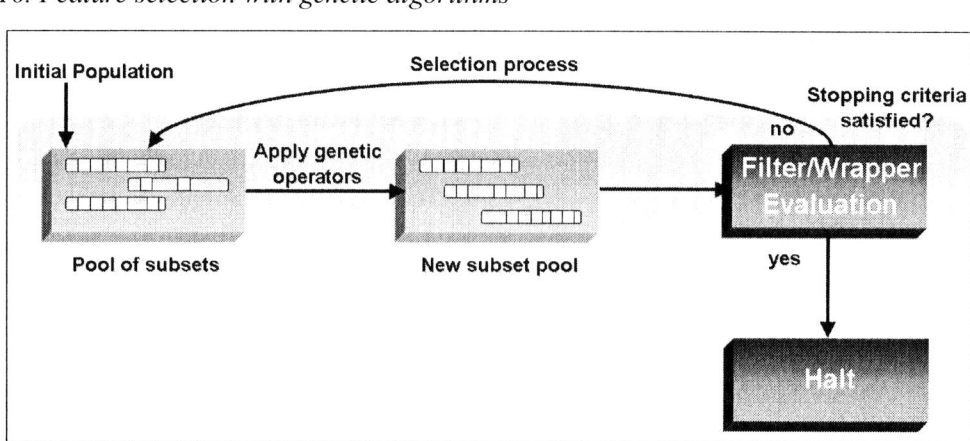

As with all feature selection approaches, GAs can get caught in local minima, missing a dataset's true minimal feature subset. Also, the fitness evaluation can be very costly, as there are many generations of many feature subsets that must be evaluated. This is particularly a problem for wrapper approaches, where classifiers are induced and evaluated for each chromosome.

The approaches reported in Bjorvand and Komorowski (1997) and Wróblewski (1995) use genetic algorithms to discover optimal or close-to-optimal reducts. Reduct candidates are encoded as bit strings, with the value in position i set if the ith attribute is present. The fitness function depends on two parameters. The first is the number of bits set. The function penalises those strings that have larger numbers of bits set, driving the process to find smaller reducts. The second is the number of classifiable objects given this candidate. The reduct should discern between as many objects as possible (ideally all of them).

Although this approach to FS is not guaranteed to find minimal subsets, it may find many subsets for any given dataset. It is also useful for situations where new objects are added to or old objects are removed from a dataset; the reducts generated previously can be used as the initial population for the new reduct-determining process. The main drawback is the time taken to compute each bit string's fitness, which is $O(a*o^2)$, where a is the number of attributes and o the number of objects in the dataset. The extent to which this hampers performance depends mainly on the population size.

Simulated Annealing-Based

Annealing is the process by which a substance is heated (usually melted) and cooled slowly in order to toughen and reduce brittleness. For example, this process is used for a metal to reach a configuration of minimum energy (a perfect, regular crystal). If the metal is annealed too quickly, this perfect organisation is unable to be achieved throughout the substance. Parts of the material will be regular, but these will be separated by boundaries where fractures are most likely to occur.

Simulated annealing (SA) (Kirkpatrick et al., 1983) is a stochastic optimization technique that is based on the computational imitation of this process of annealing. It is concerned with the change of energy (cost) of a system. In each algorithmic step, an "atom" (a feature subset in FS) is given a small random displacement, and the resulting change of energy, ΔE, is calculated. If $\Delta E \leq 0$, this new state is allowed and the process continues. However, if $\Delta E > 0$, the probability that this new state is accepted is:

$$P(\Delta E) = e^{-\frac{\Delta E}{T}} \qquad (35)$$

As the temperature, T, is lowered, the probability of accepting a state with a positive change in energy reduces. In other words, the willingness to accept a bad move decreases. The conversion of a combinatorial optimization problem into the SA framework involves the following:

1. **Concise configuration description:** The representation of the problem to be solved should be defined in a way that allows solutions to be constructed easily and evaluated quickly.
2. **Random move generator:** A suitable random transformation of the current state must be defined. Typically, the changes allowed are small, to limit the extent of search to the vicinity of the currently considered best solution. If this is not limited, the search degenerates to a random unguided exploration of the search space.
3. **Cost function definition:** The cost function (i.e., the calculation of the state's energy) should effectively combine the various criteria that are to be optimized for the problem. This function should be defined in such a

way that smaller function values indicate better solutions.
4. **Suitable annealing schedule:** As with the real-world annealing process, problems are encountered if the initial temperature is too low, or if annealing takes place too quickly. Hence, an annealing schedule must be defined that avoids these pitfalls. The schedule is usually determined experimentally.

To convert the feature selection task into this framework, a suitable representation must be used. Here, the states will be feature subsets. The random moves can be produced by randomly mutating the current state with a low probability. This may also remove features from a given feature subset, allowing the search to progress both forwards and backwards. The cost function must take into account both the evaluated subset "goodness" (by a filter evaluation function or a wrapper classifier accuracy) and also the subset size. The annealing schedule can be determined by experiment, although a good estimate may be $T(0) = |C|$ and $T(t+1) = \alpha * T(t)$, with $\alpha \geq 0.85$. Here, t is the number of iterations and α determines the rate of cooling.

The SA-based feature-selection algorithm can be seen in Figure 17. This differs slightly from the general SA algorithm in that there is a measure of local search employed at each iteration, governed by the parameter L_k. An initial solution is created, from which the next states are derived by random mutations and evaluated. The best state is remembered and used for processing in the next cycle. The chosen state may not actually be the best state encountered in this loop, due to the probability $P(\Delta E)$ that a state is chosen randomly (which will decrease over time). The temperature is decreased according to the annealing schedule, and the algorithm continues until the lowest allowed temperature has been exceeded.

Problems with this approach include how to define the annealing schedule correctly. If α is too high, the temperature will decrease slowly, allowing more frequent jumps to higher energy states, slowing convergence. However, if α is too low, the temperature decreases too quickly, and the system will converge to local minima (equivalent to brittleness in the case of metal annealing). Also, the cost-function definition is critical; there must be a balancing of the importance assigned to the different evaluation criteria involved. Biasing one over another will have the effect of directing search toward solutions that optimize that criterion only.

SimRSAR employs a simulated annealing-based feature selection mechanism to locate rough set reducts (Jensen & Shen, 2004b). The states are feature subsets, with random state mutations set to changing three features (either adding or removing them). The cost function attempts to maximize the rough set dependency (γ) whilst minimizing the subset cardinality. The cost of subset R is defined as:

$$\text{cost}(R) = \left[\frac{\gamma_C(D) - \gamma_R(D)}{\gamma_C(D)}\right]^a + \left[\frac{|R|}{|C|}\right]^b \quad (36)$$

Figure 17. Simulated annealing-based feature selection

```
SAFS(T₀, Tₘᵢₙ, α, Lₖ)
   T₀, the initial temperature;
   Tₘᵢₙ, the minimum allowed temperature;
   α, the extent of temperature decrease;
   Lₖ, the extent of local search

(1)   R ← genInitSol()
(2)   while T(t) > Tₘᵢₙ
(3)      for i=1,..., Lₖ
(4)         S ← genSol(R)
(5)         ΔE = cost(S)
(6)         if ΔE ≤ 0
(7)            M ← S
(8)         else if P(ΔE) > randNumber()
(9)            M ← S
(10)     R ← M
(11)     T(t+1) = α * T(t)
(12) output R
```

where *a* and *b* are defined in order to weight the contributions of dependency and subset size to the overall cost measure.

Ant Colony Optimization Based

Swarm intelligence (SI) is the property of a system whereby the collective behaviours of simple agents interacting locally with their environment cause coherent functional global patterns to emerge (Bonabeau, Dorigo, & Theraulez, 1999). SI provides a basis with which it is possible to explore collective (or distributed) problem solving without centralized control or the provision of a global model. One area of interest in SI is particle swarm optimization (Kennedy & Eberhart, 1995), a population-based stochastic optimization technique. Here, the system is initialised with a population of random solutions, called particles. Optima are searched for by updating generations, with particles moving through the parameter space towards the current local and global optimum particles. At each time step, the velocities of all particles are changed, depending on the current optima.

Ant colony optimization (ACO) (Bonabeau et al., 1999) is another area of interest within SI. In nature, it can be observed that real ants are capable of finding the shortest route between a food source and their nest without the use of visual information and hence, possess no global world model, adapting to changes in the environment. The deposition of pheromone is the main factor in enabling real ants to find the shortest routes over a period of time. Each ant probabilistically prefers to follow a direction rich in this chemical. The pheromone decays over time, resulting in much less pheromone on less popular paths. Given that over time, the shortest route will have the higher rate of ant traversal, this path will be reinforced and the others diminished until all ants follow the same, shortest path (the "'system'" has converged to a single solution). It is also possible that there are many equally short paths. In this situation, the rates of ant traversal over the short paths will be roughly the same, resulting in these paths being maintained while others are ignored. Additionally, if a sudden change to the environment occurs (e.g., a large obstacle appears on the shortest path), the ACO system can respond to this, and will eventually converge to a new solution. Based on this idea, artificial ants can be deployed to solve complex optimization problems via the use of artificial pheromone deposition.

ACO is particularly attractive for feature selection, as there seems to be no heuristic that can guide search to the optimal minimal subset every time. Additionally, it can be the case that ants discover the best feature combinations as they proceed throughout the search space.

ACO Framework

An ACO algorithm can be applied to any combinatorial problem as far as it is possible to define:

1. **Appropriate problem representation:** The problem can be described as a graph with a set of nodes and edges between nodes.
2. **Heuristic desirability (η) of edges:** A suitable heuristic measure of the "goodness" of paths from one node to every other connected node in the graph.
3. **Construction of feasible solutions:** A mechanism must be in place whereby possible solutions are efficiently created. This requires the definition of a suitable traversal stopping criterion to stop path construction when a solution has been reached.
4. **Pheromone updating rule:** A suitable method of updating the pheromone levels on edges is required, with a corresponding evaporation rule, typically involving the selection of the n best ants, and updating the paths they chose.
5. **Probabilistic transition rule:** The rule that determines the probability of an ant

traversing from one node in the graph to the next.

Each ant in the artificial colony maintains a memory of its history, remembering the path it has chosen so far in constructing a solution. This history can be used in the evaluation of the resulting created solution, and may also contribute to the decision process at each stage of solution construction.

Two types of information are available to ants during their graph traversal, local and global, controlled by the parameters β and α respectively. Local information is obtained through a problem-specific heuristic measure. The extent to which the measure influences an ant's decision to traverse an edge is controlled by the parameter β. This will guide ants towards paths that are likely to result in good solutions. Global knowledge is also available to ants through the deposition of artificial pheromone on the graph edges by their predecessors over time. The impact of this knowledge on an ant's traversal decision is determined by the parameter α. Good paths discovered by past ants will have a higher amount of associated pheromone. How much pheromone is deposited, and when, is dependent on the characteristics of the problem. No other local or global knowledge is available to the ants in the standard ACO model, though the inclusion of such information by extending the ACO framework has been investigated (Bonabeau et al., 1999).

Feature Selection

The feature selection task may be reformulated into an ACO-suitable problem (Jensen, 2006; Jensen & Shen, 2005). ACO requires a problem to be represented as a graph; here, nodes represent features, with the edges between them denoting the choice of the next feature. The search for the optimal feature subset is then an ant traversal through the graph, where a minimum number of nodes are visited that satisfies the traversal-stopping criterion. Figure 18 illustrates this setup: the ant is currently at node a and has a choice of which feature to add next to its path (dotted lines). It chooses feature b next, based on the transition rule, then c and then d. Upon arrival at d, the current subset $\{a,b,c,d\}$ is determined to satisfy the traversal-stopping criteria (e.g., a suitably high classification accuracy has been achieved with this subset, assuming that the selected features are used to classify certain objects). The ant terminates its traversal and outputs this feature subset as a candidate for data reduction.

A suitable heuristic desirability of traversing between features could be any subset evaluation

Figure 18. ACO problem representation for feature selection

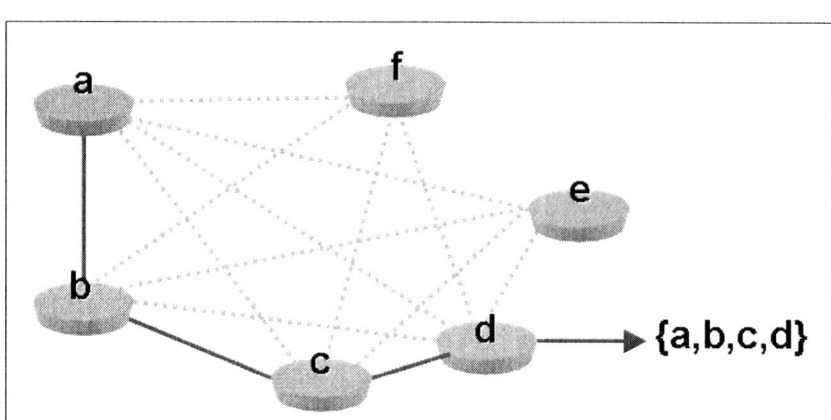

function, for example, an entropy-based measure (Quinlan, 1993) or the fuzzy-rough set dependency measure. Depending on how optimality is defined for the particular application, the pheromone may be updated accordingly. For instance, subset minimality and "goodness" are two key factors, so the pheromone update should be proportional to "goodness" and inversely proportional to size. How "goodness" is determined will also depend on the application. In some cases, this may be a heuristic evaluation of the subset; in others it may be based on the resulting classification accuracy of a classifier produced using the subset.

The heuristic desirability and pheromone factors are combined to form the so-called probabilistic transition rule, denoting the probability of an ant k at feature i choosing to move to feature j at time t:

$$p_{i,j}^k(t) = \frac{[\tau_{i,j}(t)]^\alpha \cdot [\eta_{i,j}]^\beta}{\sum_{l \in J_i^k}[\tau_{i,l}(t)]^\alpha \cdot [\eta_{i,l}]^\beta} \qquad (37)$$

where J_i^k is the set of ant k's unvisited features, $\eta_{i,j}$ is the heuristic desirability of choosing feature j when at feature i and $\tau_{i,j}(t)$ is the amount of virtual pheromone on edge (i,j). The choice of α and β is determined experimentally. Typically, several experiments are performed, varying each parameter and choosing the values that produce the best results.

Selection Process

The overall process of ACO feature selection can be seen in Figure 19. It begins by generating a number of ants, k, which are then placed randomly on the graph (i.e., each ant starts with one random feature). Alternatively, the number of ants to place on the graph may be set equal to the number of features within the data; each ant starts path construction at a different feature. From these initial positions, they traverse edges probabilistically until a traversal-stopping criterion is satisfied. The resulting subsets are

Figure 19. ACO-based feature selection overview

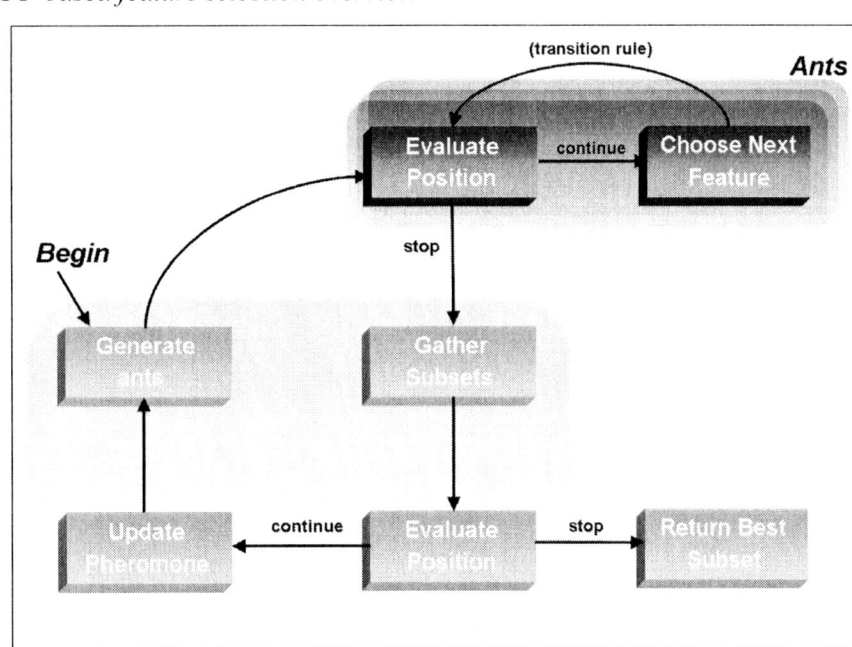

gathered and then evaluated. If an optimal subset has been found or the algorithm has executed a certain number of times, then the process halts and outputs the best feature subset encountered. If neither condition holds, then the pheromone is updated, a new set of ants are created, and the process iterates once more.

Complexity Analysis

The time complexity of the ant-based approach to feature selection is $O(IAk)$, where I is the number of iterations, A the number of original features, and k the number of ants. In the worst case, each ant selects all the features. As the heuristic is evaluated after each feature is added to the reduct candidate, this will result in A evaluations per ant. After one iteration in this scenario, Ak evaluations will have been performed. After I iterations, the heuristic will be evaluated IAk times.

Pheromone Update

Depending on how optimality is defined for the particular application, the pheromone may be updated accordingly. To tailor this mechanism to find rough set reducts, it is necessary to use the dependency measure as the stopping criterion. This means that an ant will stop building its feature subset when the dependency of the subset reaches the maximum for the dataset (the value 1 for consistent datasets). The dependency function may also be chosen as the heuristic desirability measure, but this is not necessary. In fact, it may be of more use to employ a non-rough set-related heuristic for this purpose. By using an alternative measure, such as an entropy-based heuristic, the method may avoid feature combinations that may mislead the rough set-based heuristic.

The pheromone on each edge is updated according to the following formula:

$$\tau_{i,j}(t+1) = (1-\rho).\tau_{i,j}(t) + \Delta\tau_{i,j}(t)$$

where

$$\Delta\tau_{i,j}(t) = \sum_{k=1}^{n} \frac{\gamma_{S^k}(D)}{|S^k|}$$

This is the case if the edge (i,j) has been traversed; $\Delta\tau_{i,j}(t)$ is 0 otherwise. The value ρ is a decay constant used to simulate the evaporation of the pheromone, S^k is the feature subset found by ant k. The pheromone is updated according to both the rough set measure of the goodness of the ant's feature subset and the size of the subset itself. By this definition, all ants update the pheromone. Alternative strategies may be used for this, such as allowing only the ants with the currently best feature subsets to proportionally increase the pheromone.

Results in (Jensen, 2006) show the effectiveness of the ACO-based approach for finding reducts both for crisp rough set reduction and fuzzy-rough selection. This technique regularly located optimal subsets (in terms of the subset size and corresponding dependency degree).

Particle Swarm Optimization-Based

Particle swarm optimization (PSO) is an evolutionary computation technique (Kennedy & Eberhart, 1995). The original intent was to graphically simulate the movement of bird flocking behaviour. (Shi & Eberhart, 1998) introduced the concept of inertia weight into the particle swarm optimizer to produce the standard PSO algorithm.

Standard PSO Algorithm

PSO is initialized with a population of particles. Each particle is treated as a point in an S-dimensional space. The ith particle is represented as $X_i = (x_{i1}, x_{i2}, ..., x_{iS})$. The best previous position (p_{best}, the position giving the best fitness value) of any particle is $P_i = (p_{i1}, p_{i2}, ..., p_{iS})$. The index of the global best particle is represented by g_{best}.

The velocity for particle i is $V_i = (v_{i1}, v_{i2}, ..., v_{iS})$. The particles are manipulated according to the following:

$$v_{id} = w \times v_{id} + c_1 \times rand()$$
$$\times (p_{id} - x_{id}) + c_2 \times Rand() \times (p_{gd} - x_{id}) \quad (38)$$

$$x_{id} = x_{id} + v_{id} \quad (39)$$

where w is the inertia weight. Suitable selection of the inertia weight provides a balance between global and local exploration. If a time-varying inertia weight is employed, better performance can be expected (Shi & Eberhart, 1998). The acceleration constants c_1 and c_2 represent the weighting of the stochastic acceleration terms that pull each particle toward p_{best} and g_{best} positions. Low values allow particles to roam far from target regions before being tugged back, while high values result in abrupt movement toward, or past, target regions. Particles' velocities on each dimension are limited to a maximum velocity V_{max}. If V_{max} is too small, particles may not explore sufficiently beyond locally good regions. If V_{max} is too high, particles might fly past good solutions.

The first part provides the particles with a memory capability and the ability to explore new search space areas. The second part is the "cognition" part, which represents the private thinking of the particle itself. The third part is the "social" part, which represents the collaboration among the particles. This equation is used to update the particle's velocity. The performance of each particle is then measured according to a predefined fitness function.

The PSO algorithm for feature selection can bee seen in Figure 20. Initially, a population of particles is constructed with random positions and velocities on S dimensions in the problem space. For each particle, the fitness function is evaluated. If the current particle's fitness evaluation is better than p_{best}, then this particle becomes the current best, and its position and fitness are stored. Next, the current particle's fitness is compared with the population's overall previous best fitness. If the current value is better than g_{best}, then this is set to the current particle's position, with the global best fitness updated. This position represents the

Figure 20. PSO-FS algorithm

```
PSO-FS(C, D)
  C, the set of conditional features;
  D, the set of decision features.

(1)  ∀i : X_i ← randomPosition(); V_i ← randomVelocity()
(2)  fit ← bestFit(X); globalbest ← fit; p_best ← bestPos(X);
(3)  ∀i : P_i ← X_i
(4)  while (stopping criterion not met)
(5)    for i = 1,..., S    // for each particle
(6)      if (fitness(i) > fit)   // local best
(7)        fit ← fitness(i)
(8)        p_best ← X_i
(9)      if (fitness(i) > globalbest)   // global best
(10)       globalbest ← fitness(i)
(11)       g_best ← X_i; R ← getReduct(X_i)   // convert to reduct
(12)     updateVelocity(); updatePosition()
(13) output R
```

best feature subset encountered so far, and is thus converted and stored in *R*. The velocity and position of the particle is then updated. This process loops until a stopping criterion is met, usually a sufficiently good fitness or a maximum number of iterations (generations).

Encoding

To apply PSO to rough set reduction, the particle's position is represented as binary bit strings of length *N*, where *N* is the total number of attributes. This is the same representation as that used for GA-based feature selection. Therefore, each particle position is an attribute subset.

Representation of Velocity

Each particle's velocity is represented as a positive integer, varying between 1 and V_{max}. It implies how many of the particle's bit should be changed to be the same as that of the global best position, that is, the velocity of the particle flying toward the best position. The number of different bits between two particles relates to the difference between their positions. For example, $P_{gbest} = [1,0,1,1,1,0,1,0,0,1]$, $X_i = [0,1,0,0,1,1,0,1,0,1]$. The difference between the global best and the particle's current position is $P_{gbest} - X_i = [1,-1,1,1,0,-1,1,-1,0,0]$. "1" means that, compared with the best position, this bit (feature) should be selected but it is not, decreasing classification quality. On the other hand, "-1" means that compared with the best position, this bit should not be selected but it is. Both cases will lead to a lower fitness value.

Updating Position

After the updating of velocity, a particle's position will be updated by the new velocity. If the new velocity is *V*, and the number of different bits between the current particle and g_{best} is *xg*, there exist two situations while updating the position:

1. $V \leq xg$. In such a situation, randomly change *V* bits of the particle, which are different from that of g_{best}. The particle will move toward the global best while keeping its exploration ability.
2. $V > xg$. In this case, in addition to changing all the different bits to be the same as that of g_{best}, a further (*V-xg*) bits should be randomly changed. Hence, after the particle reaches the global best position, it keeps on moving some distance toward other directions, which gives it further exploration ability.

Velocity Limit

In experimentation, the particles' velocity was initially limited to [*1, N*]. However, it was noticed that, in some cases, after several generations, swarms find good solutions (but not optimal ones), and in the following generations, g_{best} remains stationary. Hence, only a suboptimal solution is located. This indicates that the maximum velocity is too high and particles often "fly past" the optimal solution. This can be prevented by setting V_{max} as (1/3)*N and so the velocity is limited to the range [1, (1/3)*N]. Once finding a global best position, other particles will adjust their velocities and positions, searching around the best position.

Fitness Function

The following can be used as the fitness function for directing search towards optimal reducts:

$$Fitness = \alpha * \gamma_R(D) + \beta * \frac{|C| - |R|}{|C|}$$

where $\gamma_R(D)$ is the classification quality of condition attribute set *R* relative to decision *D*, |*R*| is the length of the selected feature subset, and |*C*| is the total number of features. α and β are two parameters that correspond to the importance of classification quality and subset length, with $\alpha \in [0,1]$ and $\beta = 1 - \alpha$. Setting a high α value assures that the best position is at least a real rough set reduct. The goal then is to maximize fitness values.

Setting Parameters

In the algorithm, the inertia weight decreases along with the iterations according to Shi and Eberhart (1998):

$$w = w_{max} - \frac{w_{max} - w_{min}}{iter_{max}} iter$$

where w_{max} is the initial value of the weighting coefficient, w_{min} is the final value of the weighting coefficient, $iter_{max}$ is the maximum number of iterations or generations, and $iter$ is the current iteration or generation number.

Time Complexity

Let N be the number of features and M the total objects, then the complexity of the fitness function is $O(NM^2)$. The other impact on time is the number of generation iterations; however, time is spent mainly evaluating the particles' positions.

Experimentation reported in Wang et al. (Wang, Yang, Teng, Xia, & Jensen, 2006) demonstrates the utility of PSO-based rough set feature selection. The method was compared against standard RSAR, EBR, a discernibility-based method, and a GA-based method. It regularly located better reducts, but tended to be more time consuming than RSAR and EBR, due to its nondeterministic nature.

CONCLUSION

This chapter has reviewed several techniques for feature selection based on rough set theory. Current methods tend to concentrate on alternative evaluation functions, employing rough set concepts to gauge subset suitability. These methods can be categorized into two distinct approaches: those that incorporate the degree of dependency measure (or extensions), and those that apply heuristic methods to generated discernibility matrices.

Methods based on traditional rough set theory do not have the ability to effectively manipulate continuous data. For these methods to operate, a discretization step must be carried out beforehand, which can often result in a loss of information. There are two main extensions to RST that handle this and avoid information loss: tolerance rough sets and fuzzy-rough sets. Both approaches replace crisp equivalence classes with alternatives that allow greater flexibility in handling object similarity.

Alternative search mechanisms have also been presented in this chapter, with a particular emphasis on stochastic approaches. These are of particular interest, as no perfect heuristic exists that can always guide hill-climbing approaches to optimal subsets. Methods based on GAs, simulated annealing, ant colony optimization, and particle swarm optimization were described.

Most of the effort in dependency degree-based feature selection has focussed on the use of lower approximations and positive regions in gauging subset suitability. However, there is still additional information that can be obtained through the use of upper approximations and the resulting boundary regions. The boundary region is of interest as this contains those objects whose concept membership is unknown; objects in the positive region definitely belong to a concept, and objects in the negative region do not belong. It is thought that by incorporating this additional information into the search process, better subsets should be located. Investigations into this are ongoing at the University of Wales, Aberystwyth.

An interesting direction for future work in discernibility matrix-based approaches might be the use of fuzzy propositional satisfiability. Traditionally, attributes appearing in objects are included in the discernibility matrix if their values differ. Based on these entries, a discernibility function is constructed from which prime implicants are calculated or heuristic methods employed to find reducts. Currently, there is no flexibility in this process; discrete, noise-free data must be used and attributes either appear or are absent in matrix entries. This could be extended by considering fuzzy indiscernibility, where attributes belong to the discernibility matrix with varying degrees of membership based on fuzzy similarity. From this, fuzzy propositional satisfiability (or heuristic methods) could be used to determine fuzzy reducts for the system.

REFERENCES

Bakar, A.A., Sulaiman, M.N., Othman, M., & Selamat, M.H. (2002). Propositional satisfiability algorithm to find minimal reducts for data mining. *International Journal of Computer Mathematics*, *79*(4), 379-389.

Bazan, J. (1998). A comparison of dynamic and non-dynamic rough set methods for extracting laws from decision tables. In *Rough sets in knowledge discovery* (pp. 321-365). Heidelberg: Physica-Verlag.

Bazan, J., Skowron, A., & Synak, P. (1994). Dynamic reducts as a tool for extracting laws from decision tables. In *Proceedings of the 8th Symposium on Methodologies for Intelligent Systems* (LNAI 869, pp. 346-355).

Beynon, M.J. (2000). An investigation of β-reduct selection within the variable precision rough sets model. In *Proceedings of the Second International Conference on Rough Sets and Current Trends in Computing (RSCTC 2000)* (pp. 114–122).

Beynon, M.J. (2001). Reducts within the variable precision rough sets model: A further investigation. *European Journal of Operational Research*, *134*(3), 592-605.

Bjorvand, A.T., & Komorowski, J. (1997). Practical applications of genetic algorithms for efficient reduct computation. In *Proceedings of the 15th IMACS World Congress on Scientific Computation, Modelling and Applied Mathematics, 4* (pp. 601-606).

Bonabeau, E., Dorigo, M., & Theraulez, G. (1999). *Swarm intelligence: From natural to artificial systems*. New York: Oxford University Press Inc.

Chouchoulas, A., & Shen, Q. (2001). Rough set-aided keyword reduction for text categorisation. *Applied Artificial Intelligence*, *15*(9), 843-873.

Cordón, O., del Jesus, M. J., & Herrera, F. (1999). Evolutionary approaches to the learning of fuzzy rule-based classification systems. In L.C. Jain (Ed.), *Evolution of engineering and information systems and their applications* (pp. 107-160). CRC Press.

Dash, M., & Liu, H. (1997). Feature selection for classification. *Intelligent Data Analysis*, *1*(3), 131-156.

Dubois, D., & Prade, H. (1992). Putting rough sets and fuzzy sets together. *Intelligent decision support* (pp. 203-232). Dordrecht: Kluwer Academic Publishers.

Han, J., Hu, X., & Lin, T.Y. (2004). Feature subset selection based on relative dependency between attributes. *Rough Sets and Current Trends in Computing: 4th International Conference, RSCTC 2004* (pp. 176-185). Uppsala, Sweden.

Höhle, U. (1988). Quotients with respect to similarity relations. *Fuzzy Sets and Systems*, *27*(1), 31-44.

Holland, J. (1975). *Adaptation in natural and artificial systems*. Ann Arbour: The University of Michigan Press.

Hoos, H.H., & Stützle, T. (1999). Towards a characterisation of the behaviour of stochastic local search algorithms for SAT. *Artificial Intelligence, 112,* 213-232.

Jensen, R. (2006). Performing feature selection with ACO. To appear in *Swarm intelligence and data mining.* Springer SCI book series.

Jensen, R., & Shen, Q. (2004a). Fuzzy-rough attribute reduction with application to Web categorization. *Fuzzy Sets and Systems, 141*(3), 469-485.

Jensen, R., & Shen, Q. (2004b). Semantics-preserving dimensionality reduction: Rough and fuzzy-rough based approaches. *IEEE Transactions on Knowledge and Data Engineering, 16*(12), 1457-1471.

Jensen, R., & Shen, Q. (2005). Fuzzy-rough data reduction with ant colony optimization. *Fuzzy Sets and Systems, 149*(1), 5-20.

Jensen, R., Shen, Q., & Tuson, A. (2005). Finding rough set reducts with SAT. In *Proceedings of the 10th International Conference on Rough Sets, Fuzzy Sets, Data Mining and Granular Computing* (LNAI 3641, pp. 194-203).

Jin, Y. (2000). Fuzzy modeling of high-dimensional systems: Complexity reduction and interpretability improvement. *IEEE Transactions on Fuzzy Systems, 8*(2), 212-221.

Kennedy, J., & Eberhart, R.C. (1995). Particle swarm optimization. In *Proceedings of IEEE International Conference on Neural Networks* (pp. 1942-1948). Piscataway, NJ.

Kirkpatrick, S., Gelatt Jr., C.D., & Vecchi, M.P. (1983). Optimization by simulation annealing. *Science 220*(4598), 671-680.

Koczkodaj, W. W., Orlowski, M., & Marek, V. W. (1998). Myths about rough set theory. *Communications of the ACM, 41*(11), 102-103.

Kosko, B. (1986). Fuzzy entropy and conditioning. *Information Sciences, 40*(2), 165-174.

Kryszkiewicz, M. (1994). Maintenance of reducts in the variable precision rough sets model. *ICS Research Report* 31/94, Warsaw University of Technology.

Kudo, M., & Skalansky, J. (2000). Comparison of algorithms that select features for pattern classifiers. *Pattern Recognition, 33*(1), 25-41.

Langley, P. (1994). Selection of relevant features in machine learning. In *Proceedings of the AAAI Fall Symposium on Relevance* (pp. 1-5).

Mac Parthaláin, N., Jensen, R., & Shen, Q. (2006). Fuzzy entropy-assisted fuzzy-rough feature selection. To appear in *Proceedings of the 15th International Conference on Fuzzy Systems (FUZZ-IEEE'06).*

Modrzejewski, M. (1993). Feature selection using rough sets theory. In *Proceedings of the 11th International Conference on Machine Learning* (pp. 213-226).

Nguyen, S.H., & Nguyen, H.S. (1996). Some efficient algorithms for rough set methods. In *Proceedings of the Conference of Information Processing and Management of Uncertainty in Knowledge-Based Systems* (pp. 1451-1456).

Nguyen, S.H., & Skowron, A. (1997a). Searching for relational patterns in data. In *Proceedings of the First European Symposium on Principles of Data Mining and Knowledge Discovery* (pp. 265-276). Trondheim, Norway.

Nguyen, H.S., & Skowron, A. (1997b). Boolean reasoning for feature extraction problems. In *Proceedings of the 10th International Symposium on Methodologies for Intelligent Systems* (pp. 117-126).

Øhrn, A. (1999). *Discernibility and rough sets in medicine: Tools and applications.* Department of Computer and Information Science. Trondheim, Norway, Norwegian University of Science and Technology: 239.

Pawlak, Z. (1991). *Rough sets: Theoretical aspects of reasoning about data*. Dordrecht: Kluwer Academic Publishing.

Polkowski, L. (2002). Rough sets: Mathematical foundations. *Advances in Soft Computing.* Heidelberg, Germany: Physica Verlag.

Polkowski, L., Lin, T.Y., & Tsumoto, S. (Eds). (2000). Rough set methods and applications: New developments in knowledge discovery in information systems. *Studies in Fuzziness and Soft Computing, 56.* Heidelberg, Germany: Physica-Verlag.

Quinlan, J.R. (1993). C4.5: Programs for machine learning. *The Morgan Kaufmann series in machine learning.* San Mateo, CA: Morgan Kaufmann Publishers.

Shen, Q., & Jensen, R. (2004). Selecting informative feature with fuzzy-rough sets and its application for complex systems monitoring. *Pattern Recognition, 37*(7), 1351-1363.

Shi, Y., & Eberhart, R. (1998). A modified particle swarm optimizer. In *Proceedings of the IEEE International Conference on Evolutionary Computation* (pp. 69-73), Anchorage, AK.

Siedlecki, W., & Sklansky, J. (1988). On automatic feature selection. *International Journal of Pattern Recognition and Artificial Intelligence, 2*(2), 197-220.

Siedlecki, W., & Sklansky, J. (1989). A note on genetic algorithms for large-scale feature selection. *Pattern Recognition Letters, 10*(5), 335-347.

Skowron, A., & Rauszer, C. (1992). The discernibility matrices and functions in information systems. *Intelligent Decision Support* (pp. 331-362). Dordrecht: Kluwer Academic Publishers.

Skowron, A., & Stepaniuk, J. (1996). Tolerance approximation spaces. *Fundamenta Informaticae, 27*(2), 245-253.

Ślezak, D. (1996). Approximate reducts in decision tables. In *Proceedings of the 6th International Conference on Information Processing and Management of Uncertainty in Knowledge-Based Systems (IPMU '96)* (pp. 1159-1164).

Smith, M.G., & Bull, L. (2003). Feature construction and selection using genetic programming and a genetic algorithm. In *Proceedings of the 6th European Conference on Genetic Programming (EuroGP 2003)* (pp. 229-237).

Starzyk, J.A., Nelson, D.E., & Sturtz, K. (2000). A mathematical foundation for improved reduct generation in information systems. *Journal of Knowledge and Information Systems, 2*(2), 131-146.

Stepaniuk, J. (1998). Optimizations of rough set model. *Fundamenta Informaticae, 36*(2-3), 265–283.

Thiele, H. (1998). *Fuzzy rough sets versus rough fuzzy sets - An interpretation and a comparative study using concepts of modal logics.* (Technical report no. CI- 30/98). University of Dortmund.

Wang, J., & Wang, J. (2001). Reduction algorithms based on discernibility matrix: The ordered attributes method. *Journal of Computer Science & Technology, 16*(6), 489-504.

Wang, X.Y., Yang, J., Teng, X., Xia, W., & Jensen, R. (2006). Feature selection based on rough sets and particle swarm optimization. Accepted for publication in *Pattern Recognition Letters.*

Wróblewski, J. (1995). Finding minimal reducts using genetic algorithms. In *Proceedings of the 2nd Annual Joint Conference on Information Sciences* (pp. 186-189).

Xiong, N., & Litz, L. (2002). Reduction of fuzzy control rules by means of premise learning: Method and case study. *Fuzzy Sets and Systems, 132*(2), 217-231.

Yao, Y. Y. (1998). A comparative study of fuzzy sets and rough sets. *Information Sciences, 109*(1-4), 21-47.

Zadeh, L. A. (1975). The concept of a linguistic variable and its application to approximate reasoning. *Information Sciences, 8,* 199-249, 301-357; *9,* 43-80.

Zhang, L., & Malik, S. (2002). The quest for efficient Boolean satisfiability solvers. In *Proceedings of the 18th International Conference on Automated Deduction* (pp. 295-313).

Zhang, M., & Yao, J.T. (2004). A rough sets based approach to feature selection. In *Proceedings of the 23rd International Conference of NAFIPS* (pp. 434-439). Banff, Canada.

Zhong, N., Dong, J., & Ohsuga, S. (2001). Using rough sets with heuristics for feature selection. *Journal of Intelligent Information Systems, 16*(3), 199-214.

Ziarko, W. (1993). Variable precision rough set model. *Journal of Computer and System Sciences, 46*(1), 39-59.

Chapter IV
Rough Set Analysis and Formal Concept Analysis

Yiyu Yao
University of Regina, Canada

Yaohua Chen
University of Regina, Canada

ABSTRACT

Rough-set analysis (RSA) and formal concept analysis (FCA) are two theories of intelligent data analysis. They can be compared, combined, and applied to each other. In this chapter, we review the existing studies on the comparisons and combinations of rough-set analysis and formal concept analysis and report some new results. A comparative study of two theories in a unified framework provides a better understanding of data analysis.

INTRODUCTION

Rough-set analysis (RSA) and formal concept analysis (FCA) were introduced in the early 1980s by Pawlak (1982) and Wille (1982), respectively. They can be considered as two complementary approaches for analyzing data. Many proposals have been made to compare and combine the two theories, and to apply results from one to the other.

Basic notions of rough-set analysis are the indiscernibility of objects, attribute reductions, and approximations of undefinable sets by using definable sets (Pawlak, 1982, 1991). By modeling indiscernibility as an equivalence relation, one can partition a finite universe of objects into a family of pair-wise disjoint sets called a partition. The partition provides a granulated view of the universe. An equivalence class is considered as an elementary definable set. The empty set and unions of equivalence classes are also viewed as definable sets. Consequently, an arbitrary subset of universe may not necessarily be definable. It can be approximated from below and above by a pair of maximal and minimal definable sets called upper and lower approximations. A definable set must have the same lower and upper approximations (Buszkowski, 1998). Generalized rough-set models, by considering different binary relations between objects or coverings of the universe, are proposed and studied by many researchers (Wiweger, 1989; Wu, Mi, & Zhang, 2003; Wybraniec-Skardowska, 1989; Yao, 1998,, 2003; Zakowski, 1983; Zhu, & Wang, 2003, Zhu, & Wang, 2006). In this chapter, we only focus on equivalence relations and induced definable sets.

Formal concept analysis is developed based on a formal context given by a binary relation between a set of objects and a set of attributes or properties (Wille, 1982). It provides a formal and graphical way to organize data as formal concepts. In a formal context, a pair of a set of objects and a set of attributes that uniquely associate with each other is called a formal concept (Ganter & Wille, 1999; Wille, 1982). The set of objects of a formal concept is referred to as the extension, and the set of attributes as the intension. The family of all formal concepts is a complete lattice.

Comparative examinations of rough-set analysis and formal concept analysis show that they are complementary and parallel to each other in terms of notions, issues, and methodologies (Kent, 1996; Pagliani, 1993; Qi, Wei, & Li, 2005; Wasilewski, 2005; Zhang, Yao, & Liang (Eds.), 2006). There exists a set of common notions that link the two theories together. One can immediately adopt ideas from one to the other (Yao, 2004a, 2004b; Yao, & Chen, 2006). On the one hand, the notions of formal concepts and formal concept lattices can be introduced into rough-set analysis by considering different types of formal concepts (Yao, 2004a). On the other hand, rough-set approximation operators and attribute reductions can be introduced into formal concept analysis by considering a different type of definability and classification, respectively (Hu, Sui, Lu, Wang, & Shi, 2001; Saquer, & Deogun, 1999, 2001; Yao, & Chen, 2004, 2006; Zhang, Wei, & Qi, 2005). The combination of the two theories would produce new tools for data analysis.

In this chapter, we present a critical review of the existing studies on formal concept analysis and rough-set analysis that summarizes our studies reported in several earlier papers (Chen, & Yao, 2006; Yao, 2004b; Yao, & Chen, 2005). Broadly, we can classify the existing studies into three groups. The first group may be labeled as the comparative studies (Gediga, & Düntsch, 2002; Ho, 1997; Kent, 1996; Pagliani, 1993; Qi et al., 2005; Wasilewski, 2005; Wolski, 2003, 2005; Yao, 2004a). It deals with the comparison of the two approaches with an objective to produce a more general data-analysis framework. The second group concerns the applications of the notions and ideas of formal concept analysis into rough-set analysis (Gediga & Düntsch, 2002; Wolski, 2003; Yao, 2004a). Reversely, the third group focuses on applying concepts and methods of rough-set analysis into formal concept analysis (Düntsch & Gediga, 2003; Hu et al., 2001; Pagliani & Chakraborty, 2005; Saquer & Deogun, 1999, 2001; Shao & Zhang, 2005; Wolski, 2003; Yao, 2004a; Yao & Chen, 2006; Zhang et al., 2005). Those studies lead to different types of abstract operators, concept lattices, and approximations.

In Section 2, we provide some background knowledge about formal contexts and modal-style data operators. Sections 3 and 4 briefly recall basic knowledge about formal concept analysis and rough-set analysis. In Sections 5, 6, 7, and 8, we discuss in detail the three groups of the existing studies on comparisons and combinations of rough-set analysis and formal concept analysis.

FORMAL CONTEXTS AND MODAL-STYLE DATA OPERATORS

In order to discuss issues in formal concept analysis and rough-set analysis, we consider a simple representation scheme in which a finite set of objects is described by a finite set of attributes.

Formal Contexts

Let U and V be two finite and nonempty sets. Elements of U are called objects, and elements of V are called attributes. The relationship between objects and attributes can be formally defined by a binary relation R from U to V, $R \subseteq U \times V$. The triplet (U, V, R) is called a formal context in formal concept analysis (Ganter, & Wille, 1999; Wille, 1982),, a binary information system in rough-set theory (Pawlak, 1982), and a multivalued mapping from U to V in the theory of evidence (Dempster, 1967; Shafer, 1976,1987). In this chapter, we also refer to it as a binary table. It should be noted that any multivalued table could be translated into a binary table using the scaling transformation in formal concept analysis (Buszkowski, 1998; Ganter, & Wille, 1999).

For an object $x \in U$ and an attribute $y \in V$, if $x R y$, we say that the object x has the attribute y, or alternatively, the attribute y is possessed by the object x. For an object $x \in U$, its attributes are given by the successors of x:

$$xR = \{y \in V \mid xRy\} \qquad (1)$$

For an attribute $y \in V$, the set of objects having y is given by the predecessors of y as follows:

$$Ry = \{x \in U \mid xRy\} \qquad (2)$$

We have two ways to extend the binary relation to sets of objects and attributes. For a set $A \subseteq U$, we have (Yao, 2004a):

$$A\underline{R} = \{y \in V \mid \forall x \in A (xRy)\}$$
$$= \bigcap_{x \in A} xR$$
$$A\overline{R} = \{y \in V \mid \exists x \in A (xRy)\}$$
$$= \bigcup_{x \in A} xR \qquad (3)$$

Any object $x \in A$ has all the attributes in $A\underline{R}$, and any attribute $y \in A\overline{R}$ is possessed by at least one object in A. Similarly, for a subset of attributes $B \subseteq V$, we have (Yao, 2004b):

$$\underline{R}B = \{x \in U \mid \forall y \in B(xRy)\}$$
$$= \bigcap_{y \in B} Ry,$$
$$\overline{R}B = \{x \in U \mid \exists y \in B(xRy)\}$$
$$= \bigcup_{y \in B} Ry \qquad (4)$$

Any attribute $y \in B$ is possessed by all the objects in $\underline{R}B$, and any object in $\overline{R}B$ has at least one attribute in B.

Modal-Style Data Operators

With a formal context, we can define different types of modal-style data operators (Düntsch, & Gediga, 2003; Gediga, & Düntsch, 2002; Orlowska, 1998). Each of them leads to a different type of rules summarizing the relationships between objects and attributes.

Basic set assignments: For a set of objects $A \subseteq U$ and a set of attributes $B \subseteq V$, we can define a pair of data operators, $^b: 2^U \to 2^V$ and $^b: 2^V \to 2^U$, as follows (Wong, Wang, & Yao, 1995):

$$A^b = \{y \in V \mid Ry = A\},$$
$$B^b = \{x \in U \mid xR = B\}, \qquad (5)$$

where 2^U and 2^V are the power sets of U and V, respectively. For simplicity, the same symbol is used for both mappings. The mapping ^{b} is called a basic set assignment (Yao, 2004a). A set of objects $A \subseteq U$ is called a focal set of objects if $A^b = \varnothing$, and a set of attributes $B \subseteq V$ is called a focal set of attributes if $B^b = \varnothing$ (Wong et al.,1995).

Sufficiency operators: For a set of objects $A \subseteq U$ and a set of attributes $B \subseteq V$, we can define a pair of data operators, $*: 2^U \to 2^V$ and $*: 2^V \to 2^U$, called sufficiency operators, as follows (Ganter, & Wille, 1999; Wille, 1982):

$$A^* = \{y \in V \mid \forall x \in U(x \in A \Rightarrow xRy)\}$$
$$= \{y \in V \mid A \subseteq Ry\}$$
$$= \bigcap_{x \in A} xR,$$
$$= A\underline{R},$$

$$B^* = \{x \in U \mid \forall y \in V(y \in B \Rightarrow cRy)\}$$
$$= \{x \in U \mid B \subseteq xR\}$$
$$= \bigcap_{y \in B} Ry,$$
$$= \underline{R}B. \quad (6)$$

For a set of objects A, A^* is the maximal set of attributes shared by all objects in A. For a set of attributes B, B^* is the maximal set of objects that have all attributes in B.

Necessity operators: For a set of objects $A \subseteq U$ and a set of attributes $B \subseteq V$, we can define a pair of data operators, $: 2^U \to 2^V$ and $: 2^V \to 2^U$, as follows (Düntsch, & Gediga, 2003; Gediga, & Düntsch, 2002; Yao, 2004a; Yao, 2004b):

$$A^\square = \{y \in V \mid Ry \subseteq A\},$$
$$B^\square = \{x \in V \mid xR \subseteq B\}. \quad (7)$$

By definition, an object having an attribute in A is necessarily in A. The operators are referred to as the necessity operators.

Possibility operators: For a set of objects $A \subseteq U$ and a set of attributes $B \subseteq V$, we can define a pair of data operators, $\diamond: 2^U \to 2^V$ and $\diamond: 2^V \to 2^U$, as follows (Düntsch, & Gediga, 2003; Gediga, & Düntsch, 2002; Yao, 2004a; Yao, 2004b):

$$A^\diamond = \{y \in V \mid Ry \cap A \neq \emptyset\}$$
$$= \bigcup_{z \in A} xR,$$
$$= A\overline{R},$$
$$B^\diamond = \{s \in U \mid xR \cap B \neq \emptyset\}$$
$$= \bigcup_{y \in B} Ry,$$
$$= \overline{R}B. \quad (8)$$

An object having an attribute in $A^\diamond$ is only possible in A. The operators are referred to as the possibility operators.

Connections of Modal-Style Data Operators

Since the modal-style data operators are defined based on the binary relation R, there exist connections among these modal-style data operators (Yao, 2004b). In terms of the basic set assignment, we can reexpress operators $\square$ and $\diamond$ as:

$(A).$ $A^* = \bigcup\{X^b \mid X \subseteq U, A \subseteq X\},$
$B^* = \bigcup\{Y^b \mid Y \subseteq V, B \subseteq Y\};$

$(B).$ $A^\square = \bigcup\{X^b \mid X \subseteq U, X \subseteq A\},$
$B^\square = \bigcup\{Y^b \mid B \subseteq V, Y \subseteq B\};$

$(C).$ $A^\diamond = \bigcup\{X^b \mid X \subseteq U, A \cap X \neq \emptyset\},$
$B^\diamond = \bigcup\{Y^b \mid Y \subseteq V, B \cap Y \neq \emptyset\}.$

In other words, the sufficiency, necessity, and possibility operations on a set $A \subseteq U$ can be expressed by a union of some assigned sets to the focal sets of objects that are related to A.

In a formal context (U, V, R), an equivalence relation on the universe U can be defined based on the attribute set V and is denoted as $E(V)$. That is, for two objects $x, x' \in U$,

$$xE(V)x' \Leftrightarrow xR = x'R. \quad (9)$$

The equivalence relation divides the universe U into pair-wise disjoint subsets, called equivalence classes of objects. That is, for an object $x \in U$, the equivalence class of objects containing x is given by:

$$[x] = \{x' \mid xE(V)x'\} \qquad (10)$$

Objects in $[x]$ are indistinguishable from x. The family of all equivalence classes of objects is called a partition of the universe U and denoted by $U/E(V)$. The partition $U/E(V)$ is indeed the family of the assignments to the focal sets of attributes:

$$U/E(V) = \{Y^{\flat} \neq \emptyset \mid Y \subseteq V\} \qquad (11)$$

Similarly, we can define an equivalence relation $E(U)$ on V as: for attributes $y, y' \in V$:

$$yE(U)y' \Leftrightarrow Ry = Ry' \qquad (12)$$

For an attribute $y \in V$, the equivalence classes of attributes containing y is defined by:

$$[y] = \{y' \mid yE(U)y'\}. \qquad (13)$$

The family of all equivalence classes of attributes is called a partition of the universe V, denoted by $V/E(U)$. The partition $V/E(U)$ is the family of the assignments to the focal sets of objects:

$$V/E(U) = \{X^{\flat} \neq \emptyset \mid X \subseteq U\}. \qquad (14)$$

The basic set assignments associate an equivalence class of objects with a focal set of attributes and an equivalence class of attributes with a focal set of objects. This can be easily seen from the fact that each object $x \in Y^{\flat}$ possesses exactly the set of attributes Y, and each attribute $y \in X^{\flat}$ is possessed exactly by all objects in X. Therefore, in terms of equivalence classes, we can reexpress operators $*$, $\square$ and $\Diamond$ as:

(a). $A^{*} = \bigcup\{[y] \mid [y] \in V/E(U), A \subseteq Ry\}$,
$B^{*} = \bigcup\{[x] \mid [x] \in U/E(V), B \subseteq xR\}$;

(b). $A^{\square} = \bigcup\{[y] \mid [y] \in V/E(U), Ry \subseteq A\}$,
$B^{\square} = \bigcup\{[x] \mid [x] \in U/E(V), xR \subseteq B\}$;

(c). $A^{\Diamond} = \bigcup\{[y] \mid [y] \in V/E(U), A \cap Ry \neq \emptyset\}$,
$B^{\Diamond} = \bigcup\{[x] \mid [x] \in U/E(V), B \cap xR \neq \emptyset\}$.

In other words, the sufficiency, necessity, and possibility operations on a set $A \subseteq U$ can also be expressed by a union of some equivalence classes.

FORMAL CONCEPT ANALYSIS

In formal concept analysis, one is interested in a pair of a set of objects and a set of attributes that uniquely define each other. This type of definability leads to the introduction of the notion of formal concepts.

Formal concepts: A pair (A, B), $A \subseteq U$, $B \subseteq V$, is called a formal concept of the context (U, V, R), if $A = B^*$ and $B = A^*$. Furthermore, $extent(A, B) = A$ is called the extension of the concept, and $intent(A, B) = B$ is called the intension of the concept (Ganter, & Wille, 1999; Wille, 1982).

We can interpret the unique relationship between a set of objects and a set of attributes as follows (Yao, 2004b). For $(A, B) = (B^*, A^*)$, we have:

$$\begin{aligned}
x \in A &\Leftrightarrow x \in B^{*} \\
&\Leftrightarrow B \subseteq xR \\
&\Leftrightarrow \bigwedge_{y \in B} xRy \\
\bigwedge_{x \in A} xRy &\Leftrightarrow A \subseteq Ry \\
&\Leftrightarrow y \in A^{*} \\
&\Leftrightarrow y \in B.
\end{aligned}$$

(15)

That is, the set of objects A is defined based on the set of attributes B, and vice versa.

An order relation can be defined on formal concepts based on the set inclusion relation. For two formal concepts (A_1, B_1) and (A_2, B_2), (A_1, B_1) is a subconcept of (A_2, B_2), written $(A_1, B_1) \leq (A_2, B_2)$, and (A_2, B_2) is a superconcept of (A_1, B_1), if and only if $A_1 \subseteq A_2$, or equivalently, if and only if $B_2 \subseteq B_1$ (Ganter, & Wille, 1999; Wille, 1982). In other words, a more general (specific) concept is characterized by a larger (smaller) set of objects that share a smaller (larger) set of attributes.

Concept lattice: The set of all formal concepts forms a complete lattice called a concept lattice, denoted by $L(U, V, R)$. The meet and join of the lattice are characterized by the set intersection and union (Ganter, & Wille, 1999; Wille, 1982), and can be expressed as following:

$$\bigwedge_{t \in T}(A_t, B_t) = (\bigcap_{t \in T} A_t, (\bigcap_{t \in T} B_t)^{**}),$$
$$\bigvee_{t \in T}(A_t, B_t) = ((\bigcap_{t \in T} A_t)^{**}, \bigcap_{t \in T} B_t),$$

(16)

where T is an index set and for every $t \in T$, (A_t, B_t) is a formal concept.

ROUGH-SET ANALYSIS

Rough-set analysis extends the classical set theory with two additional unary set-theoretic approximation operators (Pawlak, 1982; Yao, 1996). It is developed based on an indiscernibility (equivalence) relation on a universe of objects, and is generalized by using a binary relation (i.e., a multivalued mapping) between two universes. Various formulations of rough-set approximations have been proposed and studied (Yao, 1998, 2003).

Rough-Set Analysis on One Universe

Suppose E is an equivalence relation on U. The pair $apr = (U, E)$ is referred to as an approximation space. An approximation space induces a granulated view of the universe. Equivalence classes are interpreted as the elementary nonempty observable, measurable, or definable subsets of U. By extending the definability of equivalence classes, we assume that the empty set and unions of some equivalence classes are definable. The family of definable sets contains the empty set $\emptyset$ and is closed under set complement, intersection, and union. It is an σ-algebra $\sigma(U/E) \subseteq 2^U$ with basis U/E.

With respect to a formal context (U, V, R), a concrete equivalence relation $E(V)$ can be defined by equation (9). In this case, we have an induced approximation space $apr = (U, E(V))$. In general, for any set of attributes $W \subseteq V$, one can define an equivalence relation $E(W)$ and the corresponding approximation space $apr = (U, E(W))$.

A set of objects not in $\sigma(U/E)$ is said to be undefinable. An undefinable set must be approximated from below and above by a pair of definable sets. Conversely, the approximation operators truthfully reflect the notion of definability (Pawlak, 1982; Yao, & Chen, 2004, 2006). That is, for a set of objects $A \subseteq U$, its lower and upper approximations are equal if and only if $A \in \sigma(U/E)$, and for an undefinable set $A' \subseteq U$, its lower and upper approximations are not equal. In fact, the lower approximation for an undefinable set is a proper subset of its upper approximation.

The approximations of an undefinable set can be defined with three different formulations: subsystem-based definition, granule-based definition, and element-based definition (Yao, 1998).

Subsystem-based definition: In the subsystem-based definition, a subsystem of the power set of a universe is first constructed and the approximation operators are then defined using the subsystem.

In an approximation space $apr = (U, E)$, the subsystem is $\sigma(U/E)$, and a pair of approximation operators, $\underline{apr}, \overline{apr}: 2^U \to 2^U$, is defined by:

$$\underline{apr}(A) = \bigcup \{X \mid X \in \sigma(U/E), X \subseteq A\},$$
$$\overline{apr}(A) = \bigcap \{X \mid X \in \sigma(U/E), X \subseteq A\}. \quad (17)$$

The lower approximation $\underline{apr}(A) \in \sigma(U/E)$ is the greatest definable set contained in A, and the upper approximation $\overline{apr}(A) \in \sigma(U/E)$ is the least definable set containing A.

Granule-based definition: In the granule-based definition, equivalence classes are considered as the elementary definable sets, and approximations can be defined directly by using equivalence classes. In an approximation space $apr = (U, E)$, a pair of approximation operators, $\underline{apr}, \overline{apr}: 2^U \to 2^U$, is defined by:

$$\underline{apr}(A) = \bigcup \{[x] \mid [x] \in U/E, [x] \subseteq A\},$$
$$\overline{apr}(A) = \bigcup \{[x] \mid [x] \in U/E, A \cap [x] \neq \emptyset\} \quad (18)$$

The lower approximation is the union of equivalence classes that are subsets of A, and the upper approximation is the union of equivalence classes that have a nonempty intersection with A.

Element-based definition: In the element-based definition, the individual objects are used to construct approximations of a set of objects. In an approximation space $apr = (U, E)$, a pair of approximation operators, $\underline{apr}, \overline{apr}: 2^U \to 2^U$, is defined by:

$$\underline{apr}(A) = \bigcup \{x \mid x \in U, [x] \subseteq A\},$$
$$\overline{apr}(A) = \bigcap \{x \mid x \in U, A \cap [x] \neq \emptyset\}. \quad (19)$$

The lower approximation is the set of objects whose equivalence classes are subsets of A. The upper approximation is the set of objects whose equivalence classes have nonempty intersections with A.

The three definitions are equivalent, but with different forms and interpretations (Yao, -2003). They offer three different directions in generalizing the theory of rough sets (Yao, 1998).

The lower and upper approximation operators have the following properties: for sets of objects $A, A_1,$ and A_2:

(i). $\quad \underline{apr}(U) = U$
$\quad \overline{apr}(\emptyset) = \emptyset$

(ii). $\quad \underline{apr}(A) = (\overline{apr}(A^c))^c$,
$\quad \overline{apr}(A) = (\underline{apr}(A^c))^c$;

(iii). $\quad \underline{apr}(A_1 \cap A_2) = \underline{apr}(A_1) \cap \underline{apr}(A_2)$,
$\quad \overline{apr}(A_1 \cup A_2) = \overline{apr}(A_1) \cup \overline{apr}(A_2)$;

(iv). $\quad \underline{apr}(A) \subseteq A \subseteq \overline{apr}(A)$;

(v). $\quad \underline{apr}(\underline{apr}(A)) = \underline{apr}(A)$,
$\quad \overline{apr}(\overline{apr}(A)) = \overline{apr}(A)$;

(vi). $\quad \overline{apr}(\underline{apr}(A)) = \underline{apr}(A)$,
$\quad \underline{apr}(\overline{apr}(A)) = \overline{apr}(A)$.

Property (i) shows that the lower approximation of the empty set $\emptyset$ and the upper approximation of the universe U are themselves. Property (ii) states that the approximation operators are dual operators with respect to set complement ^{c}. Property (iii) states that the lower approximation operator is distributive over set intersection $\cap$, and the upper approximation operator is distributive over set union $\cup$. By property (iv), a set lies within its lower and upper approximations. Properties (v) and (vi) deal with the compositions of lower and upper approximation operators. The result of the composition of a sequence of lower and upper approximation operators is the same as the application of the approximation operator closest to A.

Rough-Set Analysis on Two Universes

A more general model of rough-set analysis can be formulated on two universes (Dubois, & Prade, 1990; Pei, & Xu, 2004; Wong et al., 1993, 1995; Zhang et al., 2005). A formal context (U, V, R) is called a two-universe approximation space. By the binary relation R, we have an equivalence relation $E(U)$ on V as defined by the equation 12. The induced partition $V/E(U)$ produces an σ-algebra $\sigma(V/E(U)) \subseteq 2^V$, which is the family of sets of attributes containing the empty set $\varnothing$ and unions of equivalence classes of attributes.

Similar to one universe, the lower and upper approximation operators, $\underline{apr}$, $\overline{apr}$: $2^U \to 2^V$, are defined in three different forms: subsystem-based definition, granule-based definition, and element-based definition.

Subsystem-based definition: In the subsystem-based definition, the pair of approximation operators, $\underline{apr}$, $\overline{apr}$: $2^U \to 2^V$, is defined by:

$$\underline{apr}(A) = \bigcup \{Y | Y \in \sigma(V/E(U)), \overline{R}Y \subseteq A\},$$
$$\overline{apr}(A) = \bigcap \{Y | Y \in \sigma(V/E(U)), A \subseteq \overline{R}Y\}.$$

(20)

The lower approximation is the set of attributes in which the union of their corresponding objects are contained in A. The upper approximation is the set of attributes in which the intersection of their corresponding objects contain A.

Granule-based definition: In the granule-based definition, the pair of approximation operators is defined by:

$$\underline{apr}(A) = \bigcup \{[y] | [y] \in V/E(U), Ry \subseteq A\},$$
$$\overline{apr}(A) = \bigcup \{[y] | [y] \in V/E(U), A \cap Ry \neq \varnothing\}.$$

(21)

The lower approximation is the union of equivalence classes of attributes whose corresponding sets of objects are subsets of A. The upper approximation is the union of equivalence classes of attributes whose corresponding sets of objects have nonempty intersections with A.

Element-based definition: In the element-based definition, the pair of approximation operators is defined by:

$$\underline{apr}(A) = \{y \in V | Ry \subseteq A\},$$
$$\overline{apr}(A) = \{y \in V | Ry \cap A \neq \varnothing\}.$$

(22)

The lower approximation is the set of attributes whose corresponding objects are subsets of A. The upper approximation is the set of attributes whose corresponding objects have nonempty intersections with A.

In fact, the lower and upper approximation operators can be expressed by modal-style operators, as follows (Shao, & Zhang, 2005):

$$\underline{apr}(A) = A^\square,$$
$$\overline{apr}(A) = A^\lozenge.$$

(23)

In other words, in the two-universe model, the lower approximation operator is the necessity operator, and the upper approximation operator is the possibility operator. Therefore, the lower and upper approximation operators have the following properties: for sets of objects A, A_1 and $A_2 \subseteq U$:

(i). $\underline{apr}(U) = V,$
   $\overline{apr}(\varnothing) = \varnothing;$

(ii). $\underline{apr}(A) = (\overline{apr}(A^C))^C,$
   $\overline{apr}(A) = (\underline{apr}(A^C))^C;$

(iii). $\underline{apr}(A_1 \cap A_2) = \underline{apr}(A_1) \cap \underline{apr}(A_2),$
   $\overline{apr}(A_1 \cup A_2) = \overline{apr}(A_1) \cup \overline{apr}(A_2);$

(iv). $\underline{apr}(A_1) \cup \underline{apr}(A_2) \subseteq \underline{apr}(A_1 \cup A_2)$,

$\overline{apr}(A_1 \cap A_2) \subseteq \overline{apr}(A_1) \cap \overline{apr}(A_2)$;

(v). $A_1 \subseteq A_2 \Rightarrow \underline{apr}(A_1) \subseteq (A_2)$,

$A_1 \subseteq A_2 \Rightarrow \overline{apr}(A_1) \subseteq (A_2)$.

Those properties are the counterparts of the properties of approximation operators in one universe.

Attribute Reduction

An important task of rough-set analysis is attribute reduction, which is associated with machine-learning algorithms (Bazan, Nguyen, Nguyen, Synak, & Wroblewski, J., 2000; Grzymala-Busse, 1988; Mi, Wu, & Zhang, 2003; Miao, & Wang, 1999; Qiu, Zhang, & Wu, 2005; Skowron, & Rauszer, 1991; Wang, & Wang, 2001; Wu, Zhang, Li, & Mi, 2005). For a given data table, it is possible to find a subset of the original attribute set that is the most informative and necessary (Pawlak, 1991). The unnecessary attributes can be removed from the data table without loss or with little loss of information. In other words, the number of attributes can be reduced.

For a given subset of attributes $W \subseteq V$, $W \neq \emptyset$, we can define an equivalence relation $E(W)$ on the universe U. The set W is called a reduct of V if W satisfies the following conditions (Yao, Zhao, & Wang, 2006):

R1. $U / E(V) = U / E(W)$

R2. $U / E(W - \{a\}) \neq U / E(V)$, for all $a \in W$

The condition (R1) is referred to as the jointly sufficient condition. That is, the reduced set of attributes W is sufficient to produce the same partition as the entire set V. The condition (R2) is referred to as the individually necessary condition. It states that the partition $U/E(W-\{a\})$ is not the same as the partition $U/E(V)$ when removing an attribute a from W. In other words, each individual attribute in W is necessary to keep the classification of objects in U unchanged.

There can exist many reducts for V. The intersection of all reducts of V is called the core of V. The attributes in a core of V are indispensable. In other words, without any attribute in the core of V, the partition on the universe U must be changed.

COMPARATIVE STUDIES OF ROUGH-SET ANALYSIS AND FORMAL CONCEPT ANALYSIS

Kent examined the correspondence between similar notions used in both theories, and argued that they are in fact parallel to each other in terms of basic notions, issues, and methodologies (Kent, 1996). A framework of rough concept analysis has been introduced as a synthesis of the two theories. Based on this framework, Ho developed a method of acquiring rough concepts (Ho, 1997), and Wu, Liu, and Li proposed an approach for computing accuracies of rough concepts and studied the relationships between the indiscernibility relations and accuracies of rough concepts (2004).

The notion of a formal context has been used in many studies under different names. Shafer used a compatibility relation to interpret the theory of evidence (Shafer, 1976, 1987). A compatibility relation is a binary relation between two universes, which is in fact a formal context. Wong, Wang, and Yao investigated approximation operators over two universes with respect to a compatibility relation (1993, 1995). Düntsch and Gediga referred to those operators as modal-style operators and studied a class of such operators in data analysis (Gediga, & Düntsch, 2002). The derivation operator in formal concept analysis is a sufficiency operator, and the rough-set approximation operators are the necessity and possibility operators used in modal logics.

Pagliani used a Heyting algebra structure to connect concept lattices and approximation spaces together (19). Based on the algebra structure, concept lattices, and approximation spaces can be transformed into each other. Wasilewski demonstrated that formal contexts and general approximation spaces could be mutually represented (Wasilewski, 2005). Consequently, rough-set analysis and formal concept analysis can be viewed as two related and complementary approaches for data analysis. It is shown that the extension of a formal concept is a definable set in the approximation space. Qi *et al.* argued that two theories have much in common in terms of the goals and methodologies (2005). They emphasized the basic connection and transformation between a concept lattice and a partition.

Wolski investigated Galois connections in formal concept analysis and their relations to rough-set analysis (Wolski, 2003). A logic, called S4.t, has been proposed as a good tool for approximate reasoning to reflect the formal connections between formal concept analysis and rough-set analysis (Wolski, 2005).

Yao compared the two theories based on the notions of definability, and showed that they deal with two different types of definability (Yao, 2004b):. rough-set analysis studies concepts that are defined by disjunctions of properties, formal concept analysis considers concepts that are definable by conjunctions of properties.

Based on those comparative studies, one can easily adopt ideas from one theory to another. The applications of rough-set analysis always lead to approximations and reductions in formal concept analysis. The approximation operators can result in new types of concepts and concept lattices.

CONCEPT LATTICES IN ROUGH-SET ANALYSIS

Some researchers concern the applications of notions and ideas of formal concept analysis into rough-set analysis (Gediga, & Düntsch, 2002; Wolski, 2003; Yao, 2004a). By establishing approximation operators, the necessity and possibility operators in equations (7) and (8), we construct additional concept lattices. Those lattices, their properties, their semantic interpretations, and their connections to the original concept lattice are studied extensively by Düntsch and Gediga (2002), Yao (2004a), and Wolski (2003). The results provide more insights into data analysis. One can obtain different types of inference rules regarding objects and attributes. To reflect their physical meanings, the notions of object-oriented and property-oriented concept lattices are introduced.

Object-Oriented Concept Lattice

A pair (A, B) is called an object-oriented concept if $A = B^{\diamond}$ and $B = A^{\square}$. If an object has an attribute in B then the object belongs to A. Furthermore, only objects in A have attributes in B. The set of objects A is called the extension of the concept (A, B), and the set of the attributes B is called the intension.

The family of all object-oriented concepts forms a complete lattice. The meet $\wedge$ and the join $\vee$ are defined as follows. For two object-oriented concepts (A_1, B_1) and (A_2, B_2):

$$\begin{aligned}(A_1, B_1) \wedge (A_2, B_2) &= ((B_1 \cap B_2)^{\diamond}, B_1 \cap B_2) \\ &= ((A_1 \cap A_2)^{\square\diamond}, B_1 \cap B_2), \\ (A_1, B_1) \vee (A_2, B_2) &= ((A_1 \cup A_2, (A_1 \cup A_2)^{\square}) \\ &= (A_1 \cup A_2, (B_1 \cup B_2)^{\square\diamond}).\end{aligned}$$

(24)

One can verify that the pair $((B_1 \cap B_2)^{\diamond}, B_1 \cap B_2)$ is an object-oriented concept. More specifically:

$$((B_1 \cap B_2)^{\diamond})^{\square} = B_1 \cap B_2. \qquad (25)$$

The pair $(A_1 \cup A_2, (A_1 \cup A_2)^{\square})$ is also an object-oriented concept.

For a set of objects $A \subseteq U$, we have $(A^{\Box\Diamond}) = A^\Box$, and hence, an object-oriented concept $(A^{\Box\Diamond}, A^\Box)$. For a set of attributes $B \subseteq V$, we have another object-oriented concept $(B^\Diamond, B^{\Diamond\Box})$. In the special case, for a single attribute $y \in V$, we have an object-oriented concept $(\{y\}^\Diamond, \{y\}^{\Diamond\Box}) = (Ry, (Ry)^\Box)$.

For two object-oriented concepts (A_1, B_1) and (A_2, B_2), we say that (A_1, B_1) is a subconcept of (A_2, B_2) if and only if $A_1 \subseteq A_2$, or equivalently, if and only if $B_1 \subseteq B_2$.

For a set of objects $A \subseteq U$, the pair $(A^{\Box\Diamond}, A^\Box)$ is an object-oriented concept. It follows:

$$x \in A^{\Box\Diamond} \Leftrightarrow \bigvee_{y \in A^\Box} xRy. \qquad (26)$$

The results can be extended to any object-oriented concept. For $(A, B) = (B^\Diamond, A)$, we have a rule:

$$x \in A \bigvee_{y \in B} xRy. \qquad (27)$$

That is, the set of objects A, and the set of attributes B in (A, B), uniquely determine each other.

Property-Oriented Concept Lattice

Data analysis in the object-oriented concept lattice can be carried out in a similar manner to the property-oriented concept lattice, but is focused on objects.

A pair (A, B), $A \subseteq U$, $B \subseteq V$, is called a property-oriented concept if $A = B^\Box$ and $B = A^\Diamond$. If an attribute is possessed by an object in A, then the attribute must be in B. Furthermore, only attributes B are possessed by objects in A.

The family of all property-oriented concepts forms a lattice with meet $\wedge$ and join $\vee$ defined by:

$$(A_1, B_1) \wedge (A_2, B_2) = (A_1 \cap A_2, (A_1 \cap A_2)^\Diamond)$$
$$= (A_1 \cap A_2, (B_1 \cup B_2)^{\Box\Diamond})$$
$$(A_1, B_1) \vee (A_2, B_2) = ((B_1 \cup B_2)^\Box, B_1 \cup B_2)^\Diamond)$$
$$= ((A_1 \cup A_2)^{\Diamond\Box}, B_1 \cup B_2)^\Diamond) \qquad (28)$$

For a set of objects $A \subseteq U$, we can construct a property-oriented concept $(A^{\Diamond\Box}, A^\Diamond)$.

For a set of attributes $B \subseteq V$, there is a property-oriented concept $(B^\Box, B^{\Box\Diamond})$.

For a set of attributes $B \subseteq V$, the pair $(B^{\Box\Diamond}, B^\Box)$ is a property-oriented concept. It follows:

$$\bigvee_{x \in B^\Box} xRy \Leftrightarrow y \in B^{\Box\Diamond} \qquad (29)$$

The results can be extended to any property-oriented concept. For $(A, B) = (B, A^\Diamond)$, we have a rule:

$$\bigvee_{x \in A} xRy \Leftrightarrow y \in B \qquad (30)$$

That is, the set of attributes B and the set of objects A in (A, B) uniquely determine each other.

Attribute Reductions in Concept Lattices

Some research works have been proposed to apply concepts and methods of rough-set attribute reduction into formal concept analysis (Shao, & Zhang, 2005; Zhang et al., 2005). It is considered to make the representation of implicit knowledge in formal context simpler.

Attribute reduction in a concept lattice based on a formal context can be processed through two methods. The first method directly seeks a reduct of attributes in the formal context by considering the jointly sufficient and individually necessary conditions, and then constructing the reduced concept lattice based on the reduced formal context. The second method is that a reduct of attributes is determined based on a criterion in which the reduced lattice and the original lattice show certain common features or structures. Moreover, removing any attribute in the reduct can change the structure of the concept lattice.

Reduction in a formal context: Let (U, V, R) be a formal context. The equivalence relation $E(V)$ is defined based on the attribute set V, and induces a partition of the universe $U/E(V)$. One can find out a reduct of attributes $W \subseteq V, W \neq \emptyset$ that satisfies the jointly sufficient and individual necessary conditions. Then, the formal context (U, W, R') is a reduced formal context with respect to (U, V, R), where $R' = R \cap (U \times W)$. The equivalence relation $E(W)$ defined, based on the attribute set W, produces a partition of the universe $U/E(W)$. The partition $E(W)$ is equal to the partition $E(V)$, that is,

$$U/E(W) = U/E(V) \qquad (31)$$

The concept lattice $L(U, W, R')$ derived from the reduce formal context (U, W, R') is considered as a reduced concept lattice with respect to the original concept lattice $L(U, V, R)$.

Reduction in a concept lattice: The second method of attribute reduction in concept lattice has been examined by Zhang, Wei, and Qi (2005). Let $L(U, V_1, R_1)$ and $L(U, V_2, R_2)$ be two formal concept lattices. If, for any formal concept $(A, B) \in L(U, V_2, R_2)$, there exists a formal concept $(A', B') \in L(U, V_1, R_1)$ such that $A = A'$, that is, the extensions of the formal concepts (A, B) and (A', B') are the same, then the concept lattice $L(U, V_1, R_1)$ is said to be finer than the concept lattice $L(U, V_2, R_2)$, denoted by:

$$L(U, V_1, R_1) \leq L(U, V_2, R_2) \qquad (32)$$

In other words, all the extensions of the formal concepts in the concept lattice $L(U, V_2, R_2)$ can be found in the concept lattice $L(U, V_1, R_1)$. If $L(U, V_1, R_1) \leq L(U, V_2, R_2)$ and $L(U, V_2, R_2) \leq L(U, V_1, R_1)$, then these two formal concept lattices are said to be isomorphic to each other, denoted by:

$$L(U, V_1, R_1) \cong L(U, V_2, R_2). \qquad (33)$$

For any formal context (U, V, R), there exists $L(U, V, R) \leq L(U, W, R')$, where $R' = R \cap (U \times W)$, $W \subseteq V$ and $W \neq \emptyset$. In other words, by reducing attributes from the attribute set V, the original concept lattice $L(U, V, R)$ is finer than the new concept lattice $L(U, W, R')$. W is called a consistent set of (U, V, R) if $L(U, W, R') \cong L(U, V, R)$. For a formal context (U, V, R), if there exists an attribute set $W \subseteq V, W \neq \emptyset$ and $R' = R \cap (U \times W)$ such that W satisfies following conditions:

(S1). $L(U, W, R') \cong L(U, V, R)$,

(S2). $L(U, W - \{a\}, R'') \cong L(U, V, R)$,

then W is called a reduct of (U, V, R). The intersection of all the reducts of (U, V, R) is called the core of (U, V, R). This method of attribute reduction in concept lattice based on discernibility matrix are also proposed and discussed (2005).

In fact, according to the studies by Qi et al. (2005) and the connections between modal-style operators and equivalence classes, one can infer that the two methods of attribute reductions in concept lattice are equivalent. That is, a reduct of the attribute set V obtained by one method can be achieved by another method. In other words, for a formal context (U, V, R), if a set of attributes W is a reduct of V, then two concept lattices $L(U, W, R')$ and $L(U, V, R)$ are isomorphic, that is, $L(U, W, R') \cong L(U, V, R)$, and the two partitions $U/E(V)$ and $U/E(W)$ are the same, that is, $U/E(V) = U/E(W)$.

APPROXIMATIONS IN FORMAL CONCEPT LATTICE

Many studies considered rough-set approximations in formal concept lattice (Düntsch, & Gediga, 2003; Hu et al., 2001; Saquer, & Deogun, 1999, 2001; Shao, & Zhang, 2005; Wolski, 2003). The extension of a formal concept can be viewed as a definable set of objects, although in a sense different from that of rough-set analysis (Yao, 2004a, 2004b). In fact, the extension of a formal concept is a set of indiscernible objects with respect to the intension. Based on the attributes in the intension, all objects in the extension cannot be distinguished. Furthermore, all objects in the extension share all the attributes in the intension. The collection of all the extensions, sets of objects, can be considered as a different system of definable sets (Yao, & Chen, 2004). An arbitrary set of objects may not be an extension of a formal concept. The sets of objects that are not extensions of formal concepts are regarded as undefinable sets. Therefore, in formal concept analysis, a different type of definability is proposed. An undefinable set can be approximated by extensions of formal concepts in a concept lattice.

Approximations Based on Equivalence Classes

Saquer and Deogun suggested that all concepts in a concept lattice can be considered as definable, and a set of objects can be approximated by concepts whose extensions approximate the set of objects (Saquer, & Deogun, 1999, 2001). A set of attributes can be similarly approximated by using intensions of formal concepts.

In their approach, for a given set of objects, it may be approximated by extensions of formal concepts in two steps. The classical rough-set approximations for a given set of objects are first computed. Since the lower and upper approximations of the set are not necessarily the extensions of formal concepts, they are then approximated again by using derivation operators of formal concept analysis.

At the first step, for a set of objects $A \subseteq U$, the standard lower approximation $\underline{apr}(A)$ and upper approximation $\overline{apr}(A)$ are obtained. At the second step, the lower approximation of the set of objects A is defined by the extension of the formal concept $(\underline{apr}(A)^{**}, \underline{apr}(A)^{*})$. The upper approximation of the set of objects A is defined by the extension of the formal concept $(\overline{apr}(A)^{**}, \overline{apr}(A)^{*})$. That is:

$$\underline{eapr}(A) = \underline{apr}(A)^{**},$$
$$\overline{eapr}(A) = \overline{apr}(A)^{**}. \qquad (34)$$

If $\underline{apr}(A) = \overline{apr}(A)$, we have $\underline{apr}(A)^{**} = \overline{apr}(A)^{**}$. Namely, for a definable set A, its lower and upper formal concept approximations are the same. However, the reverse implication is not true. A set of objects that has the same lower and upper approximations may not necessarily be a definable set.

Equation (35).

$$\underline{lapr}(A) = extent(\vee\{(X,Y) \mid (X,Y) \in L(U,V,R), X \subseteq A\})$$
$$= (\bigcup\{X \mid (X,Y) \in L(U,V,R), X \subseteq A\})^{**},$$
$$\overline{lapr}(A) = extent(\wedge\{(X,Y) \mid (X,Y) \in L(U,V,R), A \subseteq X\})$$
$$= (\bigcap\{X \mid X,Y) \in L(U,V,R), A \subseteq X\}.$$

Approximations Based on Lattice-Theoretic Operators

Hu *et al.* suggested an alternative formulation (2001). Instead of defining an equivalence relation, they defined a partial order on the universe of objects. For an object, its principal filter, which is the set of objects "greater than or equal to" the object and is called the partial class by Hu *et al.*, is the extension of a formal concept. The family of all principal filters is the set of join-irreducible elements of the concept lattice. Similarly, a partial order relation can be defined on the set of attributes. The family of meet-irreducible elements of the concept lattice can be constructed. The lower and upper approximations can be defined based on the families of meet-and join-irreducible elements in concept lattice. However, their formulations are slightly flawed and fail to achieve such a goal.

Yao and Chen followed the same direction of the approach proposed by Hu *et al.*, but used different formulations (Yao, & Chen, 2004, 2006).

They considered three types of lattice-theoretic formulations to define the lower and upper approximations.

Subsystem based definition: That is, for a set of objects $A \subseteq U$, its lower and upper approximations are defined by: (see equation (35)).

The lower approximation of a set of objects A is the extension of the formal concept ($\underline{lapr}(A)$, $(\underline{lapr}(A))^*$), and the upper approximation is the extension of the formal concept ($\overline{lapr}(A)$, $(\overline{lapr}(A))^*$). The concept ($\underline{lapr}(A)$, $(\underline{lapr}(A))^*$) is the supremum of those concepts whose extensions are subsets of A, and ($\overline{lapr}(A)$, $(\overline{lapr}(A))^*$) is the infimum of those concepts whose extensions are supersets of A.

Granule-based definition: The lower and upper approximations of a set of objects can be defined based on the extensions of join-irreducible and meet-irreducible concepts (Hu et al., 2001). That is, for a set of objects $A \subseteq U$, its lower and upper approximations are defined by: (see equation (36)),

Equation (36).

$$\underline{lapr}(A) = extent(\vee\{(X,Y) \mid (X,Y) \in J(L), X \subseteq A\})$$
$$= (\bigcup\{X \mid (X,Y) \in J(L), X \subseteq A\})^{**},$$
$$\overline{lapr}(A) = extent(\wedge\{(X,Y) \mid (X,Y) \in M(L), A \subseteq X\})$$
$$= (\bigcap\{X \mid X,Y) \in M(L), A \subseteq X\}.$$

Equation (37).

$$\underline{lapr}(A) = extent(\vee\{(\{x\}^{**},\{x\}^*) \mid x \in U, \{x\}^{**} \subseteq A\}),$$
$$= (\bigcup\{\{x\}^{**} \mid x \in A\})^{**}$$
$$\overline{lapr}(A) = extent(\wedge\{(\{y\}^*,\{y\}) \mid y \in V, A \subseteq \{y\}^*\}),$$
$$= (\bigcap\{\{y\}^* \mid y \in V, A \subseteq \{y\}^*\}.$$

where $J(L)$ and $M(L)$ are represented as the family of all join-irreducible formal concepts and the family of all meet-irreducible formal concepts, respectively.

Elementary-based definition: Ganter and Wille have shown that a formal concept in a concept lattice can be expressed by the join of object concepts in which the object is included in the extension of the formal concept (Ganter, & Wille, 1999). A formal concept can also be expressed by the meet of attribute concepts in which the attribute is included in the intension of the formal concept. Therefore, the lower and upper approximations of a set of objects can be defined based on the extensions of object and attribute concepts. That is, for a set of objects $A \subseteq U$, its lower and upper approximations are defined by object and attribute concepts: (see equation (37)).

By comparing with the standard rough-set approximations, one can observe two problems of the approximation operators defined by using lattice-theoretic operators. The lower approximation of a set of objects A is not necessarily a subset of A. Although a set of objects A is undefinable, that is, A is not the extension of a formal concept, its lower and upper approximations may be the same. In order to avoid these shortcomings, we present another formulation by using set-theoretic operators.

Approximations Based on Set-Theoretic Operators

Yao and Chen proposed an approach to solve the problems in lattice-theoretic approaches (Yao, & Chen, 2004, 2006). The extension of a formal concept is a definable set of objects. A system of definable sets can be derived from a concept lattice. For a formal concept lattice $L(U, V, R)$, the family of all extensions is denoted by:

$$EXT(L) = \{extent(X,Y) \mid (X,Y) \in L(U,V,R)\}. \tag{38}$$

The system $EXT(L)$ contains the entire set U and is closed under intersection. Thus, $EXT(L)$ is a closure system (Cohn, 1965).

The lower approximation is a maximal set in $EXT(L)$ that are subsets of A, and the upper approximation is a minimal set in $EXT(L)$ that are supersets of A. That is, for a set of objects $A \subseteq U$, its upper approximation is defined by:

$$\overline{sapr}(A) = \bigcap \{X \mid X \in EXT(L), A \subseteq X\}, \tag{39}$$

and its lower approximation is a family of sets

$$\underline{sapr}(A) = \{X \mid X \in EXT(L), X \subseteq A, \\ \forall X' \in EXT(L)(X \subset X' \Rightarrow X' \not\subseteq A)\} \tag{40}$$

An upper approximation is unique, but the maximal set contained in A is generally not unique.

The upper approximation $\overline{sapr}(A)$ is the same as $\overline{lapr}(A)$, namely, $\overline{sapr}(A) = \overline{lapr}(A)$. However, the lower approximation is different. An important feature is that a set can be approximated from below by several definable sets of objects. In general, for $A' \in \underline{sapr}(A)$, we have $A' \subseteq \underline{lapr}(A)$. Thus, in a concept lattice $L(U, V, R)$, for a subset of the universe of objects $A \subseteq U$, $\overline{sapr}(A) = A$ and $\underline{sapr}(A) = \{A\}$, if and only if A is an extension of a concept.

The lower approximation offers more insights into the notion of approximations. In some situations, the union of a family of definable sets is not necessarily a definable set. It may not be reasonable to insist on a unique approximation. The approximation of a set by a family of sets may provide a better characterization of the set.

Approximations Based on Modal-Style Operators

Some researchers used two different systems of concepts to approximate a set of objects or a set of attributes (Düntsch, & Gediga, 2003;, Pagliani, & Chakraborty, 2005; Pei, & Xu, 2004; Shao, & Zhang, 2005; Wolski, 2003).

Another class of approximation operators can be derived by the combination of operators $\square$ and $\lozenge$. The combined operators $\square\lozenge$ and $\lozenge\square$ have the following important properties (Düntsch, & Gediga, 2003):

1. $\square\lozenge$ is a closure operator on U and V,

2. $\square\lozenge$ and $\lozenge\square$ are dual to each other,

3. $\lozenge\square$ is an interior operator on U and V.

Based on those properties, the lower and upper approximations of a set of objects and a set of attributes can be defined, respectively, as follows (Düntsch, & Gediga, 2003; Pagliani, & Chakraborty, 2005; Shao, & Zhang, 2005; Wolski, 2003):

$$\overline{rapr}(A) = A^{\square\lozenge}, \overline{rapr}(B) = B^{\square\lozenge}, \quad (41)$$

and

$$\overline{rapr}(A) = A^{\lozenge\square}, \overline{rapr}(B) = B^{\lozenge\square}.$$

(Wu et al., 2003)

The operators $\square\lozenge$ and $\lozenge\square$ and the corresponding rough-set approximations have been used and studied by many authors, for example, Düntsch and Gediga (2003), Pagliani (1993), Pagliani and Chakraborty (2005), Pei and Xu (2004), Shao and Zhang (2005) and Wolski (2003), Wolski, (2005).

If a set of objects A equals to its lower approximation $\underline{rapr}(A)$, we say that A is a definable set of objects in the object-oriented concept lattice denoted by $L_o(U, V, R)$. If the set of objects A equals to its upper approximation $\overline{rapr}(A)$, we say that A is a definable set of objects in the property-oriented concept lattice denoted by $L_p(U, V, R)$. The lower and upper approximations of a set of objects are equal if and only if the set of objects is a definable set of objects in both lattices $L_o(U, V, R)$ and $L_p(U, V, R)$. Similarly, the lower and upper approximations of a set of attributes are equal if and only if the set of attributes is a definable set of attributes in both lattices $L_o(U, V, R)$ and $L_p(U, V, R)$.

CONCLUSION

This review of rough-set analysis and formal concept analysis provides a solid base for further research. Rough-set analysis emphasizes the indiscernibility relation between objects, attribute reductions, and approximations of undefinable sets using definable sets. Formal concept analysis deals with unique association between objects and attributes and concept lattices. They capture different aspects of data and represent different types of knowledge embedded in data sets. The introduction of the notion of reductions and approximations into formal concept analysis and the notion of concept lattice into rough-set analysis combines the two theories. The studies describe a particular characteristic of data, improve our understanding of data, and produce new tools for data analysis.

REFERENCES

Bazan, J.G., Nguyen, H.S., Nguyen, S.H., Synak, P., & Wroblewski, J. (2000). Rough set algorithms in classification problem. In L. Polkowski, S. Tsumoto, & T.Y. Lin (Eds.), *Rough set methods and applications* (pp. 49-88).

Birkhoff, G. (1967). *Lattice theory* (3rd ed.). Providence, RI: American Mathematical Society Colloquium Publications.

Buszkowski, W. (1998). Approximation spaces and definability for incomplete information systems. In *Proceedings of 1st International Conference on Rough Sets and Current Trends in Computing (RSCTC'98)* (pp. 115-122).

Chen, Y.H., & Yao, Y.Y. (2006). Multiview intelligent data analysis based on granular computing. In *Proceedings of IEEE International Conference on Granular Computing (Grc'06)* (pp. 281-286).

Cohn, P.M. (1965). *Universal algebra*. New York: Harper and Row Publishers.

Dempster, A.P. (1967). Upper and lower probabilities induced by a multivalued mapping. *Annals of Mathematical Statistics, 38*, 325-339.

Dubois, D., & Prade, H. (1990). Rough fuzzy sets and fuzzy rough sets. *International Journal of General Systems, 17*, 191-209.

Düntsch, I., & Gediga, G. (2003). Approximation operators in qualitative data analysis. In H. de Swart, E. Orlowska, G. Schmidt, & M. Roubens (Eds.), *Theory and application of relational structures as knowledge instruments* (pp. 216-233). Heidelberg: Springer.

Ganter, B. & Wille, R. (1999). *Formal concept analysis: Mathematical foundations*. New York: Springer-Verlag.

Gediga, G., & Düntsch, I. (2002). Modal-style operators in qualitative data analysis. In *Proceedings of IEEE International Conference on Data Mining* (pp. 155-162).

Grzymala-Busse, J. (1988). Knowledge acquisition under uncertainty—a rough set approach, *Journal of Intelligent and Robotics Systems, 1*, 3-16.

Ho, T.B. (1997). Acquiring concept approximations in the framework of rough concept analysis. In *Proceedings of 7th European-Japanese Conference on Information Modelling and Knowledge Bases* (pp. 186-195).

Hu, K., Sui, Y., Lu, Y., Wang, J., & Shi, C. (2001). Concept approximation in concept lattice. In *Proceedings of 5th Pacific-Asia Conference on Knowledge Discovery and Data Mining (PAKDD'01)* (pp. 167-173).

Kent, R.E. (1996). Rough concept analysis. *Fundamenta Informaticae, 27*, 169-181.

Mi, J.S., Wu, W.Z. & Zhang, W.X. (2003). Approaches to approximation reducts in inconsistent decision tables. In *Proceedings of 9th International Conference on Rough Sets, Fuzzy Sets, Data Mining, and Granular Computing (RSFDGrC'03)* (pp. 283-286).

Mi, J.S., Wu, W.Z. & Zhang, W.X. (2004). Approaches to knowledge reduction based on variable precision rough set model, *Information Sciences, 159*, 255-272.

Miao, D., & Wang, J. (1999). An information representation of the concepts and operations in rough set theory, *Journal of Software, 10*, 113-116.

Orlowska, E. (1998). Introduction: What you always wanted to know about rough sets. In E. Orlowska (Ed.), *Incomplete information: Rough set analysis*. New York: Physica-Verlag Heidelberg.

Pagliani, P. (1993). From concept lattices to approximation spaces: Algebraic structures of some spaces of partial objects. *Fundamenta Informaticae, 18*, 1-25.

Pagliani, P., & Chakraborty, M.K. (2005). Information quanta and approximation spaces. I: Nonclassical approximation operators. In *Proceedings of IEEE International Conference on Granular Computing* (pp. 605-610).

Pawlak, Z. (1982). Rough sets, *International Journal of Computer and Information Sciences, 11*, 341-356.

Pawlak, Z. (1991). *Rough sets—Theoretical aspects of reasoning about data*. Boston; Dordrecht: Kluwer Publishers.

Pei, D.W., & Xu, Z.B. (2004). Rough set models on two universes. *International Journal of General Systems, 33*, 569-581.

Qi, J.J., Wei, L., & Li, Z.Z. (2005). A partitional view of concept lattice. In *Proceedings of 10th International Conference on Rough Sets, Fuzzy Sets, Data Mining, and Granular Computing (RSFDGrC'05), Part I* (pp. 74-83).

Qiu, G. F., Zhang, W. X. & Wu, W. Z. (2005). Characterizations of attributes in generalized approximation representation spaces. In *Proceedings of 10th International Conference on Rough Sets, Fuzzy Sets, Data Mining, and Granular Computing (RSFDGrC'05)* (pp. 84-93).

Saquer, J., & Deogun, J. (1999). Formal rough concept analysis. In *Proceedings of 7th International Workshop on Rough Sets, Fuzzy Sets, Data Mining, and Granular-Soft Computing (RSFDGrC'99)* (pp. 91-99).

Saquer, J., & Deogun, J. (2001). Concept approximations based on rough sets and similarity measures. *International Journal of Applied Mathematics and Computer Science, 11*, 655-674.

Shafer, G. (1976). *A mathematical theory of evidence*. Princeton: Princeton University Press.

Shafer, G. (1987). Belief functions and possibility measures. In J.C. Bezdek (Ed.), *Analysis of fuzzy information, (Vol. 1) Mathematics and logi* (pp. 51-84). Boca Raton: CRC Press.

Shao, M.W. & Zhang, W. X. (2005). Approximation in formal concept analysis. In *Proceedings of 10th International Conference on Rough Sets, Fuzzy Sets, Data Mining, and Granular Computing (RSFDGrC'05), Part I* (pp. 43-53).

Skowron, A., & Rauszer, C. (1991). The discernibility matrices and functions in information systems. In R. Slowinski (Ed.), *Intelligent decision support handbook of applications and advance of the rough set theory* (pp. 331-362).

Slezak, D. (2000). Various approaches to reasoning with frequency based decision reducts: A survey. In L. Polkowski, S. Tsumoto, T. Y. Lin (Eds.), *Rough set methods and applications* (pp. 235-285).

Wang, G., Yu, H., & Yang, D. (2002). Decision table reduction based on conditional information entropy, *Chinese Journal of Computers, 25*, 759-766.

Wang, J., & Wang, J. (2001). Reduction algorithms based on discernibility matrix: The ordered attributes method. *Journal of Computer Science and Technology, 16*, 489-504.

Wasilewski, P. (2005). Concept lattices vs. approximation spaces. In *Proceedings of 10th International Conference on Rough Sets, Fuzzy Sets, Data Mining, and Granular Computing (RSFDGrC'05), Part I* (pp. 114-123).

Wille, R. (1982). Restructuring lattice theory: An approach based on hierarchies of concepts. In I. Rival (Ed.), *Ordered set* (pp. 445-470). Dordecht; Boston: Reidel.

Wiweger, A. (1989). On topological rough sets. *Bulletin of the Polish Academy of Sciences, Mathematics, 37*, 89-93.

Wolski, M. (2003). Galois connections and data analysis. *Fundamenta Informaticae CSP*, 1-15.

Wolski, M. (2005). Formal concept analysis and rough set theory from the perspective of finite topological approximations. *Journal of Transactions on Rough Sets*, III (LNCS 3400), 230-243.

Wong, S.K.M., Wang, L.S., & Yao, Y.Y. (1993). Interval structure: A framework for representing uncertain information. In *Proceedings of 8th Conference on Uncertainty in Artificial Intelligence* (pp. 336-343).

Wong, S.K.M., Wang, L.S., & Yao, Y.Y. (1995). On modeling uncertainty with interval structures. *Computational Intelligence, 11*, 406-426.

Wu, Q., Liu, Z.T., & Li, Y. (2004). Rough formal concepts and their accuracies. In *Proceedings of 2004 International Conference on Services Computing (SCC'04)* (pp. 445-448).

Wu, W.Z., Mi, J.S., & Zhang, W.X. (2003). Generalized fuzzy rough sets. *Information Sciences, 151*, 263-282.

Wu, W.Z., Zhang, M., Li, H.Z., & Mi, Z.H. (2005). Knowledge reduction in random information systems via Dempster-Shafer theory of evidence. *Information Sciences, 174*, 143-164.

Wybraniec-Skardowska, U. (1989). On a generalization of approximation space. *Bulletin of the Polish Academy of Sciences, Mathematics, 37*, 51-61.

Yao, Y.Y. (1996). Two views of the theory of rough sets in finite universe. *International Journal of Approximate Reasoning, 15*, 291-317.

Yao, Y.Y. (1998a). Generalized rough set models. In L. Polkowski & A. Skowron (Eds.), *Rough sets in knowledge discovery* (pp. 286-318). Heidelberg: Physica-Verlag.

Yao, Y.Y. (1998b). On generalizing Pawlak approximation operators. In *Proceedings of 1st International Conference on Rough Sets and Current Trends in Computing (RSCTC'98)* (pp. 298-307).

Yao, Y.Y. (1998c). Constructive and algebraic methods of the theory of rough sets. *Information Sciences, 109*, 21-47.

Yao, Y.Y. (2003). On generalizing rough set theory. In *Proceedings of 9th International Conference on Rough Sets, Fuzzy Sets, Data Mining, and Granular Computing (RSFDGrC'03)* (pp. 44-51).

Yao, Y.Y. (2004a). A comparative study of formal concept analysis and rough set theory in data analysis. In W. Zakowski (Ed.), *Proceedings of 3rd International Conference on Rough Sets and Current Trends in Computing (RSCTC'04)* (1983)-68.

Yao, Y.Y. (2004b). Concept lattices in rough set theory. In *Proceedings of 23rd International Meeting of the North American Fuzzy Information Processing Society (NAFIPS'04)* (pp. 796-801).

Yao, Y.Y., & Chen, Y.H. (2004). Rough set approximations in formal concept analysis. In *Proceedings of 23rd International Meeting of the North American Fuzzy Information Processing Society (NAFIPS'04)* (pp. 73-78).

Yao, Y.Y., & Chen, Y.H. (2005). Subsystem based generalizations of rough set approximations. In *Proceedings of 15th International Symposium on Methodologies for Intelligent Systems (ISMIS'05)* (pp. 210-218).

Yao, Y.Y., & Chen, Y.H. (2006). Rough set approximations in formal concept analysis. *Journal of Transactions on Rough Sets, V, LNCS 4100*, 285-305.

Yao, Y.Y., & Lin, T.Y. (1996). Generalization of rough sets using modal logic. *International Journal of Intelligent Automation and Soft Computing, 2*, 103-120.

Yao, Y.Y., Zhao, Y., & Wang, J. (2006). On reduct construction algorithms. In *Proceedings of 1st International Conference on Rough Sets and Knowledge Technology (RSKT'06)* (pp. 297-304).

Zadeh, L.A. (2003). Fuzzy logic as a basis for a theory of hierarchical definability (THD). In *Proceedings of 33rd International Symposium on Multiple-Valued Logic (ISMVL'03)* (pp. 3-4).

Zakowski, W. (1983). Approximations in the space (U, Π). *Demonstratio Mathematica, XVI*, 761-769.

Zhang, W.X., Wei, L., & Qi, J.J. (2005). Attribute reduction in concept lattice based on discernibility matrix. In *Proceedings of 10th International Conference on Rough Sets, Fuzzy Sets, Data Mining, and Granular Computing (RSFDGrC'05), Part II* (pp. 517-165).

Zhang, W.X., Yao, Y.Y., & Liang, Y. (Eds.). (2006). *Rough set and concept lattice.* Xi'An: Xi'An Jiao-Tong University Press.

Zhu, W., & Wang, F.Y. (2003). Reduction and axiomization of covering generalized rough sets. *Information Sciences, 152*, 217-230.

Zhu, W., & Wang, F.Y. (2006). Covering based granular computing for conflict analysis. In *Proceedings of IEEE International Conference on Intelligence and Security Informatics (ISI'06)* (pp. 566-571).

Section II
Current Trends and Models

Chapter V
Rough Sets:
A Versatile Theory for Approaches to Uncertainty Management in Databases

Theresa Beaubouef
Southeastern Louisiana University, USA

Frederick E. Petry
Naval Research Laboratory, USA

ABSTRACT

This chapter discusses ways in which rough-set theory can enhance databases by allowing for the management of uncertainty. Rough sets can be integrated into an underlying database model, relational or object oriented, and also used in the design and uerying of databases, because rough-sets are a versatile theory, theories. The authors discuss the rough relational databases model, the rough object-oriented database model, and fuzzy set and intuitionistic set extensions to each of these models. Comparisons and benefits of the various approaches are discussed, illustrating the usefulness and versatility of rough sets for uncertainty management in databases.

INTRODUCTION

Rough-set theory has become well established since first introduced by Pawlak in the 1970s. It is based on two simple concepts: indiscernibility and approximation regions. Rough-set theory is a formal theory, mathematically sound. It has been applied to several areas of research such as logic and knowledge discovery, and has been implemented in various applications in the real world. Because of rough sets' ability to define uncertain things in terms of certain, definable things, it is a natural mechanism for integrating real-world uncertainty in computerized databases. Moreover, other uncertainty-management techniques may be combined with rough sets to offer even greater uncertainty management in databases. This chapter discusses how rough-sets theory can be applied to several areas of databases including design, modeling, and querying. Both relational and object-oriented databases benefit from rough-set techniques, and when combined with fuzzy or intuitionistic sets, these databases are very rich in the modeling of uncertainty for real-world enterprises.

BACKGROUND

Rough-set theory, (Pawlak, 1984, 1991) is a technique for dealing with uncertainty. The following concepts are necessary for rough sets:

- U is the *universe*, which cannot be empty,
- R is the *indiscernibility relation*, or equivalence relation.
- $A = (U, R)$, an ordered pair, is called an *approximation space*.
- $[x]_R$ denotes the equivalence class of R containing x, for any element x of U.
- *Elementary sets* in A—the equivalence classes of R.
- *Definable set* in A—any finite union of elementary sets in A.

A given approximation space defined on some universe U has an equivalence relation R imposed upon it, partitioning U into equivalence classes called elementary sets that may be used to define other sets in A. Given that $X \subseteq U$, X can be defined in terms of the definable sets in A by:

lower approximation of X in A is the set
$\underline{R}X = \{x \in U \mid [x]_R \subseteq X\}$

upper approximation of X in A is the set
$\overline{R}X = \{x \in U \mid [x]_R \cap X \neq \emptyset\}$.

The major rough-set concepts of interest are the use of an indiscernibility relation to partition domains into equivalence classes and the concept of lower and upper approximation regions to allow the distinction between certain and possible, or partial, inclusion in a rough set. Indiscernibility is the inability to distinguish between two or more values. For example, the average person describing the color of a suspect's hair may say that it is "brown," when it actually is dark brown. As it turns out, "brown" is probably good enough for helping the police identify the suspect. However, a beautician who specializes in hair color will find it important to discern between the various shades of brown. Indiscernibility can also arise from lack of precision in measurement, limitations of computational representation, or the granularity or resolution of the sampling or observations

ROUGH SETS IN DATABASE DESIGN

Beaubouef and Petry (2004a, 2005a) introduce a rough-set design methodology for databases. Conceptual modeling is accomplished through rough entity-relationship modeling and then for relational databases, rough normalization is discussed. The process of normalization makes use of the concept of rough functional dependencies.

We must first design a database using some type of semantic model. We use a variation of the entity-relationship diagram that we call a fuzzy-rough E-R diagram. This diagram is similar to the standard E-R diagram in that entity types are depicted in rectangles, relationships with diamonds, and attributes with ovals. However, in the fuzzy-rough model, it is understood that membership values exist for all instances of entity types and relationships. Attributes that allow values where we want to be able to define equivalences are denoted with an asterisk (*) above the oval. These values are defined in the indiscernibility relation, which is not actually part of the database design, but inherent in the fuzzy-rough model.

Our fuzzy-rough E-R model is similar to the second and third levels of fuzziness defined by Zvieli and Chen (1986). However, in our model, all entity and relationship occurrences (second level) are of the fuzzy type so we do not mark an "*f*" beside each one. Zvieli and Chen's third level considers attributes that may be fuzzy, which they designate by triangles instead of ovals in the E-R diagram. We do not introduce fuzziness at the attribute level of our model in this paper, only roughness, or indiscernibility, and denote those attribute with the "*".

ROUGH SETS APPLIED TO THE RELATIONAL DATABASE MODEL

The rough relational database model (Beaubouef, Petry, & Buckles, 1995) is a logical database model where rough sets are integrated into the traditional relational database model. In this model, domains are partitioned into equivalence classes. The database is a nonfirst normal form extension of the traditional relational database, and values for attributes may be a value representing an equivalence class or a set of values. Each relation contains a rough set of tuples. A tuple can belong to the lower approximation region representing certain data, or to the boundary region of the rough set of tuples, representing uncertain data. In this model, rough relational operators are defined to operate on the rough relational database.

Every attribute domain is partitioned by some equivalence relation designated by the database designer or user. Within each domain, a group of values that is considered indiscernible forms an equivalence class. The query mechanism uses class equivalence rather than value equality in retrievals, making the exact wording of a query less critical.

Recall is also improved in the rough relational database. A rough relation will be seen shortly to represent imprecision by the use of the indiscernibility relationship among the domain values of the attributes. So rough relations provide *possible* matches to the query in addition to the *certain* matches that are obtained in the standard relational database. This is accomplished by using set containment, in addition to equivalence of attributes, in the calculation of lower and upper approximation regions of the query result.

The rough relational database has several features in common with the ordinary relational database. Both models represent data as a collection of *relations* containing *tuples*. These relations are sets. The tuples of a relation are its elements, and like the elements of sets in general, are unordered and nonduplicated. A tuple t_i takes the form $(d_{i1}, d_{i2}, ..., d_{im})$, where d_{ij} is a *domain value* of a particular domain set D_j in the ordinary relational database, $d_{ij} \in D_j$. In the rough relational database, however, as in other nonfirst normal form extensions to the relational model (Makinouchi, 1977; Roth, Korth, & Batory, 1987) $d_{ij} \subseteq D_j$, and although it is not required that d_{ij} be a singleton, $d_{ij} \neq \varnothing$. Let $P(D_i)$ denote the powerset$(D_i) - \varnothing$.

Definition. A *rough relation* R is a subset of the set cross product $P(D_1) \times P(D_2) \times \cdots \times P(D_m)$.

A rough tuple **t** is any member of R, which implies that it is also a member of $P(D_1) \times P(D_2) \times \cdots \times P(D_m)$. If t_i is some arbitrary tuple, then $t_i = (d_{i1}, d_{i2}, ..., d_{im})$, where $d_{ij} \subseteq D_j$. A tuple in this model differs from that of ordinary databases in that the tuple components may be sets of domain values rather than single values. For notational convenience, the set braces are omitted from singletons.

Definition. An *interpretation* $\alpha = (a_1, a_2, ..., a_m)$ of a rough tuple $t_i = (d_{i1}, d_{i2}, ..., d_{im})$ is any value assignment such that $a_j \in d_{ij}$ for all j.

The interpretation space is the cross product $D_1 \times D_2 \times \cdots \times D_m$, but is limited for a given relation R to the set of those tuples that are valid according to the underlying semantics of R. In an ordinary relational database, because domain values are atomic, there is only one possible interpretation for each tuple t_i, the tuple itself. In the rough relational database, this is not always the case.

Let $[d_{xy}]$ denote the equivalence class to which d_{xy} belongs. When d_{xy} is a set of values, the equivalence class is formed by taking the union of equivalence classes of members of the set; if $d_{xy} = \{c_1, c_2, ..., c_n\}$, then $[d_{xy}] = [c_1] \cup [c_2] \cup ... \cup [c_n]$.

Definition. Tuples $t_i = (d_{i1}, d_{i2}, ..., d_{im})$ and $t_k = (d_{k1}, d_{k2}, ..., d_{km})$ are *redundant* if $[d_{ij}] = [d_{kj}]$ for all $j = 1, ..., m$.

In rough relations, there are no redundant tuples. The merging process used in relational database operations removes duplicate tuples since duplicates are not allowed in sets, the structure upon which the relational model is based. If a tuple is a member of the lower approximation region, it is, by definition, also a member of the upper approximation region. So, where operations result in redundant tuples, one in the lower and another in the boundary region of the upper approximation region, the less certain tuple is the one eliminated, retaining the one in the lower approximation region. It is possible for more than one tuple to have the same interpretation. The very nature of roughness in using both lower and upper approximations precludes the restriction of uniqueness in interpretations.

Indiscernibility can be represented in the rough relational database by an additional relation. The tuples of this relation represent all of the possible singleton values d_{ij} for every domain D_j. Each tuple also contains an arbitrary indiscernibility identifier that associates the value d_{ij} with the equivalence class to which it belongs. This indiscernibility relation is an integral part of the rough relational database. All database retrieval operations implicitly access the indiscernibility relation in addition to these rough relations expressed in the query.

In addition to indiscernibility, the rough relational database must incorporate lower and upper approximations into the querying in order to retrieve a rough set. We can view rough querying as follows. First, retrieve the elements of the lower approximation as previously described, utilizing the notion of indiscernibility. Next, retrieve the elements of the upper approximation, also utilizing the indiscernibility relation, and return those tuples that are not also in the lower approximation. The lower approximation includes all tuples whose individual attribute values are equivalent to those expressed by the query. The upper approximation is based on set containment of values.

Rough Relational Operators

There are two basic types of relational operators. The first type arises from the fact that relations are considered sets of tuples. Therefore, operations that can be applied to sets also apply to relations. The most useful of these for database purposes are *set difference, union,* and *intersection*. Operators that do not come from set theory, but that are useful for retrieval of relational data, are *select, project,* and *join*.

In the rough relational database, relations are rough sets as opposed to ordinary sets. Therefore, new rough operators (-, $\cup$, $\cap$, $\times$, σ, π, $*$), which are comparable to the standard relational operators, must be developed for the rough relational database. Moreover, a mechanism must exist within the database to mark tuples of a rough relation as belonging to the lower or upper approximation of that rough relation. Because the definitions for the rough relational operators that follow are independent of any implementation details, only issues related to the determination of the approximation area to which a tuple belongs will be discussed.

The definitions for the set operations for rough relations are comparable to those defined for ordinary relations in the standard relational database model. These binary operations require that the argument relations be "union compatible." Two relations, $X(A_1, A_2, ..., A_n)$ and $Y(B_1, B_2, ..., B_n)$, are union compatible if they have the same number of attributes in their relation schemas and if the domain of A_i is equal to the domain of B_i for all $i = 1, n$.

Rough Difference

The relational difference operator is a binary operator that returns those elements of the first relation that are not elements of the second relation. Let X and Y be two union-compatible rough relations.

Definition. The *rough difference*, **X - Y**, between X and Y is a rough relation, T, where:

$\underline{R}T = \{t \in \underline{R}X \mid t \notin \underline{R}Y\}$ and $\overline{R}T = \{t \in \overline{R}X \mid t \notin \overline{R}Y\}$.

The lower approximation of $T = X - Y$ contains those tuples belonging to the lower approximation of X that are not redundant with a tuple in the lower approximation of Y. The upper approximation of the rough relation, T, contains those tuples in the upper approximation of X that are not included in the upper approximation of Y.

For example, consider the sample relations, X and Y, that contain the following tuples where tuples of the lower approximation region are denoted with an *:

X = (Smith, Wetland)*
(Jones, Forest)*
(Crown, Wetland)
(Crown, Forest)

Y = (Smith, Wetland)*
(Crown, Wetland)*
(Jones, Forest)
(Jones, Commercial)

Then, the difference $X - Y$ contains the tuples (Jones, Forest)* and (Crown, Forest).

Other set operators are defined in a similar manner.

Rough Union

Let X and Y be two union compatible rough relations.

Definition. The *rough union* of X and Y, $X \cup Y$ is a rough relation, T, where:

$\underline{R}T = \{t \in \underline{R}X \cup \underline{R}Y\}$ and $\overline{R}T = \{t \in \overline{R}X \cup \overline{R}Y\}$.

The lower approximation of the resulting rough relation T contains those tuples that are a member of either or both of the lower approximations of X and Y, and the upper approximation of T contains tuples that belong to either or both of the upper approximations of X and Y.

Rough Intersection

Rough intersection is defined similarly.

Definition. The *rough intersection* of X and Y, $X \cap Y$ is a rough relation T where:

$\underline{R}T = \{t \in \underline{R}X \cap \underline{R}Y\}$ and $\overline{R}T = \{t \in \overline{R}X \cap \overline{R}Y\}$.

In rough intersection, comparison of tuple values is based on redundancy, as opposed to the standard relational model, which bases comparisons on equality of values. The lower approximation of the resulting rough relation T contains those tuples of the lower approximation of X that have corresponding redundant tuples in the lower approximation of Y, and the upper approximation of T contains tuples of the upper approximation of X that have redundant tuples in the upper approximation of Y.

We now define the rough relational select, project, and join operations.

Rough Selection

The select operator for the rough relational database model, σ, is a unary operator that takes a rough relation X as its argument, and returns a rough relation containing a subset of the tuples of X, selected on the basis of values for one or more specified attributes. The operation $\sigma_{A=a}(X)$, for example, returns those tuples in X where the value for attribute A is equivalent to the value **a**, or more precisely, a member of the equivalence class [**a**].

Let R be a relation schema, X, a rough relation on that schema; A, an attribute in R; and **a** = $\{a_i\}$, where $a_i, b_j \in \text{dom}(A)$. $\cup_x$ denotes "the union over all x"; $t(A)$ denotes a tuple's value for attribute A.

Definition. The *rough selection*, $\sigma_{A=a}(X)$, of tuples from X is a rough relation Y having the same schema as X and where:

$$\underline{R}Y = \{t \in X \mid \cup_i [a_i] = \cup_j [b_j]\},\ a_i \in \mathbf{a},\ b_j \in t(A)$$

$$\overline{R}Y = \{t \in X \mid \cup_i [a_i] \subseteq \cup_j [b_j]\},\ a_i \in \mathbf{a},\ b_j \in t(A).$$

The lower approximation of $Y = \sigma_{A=a}(X)$ contains those tuples where the value of attribute A for that tuple is indiscernible from the members of **a**, as indicated in the select condition. The upper approximation contains those tuples where the members of **a** form a subset of the values of attribute A for that tuple.

Rough Projection

Project is a unary operator that takes a rough relation as its argument and returns a rough relation containing a subset of the columns of the original relation. Let X be a rough relation with schema A, and let B be a subset of A. The rough projection of X onto schema B is a rough relation Y obtained by omitting the columns of X that correspond to attributes in $A - B$, and removing redundant tuples.

Definition. The *rough projection* of X onto B, $\pi_B(X)$, is a relation Y with schema $Y(B)$ where:

$$Y(B) = \{t(B) \mid t \in X\}.$$

Each $t(B)$ is a tuple retaining only those attributes in the requested set B. Additionally, the rough project must maintain which tuples belong to the lower approximation and which belong to the upper approximation. When comparing tuples for redundancy, if redundant tuples both belong to the lower approximation or both belong to the upper approximation, either can be deleted. In cases where that one tuple is from the lower approximation and the other from the upper approximation, the tuple from the lower approximation is retained.

Rough Join

The join operator is a binary operator that takes related tuples from two relations and combines them into single tuples of the resulting relation. It uses common attributes to combine the two relations into one, usually larger, relation. Let $X(A_1, A_2, ..., A_m)$ and $Y(B_1, B_2, ..., B_n)$ be rough relations with m and n attributes, respectively, and let $AB = C$, the schema of the resulting rough relation T.

Definition. The *rough join*, $X \bowtie_{<\text{JOIN CONDITION}>} Y$, of two relations X and Y, is a relation $T(C_1, C_2, ..., C_{m+n})$ where:

$$T = \{t \mid \exists\, t_X \in X,\, t_Y \in Y \text{ for } t_X = t(A),\, t_Y = t(B)\},$$

and where

$t_X(A \cap B) = t_Y(A \cap B)$, for $\underline{R}T$
$t_X(A \cap B) \subseteq t_Y(A \cap B)$ or $t_Y(A \cap B) \subseteq t_X(A \cap B)$, for $\overline{R}T$

<JOIN CONDITION> is a conjunction of one or more conditions of the form **A = B**.

Properties of the rough relational operators can be found in Beaubouef et.al. (1995). In that paper, the authors show that properties of relational databases and for rough set theory hold in the rough relational database model. Rough-set techniques can also be used for the rough querying of crisp data in relational databases (Beaubouef & Petry, 1994). An ordinary relational database can remain unchanged, and yet a query mechanism built for the model where rough predicates return both certain and possible results to a query.

THE FUZZY ROUGH RELATIONAL DATABASE MODEL

The rough relational model can be extended to incorporate fuzzy set (Zadeh, 1965) uncertainty as well (Beaubouef & Petry, 2000). In the fuzzy rough relational model, the tuples of the boundary, or uncertain region, of a fuzzy rough relation have fuzzy membership values associated with them. This fuzzy membership value, between zero and one, represents the degree of uncertainty of tuples in the uncertain region, whereas in the rough relational database, all tuples in this region are equally uncertain.

It has been shown in Wygralak (1989) that rough sets can be expressed by a fuzzy membership function $\mu \rightarrow \{0, 0.5, 1\}$ to represent the negative, boundary, and positive regions. In this model, all elements of the lower approximation, or positive region, have a membership value of one. Those elements of the upper approximation that are not also in the lower approximation region, that is, those elements of the boundary region, are assigned a membership value of 0.5. Elements not belonging to the rough set have a membership value of zero. Rough-set definitions of union and intersection were modified in Beaubouef and Petry (2000) so that the fuzzy model would satisfy all the properties of rough sets. This allowed a rough set to be expressed as a fuzzy set.

Fuzziness is integrated into the rough relational database model not as a means for expressing rough relations in an alternate manner, but to quantify levels of roughness in boundary region areas through the use of fuzzy membership values. Therefore, the fuzzy rough set should not require membership values of elements of the boundary region to equal 0.5, but allow them to take on any values between zero and one, not including zero and one. Additionally, the union and intersection operators for fuzzy rough sets are comparable to those for ordinary fuzzy sets, where MIN and MAX are used to obtain membership values of redundant elements.

Let U be a *universe*, X a rough set in U.

Definition. A *fuzzy rough set* Y in U is a membership function $\mu_Y(x)$ that associates a grade of membership from the interval [0,1] with every element of U where:

$\mu_Y(\underline{R}X) = 1$
$\mu_Y(U - X) = 0$
$0 < \mu_Y(X - \underline{R}X) < 1$.

All elements of the positive region have a membership value of one and elements of the boundary region have a membership value between zero and one.

Definition. The *union* of two fuzzy rough sets A and B is a fuzzy rough set C where:

$C = \{x \mid x \in A \text{ OR } x \in B\}$
$\mu_C(x) = \text{MAX}[\mu_A(x), \mu_B(x)]$.

Definition. The *intersection* of two fuzzy rough sets A and B is a fuzzy rough set C where:

$C = \{x \mid x \in A \text{ AND } x \in B\}$
$\mu_C(x) = \text{MIN}[\mu_A(x), \mu_B(x)]$.

The fuzzy rough relational database extends the rough model by including a membership value, $d_{i\mu} \in D_\mu$, where D_μ is the interval [0,1], the domain for fuzzy membership values, in the tuple of a fuzzy rough relation R.

Definition. A *fuzzy rough relation R* is a subset of the set cross product $P(D_1) \times P(D_2) \times \cdots \times P(D_m) \times D_\mu$.

For a specific relation, R, membership is determined semantically.

A *fuzzy rough tuple* **t** is any member of R. If $\mathbf{t}_i$ is some arbitrary tuple, then $\mathbf{t}_i = (d_{i1}, d_{i2}, ..., d_{im}, d_{i\mu})$ where $d_{ij} \subseteq D_j$ and $d_{i\mu} \in D_\mu$. Interpretation and redundancy of fuzzy rough tuples are defined in a similar manner as in the rough relational database.

Definition. Tuples $\mathbf{t}_i = (d_{i1}, d_{i2}, ..., d_{in}, d_{i\mu})$ and $\mathbf{t}_k = (d_{k1}, d_{k2}, ..., d_{kn}, d_{k\mu})$ are *redundant* if $[d_{ij}] = [d_{kj}]$ for all $j = 1, ..., n$.

If a relation contains only those tuples of a lower approximation, that is, those tuples having a μ value equal to one, the interpretation α of a tuple is unique. This follows immediately from the definition of redundancy. In fuzzy rough relations, there are no redundant tuples. The merging process used in relational database operations removes duplicate tuples since duplicates are not allowed in sets, the structure upon which the relational model is based.

Tuples may be redundant in all values except μ. As in the union of fuzzy rough sets, where the maximum membership value of an element is retained, it is the convention of the fuzzy rough relational database to retain the tuple having the higher μ value when removing redundant tuples during merging. If we are supplied with identical data from two sources, one certain and the other uncertain, we would want to retain the data that is certain, avoiding loss of information.

Recall that the rough relational database is in nonfirst normal form; there are some attribute values that are sets. Another definition, which will be used for upper approximation tuples, is necessary for some of the alternate definitions of operators to be presented. This definition captures redundancy between elements of attribute values that are sets:

Definition. Two subtuples $X = (d_{x1}, d_{x2}, ..., d_{xm})$ and $Y = (d_{y1}, d_{y2}, ..., d_{ym})$ are *roughly-redundant*, $\approx_R$, if for some $[p] \subseteq [d_{xj}]$ and $[q] \subseteq [d_{yj}]$, $[p] = [q]$ for all $j = 1, ..., m$.

In order for any database to be useful, a mechanism for operating on the basic elements and retrieving specified data must be provided. The concepts of redundancy and merging play a key role in the operations defined next.

Fuzzy Rough Relational Operators

We previously defined several operators for the rough relational algebra. We now define similar operators for the fuzzy rough relational database. Recall that for all of these operators, the indiscernibility relation is used for equivalence of attribute values rather than equality of values.

Fuzzy Rough Difference

The fuzzy rough relational difference operator is very much like the ordinary difference operator in relational databases and in sets in general.

In the fuzzy rough relational database, the difference operator is applied to two fuzzy rough relations and, as in the rough relational database, indiscernibility, rather than equality of attribute values, is used in the elimination of redundant tuples. Hence, the difference operator is somewhat more complex. Let X and Y be two union-compatible fuzzy rough relations.

Definition. The *fuzzy rough difference*, $\mathbf{X} - \mathbf{Y}$, between X and Y, is a fuzzy rough relation T where:

$$T = \{t(d_1, ..., d_n, \mu_i) \in X \mid t(d_1, ..., d_n, \mu_i) \notin Y\} \cup \{t(d_1, ..., d_n, \mu_i) \in X \mid t(d_1, ..., d_n, \mu_j) \in Y \text{ and } \mu_i > \mu_j\}$$

The resulting fuzzy rough relation contains all those tuples that are in the lower approximation of X, but not redundant with a tuple in the lower approximation of Y. It also contains those tuples belonging to upper approximation regions of both X and Y, but which have a higher μ value in X than in Y. For example, let X contain the tuple (*Wetland*, 1.0) and Y contain the tuple (*Wetland*, .02). It would not be desirable to subtract out certain information with possible information, so $X - Y$ yields (*Wetland*, 1.0).

Fuzzy Rough Union

Because relations in databases are considered as sets, the union operator can be applied to any two union-compatible relations to result in a third relation that has as its tuples all the tuples contained in either or both of the two original relations. The union operator can be extended to apply to fuzzy rough relations. Let X and Y be two union compatible fuzzy rough relations.

Definition. The *fuzzy rough union* of X and Y, $\mathbf{X} \cup \mathbf{Y}$ is a fuzzy rough relation T where:

$$T = \{t \mid t \in X \text{ OR } t \in Y\} \text{ and } \mu_T(t) = \text{MAX}[\mu_X(t), \mu_Y(t)].$$

The resulting relation T contains all tuples in either X or Y or both, merged together, and having redundant tuples removed. If X contains a tuple that is redundant with a tuple in Y except for the μ value, the merging process will retain only that tuple with the higher μ value.

Fuzzy Rough Intersection

The fuzzy rough intersection, another binary operator on fuzzy rough relations can be defined similarly.

Definition. The *fuzzy rough intersection* of X and Y, $\mathbf{X} \cap \mathbf{Y}$ is a fuzzy rough relation T where:

$$T = \{t \mid t \in X \text{ AND } t \in Y\} \text{ and } \mu_T(t) = \text{MIN}[\mu_X(t), \mu_Y(t)].$$

In intersection, the MIN operator is used in the merging of equivalent tuples having different μ values, and the result contains all tuples that are members of both of the original fuzzy rough relations.

Fuzzy Rough Selection

As was defined for rough selection, σ, is a unary operator that takes a fuzzy rough relation X as its argument, and returns a fuzzy rough relation containing a subset of the tuples of X, selected on the basis of values for a specified attribute.

Let R be a relation schema, X a fuzzy rough relation on that schema, A an attribute in R, $\mathbf{a} = \{a_i\}$ and $\mathbf{b} = \{b_j\}$, where $a_i, b_j \in \text{dom}(A)$, and $\cup_x$ is interpreted as "the union over all x."

Definition. The *fuzzy rough select*, $\sigma_{A=a}(X)$, of tuples from X is a fuzzy rough relation Y having the same schema as X and where:

$$Y = \{t \in X \mid \cup_i[a_i] \subseteq \cup_j[b_j]\},$$

where $a_i \in \mathbf{a}$, $b_j \in t(A)$ and where membership values for tuples are calculated by multiplying the original membership value by:

card(**a**)/card(**b**)

where *card(x)* returns the cardinality, or number of elements, in *x*.

Assume we want to retrieve those elements where **LandClass** = "Wetland" from the following fuzzy rough tuples:

(Wetland, 1)
({Wetland, Pasture, Forest}, .7)
(Urban, 1)
({Wetland, Cropland}, .9)

The result of the selection is the following:

(Wetland, 1)
({Wetland, Pasture, Forest}, .23)
({Wetland, Cropland}, .45)

where the μ for the second tuple is 0.7/3 and for the third tuple is 0.9/2, since these tuples' cardinalities are respectively 3 and 2.

Fuzzy Rough Project

Similarly to rough project, the fuzzy rough projection of *X* (a fuzzy rough relation with schema *A*) onto *B* is a fuzzy rough relation *Y* obtained by omitting the columns of *X* that correspond to attributes in *A − B*, and removing redundant tuples. Recall the definition of redundancy accounts for indiscernibility, which is central to the rough, sets theory, and that higher μ values have priority over lower ones.

Definition. The *fuzzy rough projection* of *X* onto B, $\pi_B(X)$, is a fuzzy rough relation *Y* with schema Y(B) where:

$Y(B) = \{t(B) \mid t \in X\}$.

Fuzzy Rough Join

Let $X(A_1, A_2, ..., A_m)$ and $Y(B_1, B_2, ..., B_n)$ be fuzzy rough relations with **m** and **n** attributes, respectively, and $AB = C$, the schema of the resulting fuzzy rough relation *T*.

Definition. The *fuzzy rough join*, $X \bowtie_{<JOIN\ CONDITION>} Y$, of two relations *X* and *Y*, is a relation $T(C_1, C_2, ..., C_{m+n})$ where:

$T = \{t \mid \exists\ t_X \in X, t_Y \in Y\ \text{for}\ t_X = t(A), t_Y = t(B)\}$,

and where

$$t_X(A \cap B) = t_Y(A \cap B), \mu=1 \quad (1)$$

$$t_X(A \cap B) \subseteq t_Y(A \cap B)\ \text{or}\ t_Y(A \cap B) \subseteq t_X(A \cap B),$$
$$\mu = MIN(\mu_X, \mu_Y) \quad (2)$$

<JOIN CONDITION> is a conjunction of one or more conditions of the form **A = B**.

Only those tuples that resulted from the "joining" of tuples that were both in lower approximations in the original relations belong to the lower approximation of the resulting fuzzy rough relation. All other "joined" tuples belong to the upper approximation only (the boundary region), and have membership values less than one. The fuzzy membership value of the resultant tuple is simply calculated, as in Buckles and Petry (1982), by taking the minimum of the membership values of the original tuples. Taking the minimum value also follows the logic of Ola and Ozsoyoglu (1993), where in joins of tuples with different levels of information uncertainty, the resultant tuple can have no greater certainty than that of its least certain component.

INTUITIONISTIC SETS

To further extend the fuzzy rough relational database, intuitionistic sets (Atanassov, 1986, 1999) can be used. An intuitionistic set (intuitionistic fuzzy set) is a generalization of the traditional fuzzy set introduced by Zadeh (1965). In intuitionistic sets, there are two membership functions: membership and nonmembership. Each of these must be between zero and one, and, in addition, their sum must be between zero and one.

Let set X be fixed. An intuitionistic set A is defined by the following:

$A = \{ <x, \mu_A(x), v_A(x)> \mid x \in X\}$, and where $\mu_A: X \to [0,1]$, and $v_A: X \to [0,1]$.

The degree of membership of element $x \in X$ to the set A is denoted by $\mu_A(x)$, and the degree of nonmembership of element $x \in X$ to the set A is denoted by $v_A(x)$. A is a subset of X.

Additionally, for all $x \in X$,

$0 \leq \mu_A(x) + v_A(x) \leq 1$.

A hesitation margin:

$\pi_A(x) = 1 - (\mu_A(x) + v_A(x))$,

expresses a degree of uncertainty about whether x belongs to X or not, or uncertainty about the membership degree. This hesitancy may cater toward membership or nonmembership.

Example

A person may be *happy* or *unhappy* in traditional logic. In the two-valued logic, there are only two choices. There is no continuum between happy and unhappy; nor is there any uncertainty involved.

In rough sets, many things may be considered in the realm of happiness and unhappiness, and some of them will be grouped together in equivalence classes. Some of these classes are entirely included in the set *happy*: [overjoyed, ecstatic], [glad] or [happy], for example. Some are not in the rough set *happy* at all [upset, angry] or [furious], [sad, unhappy], for example. Lastly, there are some that involve uncertainty about the belonging to the rough set *happy*. These may include such equivalence classes as [satisfied, content], or [nonchalant, indifferent]. These would belong to the boundary, or uncertain region of the rough set.

In fuzzy sets, a person could be happy to a certain degree. The degree of membership of an element to the fuzzy set of happy is represented by a membership value between zero and one. For example, one could be happy to a degree of .8. This implies unhappiness to a degree of .2. However, a person could be happy to a degree of .8, but not unhappy at all, or at least not to that extent. This cannot be represented in fuzzy sets.

In intuitionistic fuzzy sets, however, there are measures for both the degree of membership and the degree of nonmembership. A person could be happy to a degree of .8, but only unhappy to a degree of .05, resulting in a hesitancy of .15. This two-sided fuzziness in the intuitionistic set provides greater management of uncertainty for many real-world cases.

INTUITIONISTIC ROUGH SETS

In this section, we introduce the intuitionistic rough set that incorporates the beneficial properties of both rough set and intuitionistic set techniques. Intuitionistic rough sets are generalizations of fuzzy rough sets that give more information about the uncertain, or boundary region. They follow the definitions for partitioning of the universe into equivalence classes as in traditional rough sets, but instead of having a simple boundary region, there are basically two boundaries formed from the membership and nonmembership functions.

Let U be a *universe*, Y a rough set in U, defined on some approximation space that partitions U into equivalence classes.

Definition. An *intuitionistic rough set* Y in U is $<Y, \mu_Y(x), v_Y(x)>$, where $\mu_Y(x)$ is a membership function that associates a grade of membership from the interval [0,1] with every element (equivalence class) of U, and $v_Y(x)$ associates a degree of nonmembership from the interval [0,1] with every element (equivalence class) of U, where:

$$0 \leq \mu_Y(x) + v_Y(x) \leq 1,$$

where x denotes the equivalence class containing x.

A hesitation margin:

$$\pi_Y(x) = 1 - (\mu_Y(x) + v_Y(x)),$$

Consider the following special cases $<\mu, v>$ for some element of Y:

<1, 0> Denotes total membership. This corresponds to elements found in $\underline{R}Y$.
<0, 1> Denotes elements that do not belong to Y. Same as $U - \overline{R}Y$.
<.5, .5> Corresponds to traditional rough-set boundary region.
<p, 1-p> Corresponds to fuzzy rough set in that there is a single boundary. In this case, we assume that any degree of membership has a corresponding complementary degree of nonmembership.
<p, 0> Corresponds to fuzzy rough set. In this case, there is no complement to what p shows membership in.
<0, q> This case cannot be modeled by fuzzy rough sets. It denotes things that are not a member of $\underline{R}Y$ or $\overline{R}Y$. It falls somewhere in the region $U - \overline{R}Y$.
<x, y> Intuitionistic set general case, uncertain double boundary, one for membership, and one for nonmembership.

Let Y' denote the complement of Y. Then the intuitionistic set having $<\mu_Y(x), \mu_{Y'}(x)>$ is the same as fuzzy rough set.

The last two cases shown, $<0, q>$ and $<x, y>$, cannot be represented by fuzzy sets, rough sets, or fuzzy rough sets. These are the situations that show that intuitionistic rough sets provide greater uncertainty management than the others alone. Note, however, that with the intuitionistic set, we do not lose the information about uncertainty provided by other set theories, since from the first few cases, we see that they are special cases of the intuitionistic rough set.

We may also perform operations on the intuitionistic rough sets such as union and intersection. As with fuzzy rough sets, the definition of these operators is necessary for applications in the intuitionistic rough relational database model.

Definition. The *union* of two intuitionistic rough sets, A and B, is an intuitionistic rough set C where:

$$C = \{x \mid x \in A \text{ OR } x \in B\}$$
$$\mu_C(x) = \text{MAX}[\mu_A(x), \mu_B(x)], \quad v_C(x) = \text{MIN}[v_A(x), v_B(x)].$$

Definition. The *intersection* of two intuitionistic rough sets, A and B, is an intuitionistic rough set C where:

$$C = \{x \mid x \in A \text{ AND } x \in B\}$$
$$\mu_C(x) = \text{MIN}[\mu_A(x), \mu_B(x)], \quad v_C(x) = \text{MAX}[v_A(x), v_B(x)].$$

In this section, we defined intuitionistic rough sets and compared them with rough sets and fuzzy sets. Although there are several various ways of combining rough and fuzzy sets, we focused on those fuzzy rough sets, as defined in Beaubouef and Petry (2000, 2002) and used for fuzzy rough databases, since our intuitionistic rough relational database model follows from this. The intuitionistic rough relational database model will have an advantage over the rough and fuzzy rough database models in that the nonmembership uncertainty of intuitionistic set theory will also play a role, providing even greater uncertainty management than the original models.

INTUITIONISTIC ROUGH RELATIONAL DATABASE MODEL

The intuitionistic rough relational database extends our previous description of the fuzzy rough relational database. Here an *intuitionistic rough tuple* t_i takes the form $(d_{i1}, d_{i2}, ..., d_{im}, d_{i\mu}, d_{iv})$, where d_{ij} is a *domain value* as before, and $d_{i\mu} \in D_\mu$, where D_μ is the interval [0,1], the domain for intuitionistic membership values, and D_v is the interval [0,1], the domain for intuitionistic nonmembership values.

Definition. An intuitionistic *rough relation* R is a subset of the set cross product $P(D_1) \times P(D_2) \times \cdots \times P(D_m) \times D_\mu \times D_v$.

For a specific relation, R, membership is determined semantically. Given that D_1 is the set of names of nuclear/chemical plants, D_2 is the set of locations, and assuming that RIVERB is the only nuclear power plant that is located in VENTRESS,

(RIVERB, VENTRESS, 1, 0)
(RIVERB, OSCAR, .7, .3)
(RIVERB, ADDIS, 1, 0)
(CHEMO, VENTRESS, .3, .2)

are all elements of $P(D_1) \times P(D_2) \times D_\mu \times D_v$. However, only the element (*RIVERB, VENTRESS, 1, 0*) of those listed previously is a member of the relation $R(PLANT, LOCATION, \mu, v)$ that associates each plant with the town or community in which it is located.

Definition. An *interpretation* $\alpha = (a_1, a_2, ..., a_m, a_\mu, a_v)$ of an intuitionistic rough tuple $t_i = (d_{i1}, d_{i2}, ..., d_{im}, d_{i\mu}, d_{iv})$ is any value assignment such that $a_j \in d_{ij}$ for all j.

As before, the interpretation space is the cross product $D_1 \times D_2 \times \cdots \times D_m \times D_\mu \times D_v$ but for a given relation R consists only of the set of those tuples that are valid according to the underlying semantics of R. Again there are, in general, more than one possible interpretation for each tuple t_i.

Let $[d_{xy}]$ denote the equivalence class to which d_{xy} belongs. When d_{xy} is a set of values, the equivalence class is formed by taking the union of equivalence classes of members of the set; if $d_{xy} = \{c_1, c_2, ..., c_n\}$, then $[d_{xy}] = [c_1] \cup [c_2] \cup ... \cup [c_n]$.

Definition. Tuples $t_i = (d_{i1}, d_{i2}, ..., d_{in}, d_{i\mu}, d_{iv})$ and $t_k = (d_{k1}, d_{k2}, ..., d_{kn}, d_{k\mu}, d_{kv})$ are *redundant* if $[d_{ij}] = [d_{kj}]$ for all $j = 1, ..., n$.

Definition. Two subtuples $X = (d_{x1}, d_{x2}, ..., d_{xm})$ and $Y = (d_{y1}, d_{y2}, ..., d_{ym})$ are *roughly-redundant*, $\approx_R$, if for some $[p] \subseteq [d_{xj}]$ and $[q] \subseteq [d_{yj}]$, $[p] = [q]$ for all $j = 1, ..., m$.

Interpretations are unique for tuples in relations containing only the lower approximation region, since there is no redundancy. Tuples may be redundant in all values except μ and v. As in the union of intuitionistic rough sets, where the maximum membership value of an element is retained, it is the convention of the intuitionistic rough relational database to retain the tuple having the higher μ value when removing redundant tuples during merging. If the μ values are equal but the v values unequal, we retain that tuple having the lower v value. For identical data from two sources, one certain and the other uncertain, we would want to retain the data that is certain, avoiding loss of information.

Intuitionistic Rough Relational Operators

In prior sections, we defined several operators for the rough relational algebra, and then described the expressive power of the fuzzy rough versions of these operators in the fuzzy rough relational database model. We now review similar operators for the intuitionistic rough relational database.

Intuitionistic Rough Difference

The intuitionistic rough relational-difference operator, like the ordinary difference operator in relational databases, is a binary operator that returns those elements of the first argument that are not contained in the second argument. It is applied to two intuitionistic rough relations and, as in the rough relational database, indiscernibility, rather than equality of attribute values, is used in the elimination of redundant tuples. Let X and Y be two union compatible intuitionistic rough relations.

Definition. The *intuitionistic rough difference*, $\mathbf{X} - \mathbf{Y}$, between X and Y is a intuitionistic rough relation T where:

$T = \{t(d_1, ..., d_n, \mu_i, v_i) \in X \mid t(d_1, ..., d_n, \mu_j, v_j) \notin Y\} \cup \{t(d_1, ..., d_n, \mu_i, v_i) \in X \mid t(d_1, ..., d_n, \mu_j, v_j) \in Y \text{ and } \mu_i > \mu_j\} \cup \{t(d_1, ..., d_n, \mu_i, v_i) \in X \mid t(d_1, ..., d_n, \mu_j, v_j) \in Y \text{ and } \mu_i = \mu_j \text{ and } v_i < v_j\}$

The resulting intuitionistic rough relation contains all those tuples that are in the lower approximation of X, but not redundant with a tuple in the lower approximation of Y. It also contains those tuples belonging to uncertain regions of both X and Y, but which have a higher μ value in X than in Y or equal μ values and lower v values.

Intuitionistic Rough Union

The intuitionistic rough union operator is a binary operator applied to two union compatible intuitionistic rough relations that results in a third relation that has as its tuples all the tuples contained in either or both of the two original relations. Let X and Y be two union compatible intuitionistic rough relations.

Definition. The *intuitionistic rough union* of X and Y, $\mathbf{X} \cup \mathbf{Y}$ is a intuitionistic rough relation T where:

$T = \{t \mid t \in X \text{ OR } t \in Y\}$ and $\mu_T(t) = \text{MAX}[\mu_X(t), \mu_Y(t)]$, and if $\mu_X(t) = \mu_Y(t)$, $v_T(t) = \text{MIN}[v_X(t), v_Y(t)]$.

The resulting relation T contains all tuples in either X or Y or both, merged together and having redundant tuples removed. If X contains a tuple that is redundant with a tuple in Y except for the μ value, the merging process will retain only that tuple with the higher μ value. Those tuples redundant in all values except v will retain the tuple having the lower v value.

Intuitionistic Rough Intersection

The intuitionistic rough intersection, another binary operator on intuitionistic rough relations, can be defined similarly.

Definition. The *intuitionistic rough intersection* of X and Y, $\mathbf{X} \cap \mathbf{Y}$ is an intuitionistic rough relation T where:

$T = \{t \mid t \in X \text{ AND } t \in Y\}$ and $\mu_T(t) = \text{MIN}[\mu_X(t), \mu_Y(t)]$ and if $\mu_X(t) = \mu_Y(t)$, $v_T(t) = \text{MAX}[v_X(t), v_Y(t)]$

In intersection, the MIN operator is used in the merging of equivalent tuples having different μ values, and the result contains all tuples that are members of both of the original intuitionistic rough relations. For like μ values in redundant tuples, the v values are compared and the tuple having the higher will be retained.

Intuitionistic Rough Select

The definitions of the select operator for the intuitionistic rough relational database model, σ, follows the same format as for the rough relational database model.

Definition. The *intuitionistic rough selection*, $\sigma_{A=a}(X)$, of tuples from X is an intuitionistic rough relation Y having the same schema as X and where:

$$Y = \{t \in X \mid \cup_i [a_i] \subseteq \cup_j [b_j]\},$$

where $a_i \in \mathbf{a}$, $b_j \in t(A)$, and where membership values for tuples are calculated by multiplying the original membership value μ by:

card($\mathbf{a}$)/card($\mathbf{b}$)

where $card(x)$ returns the cardinality, or number of elements, in x. The nonmembership value v remains the same as in the intuitionistic rough relation X, since the result of performing this operation does not give us any additional information about nonmembership.

Intuitionistic Rough Project

Again we follow the format of the rough relational database model for the project operator here. Recall that higher μ values have priority over lower ones in the removal of redundant tuples. For like μ values, the v values are examined, and the tuple having lower v value is retained.

Definition. The *intuitionistic rough projection* of X onto B, $\pi_B(\mathbf{X})$, is an intuitionistic rough relation Y with schema $Y(B)$ where:

$$Y(B) = \{t(B) \mid t \in X\}.$$

Intuitionistic Rough Join

Join here follows the format of the fuzzy rough relational database model, but must account for both of the membership values that are included in the intuitionistic rough relational tuples.

Definition. The *intuitionistic rough join*, $\mathbf{X} \bowtie_{<\text{JOIN CONDITION}>} \mathbf{Y}$, of two relations X and Y, is a relation $T(C_1, C_2, ..., C_{m+n})$ where:

$$T = \{t \mid \exists\, t_X \in X, t_Y \in Y \text{ for } t_X = t(A), t_Y = t(B)\},$$

and where

$$t_X(A \cap B) = t_Y(A \cap B), \mu = 1 \quad (1)$$

$$t_X(A \cap B) \subseteq t_Y(A \cap B) \text{ or } t_Y(A \cap B) \subseteq t_X(A \cap B), \mu = \text{MIN}(\mu_X, \mu_Y), \text{ if } \mu_X = \mu_Y, v = \text{MAX}(v_X, v_Y) \quad (2)$$

<JOIN CONDITION> is a conjunction of one or more conditions of the form $\mathbf{A} = \mathbf{B}$.

Membership values of joined tuples are computed as for the fuzzy rough relational database join operation. For equal membership values, the maximum nonmembership value is retained.

ROUGH SETS AND THE OBJECT-ORIENTED DATABASE MODEL

Object-oriented (OO) design and programming have become widely utilized partly due to its ability to model complex objects that can more realistically model a real-world enterprise. The concepts of *classes* and *inheritance* allow for code reuse through specialization and generalization. A class *inherits* data and behavior from classes at higher levels in the class hierarchy. These promote code reuse, which can save development time and reduce errors. These benefits also apply to the object-oriented database model.

Indiscernibility is the inability to distinguish between two or more values. It can arise from lack of precision in measurement, limitations of computational representation, or the granularity or resolution of the sampling or observations. In the rough-set object-oriented database, indiscernibility is managed through classes. Every attribute domain is implemented as a class hierarchy, with the lowest elements of the hierarchy representing the equivalence classes based on the finest possible partitioning for the domain as it pertains to the application. An approach integrating rough-set concepts into an OO model, and how changing the granularity of the partitioning affects query results, is described in Beaubouef and Petry (2002).

Generalized Object-Oriented Database framework

In this section, we will develop the rough object-oriented database model. We first describe the formal framework and type definitions for generalized object-oriented databases, proposed by DeTré and De Caluwe (1999), that conforms to the standards set forth by the Object Database Management Group (Catell, Barry, Bartels, Berler, Eastman, Gamerman, Jordan, Springer, Strickland, & Wade, 1997). We extend this framework, however, to allow for rough-set indiscernibility and approximation regions for the representation of uncertainty, as we have previously described for relational databases. The rough-object database scheme is formally defined by the following type system and constraints.

The type system, ts, contains literal types $T_{literal}$, which can be a base type, a collection literal type, or a structured literal type. It also contains T_{object}, which specifies object types, and $T_{reference}$, the set of specifications for reference types. In the type system, each domain $dom_{ts} \in D_{ts}$, the set of domains. This domain set, along with a set of operators O_{ts} and a set of axioms A_{ts}, capture the semantics of the type specification. The type system is then defined based on these type specifications, the set of all programs P, and the implementation function mapping each type specification for a domain onto a subset of the powerset of P that contains all the implementations for the type system.

We are particularly interested in object types, and may specify a class t of object types as:

Class $id(id_1:s_1; ...; id_n:s_n)$ or

Class $id: \overline{id}_1, ..., \overline{id}_n(id_1:s_1; ...; id_n:s_n)$

where id, an identifier, names an object type, $\{ \overline{id}_i \mid 1 \leq i \leq m \}$ is a finite set of identifiers denoting parent types of t, and $\{ id_i:s_i \mid 1 \leq i \leq n \}$ is the finite set of characteristics specified for object type t within its syntax. This set includes all the attributes, relationships, and method signatures for the object type. The identifier for a characteristic is id_i and the specification is s_i for each of the $id_i:s_i$.

Consider a geographical information system (GIS) that stores spatial data concerning water and landforms, structures, and other geographic information. If we have simple types defined for string, set, geo, integer, and so forth, we can specify an object type

Class *ManMadeFeature* (
Location: geo;
Name: string;
Height: integer;
Material: Set(string));

Some example instances of the object type *ManMadeFeature* might include:

[oid1, Ø, *ManMadeFeature*, Struct(0289445, "WWK radio tower," 60, Set(steel, plastic, aluminum))]

or

[oid2, Ø, *ManMadeFeature*, Struct(01122345, "Madison water tower," 30, Set(steel, water, iron))],

following the definition of instance of an object type DeTré and De Caluwe (1999), the quadruple o = [oid, N, t, v], consisting of a unique object identifier, a possibly empty set of object names, the name of the object type, and for all attributes, the values ($v_i \in dom_{sf}$) for that attribute, which represent the state of the object. The object type *t* is an instance of the type system *ts*, and is formally defined in terms of the type system and its implementation function t = [*ts*, f_{impl}^{type} (*ts*)].

Rough-Set Object-Oriented Database

In the rough-set object-oriented database, indiscernibility is managed through classes. Every domain is implemented as a class hierarchy, with the lowest elements of the hierarchy representing the equivalence classes based on the finest possible partitioning for the domain as it pertains to the application. Consider, for example, a geographic information system, where objects have an attribute called *landClass*. There are many different classifications for land area features, such as those covered by forests, pastures, urban area, or some type of water, for example. An example of part of the hierarchy related to water is depicted in Figure 1.

Ignoring the nonwater parts of the *landClass* domain, and focusing on the water-related parts, we see that the domain set:

$dom_{landClass}$ = {creek, brook, stream, branch, river, lake, pond, waterhole, slough}

can be partitioned in several different ways. One partitioning that represents the finest partitioning (more, but smaller, equivalence classes) is given by:

R^1 = {[creek, brook, stream], [branch], [river], [lake], [pond, waterhole], [slough]}.

This is evidenced by the lowest level of the hierarchy.

An object type (domain class) for landClass may be defined as:

Class landClass (
numEquivClass: integer;
name: string;
indiscernibility: Set(Ref(equivClass)))

At this lowest level, each landClass object has only one reference in its attribute for indiscernibility, the object identifier for the particular equivalence class. These reference individual equivalence class objects defined by:

Class equivClass(
element: Set(string);

Figure 1. Part of the class hierarchy of landClass features involving water

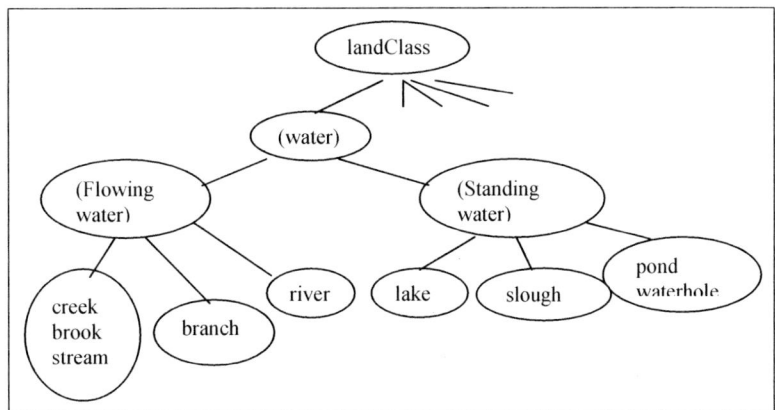

N: integer;
Name: string).

In this case, we have six separate equivalence classes, three of which are shown below:

[oid56, Ø, *equivClass*, Struct(Set("creek," "brook," "stream"), 3, "creek")]

[oid57, Ø, *equivClass*, Struct(Set("branch"), 1, "branch")]

[oid61, Ø, *equivClass*, Struct(Set("pond," "waterhole"), 2, "pond")]

Note that the name of the class can be set equal to any of the values within the class.

Let the other water classes be similarly defined, with oid58 denoting the "river" class, oid59 denoting "lake", and oid60 denoting "slough." If we want to change the partitioning, such that our application only distinguishes between flowing and standing water, for example, our equivalence classes become:

R^2 = {[creek, brook, river, stream, branch], [lake, pond, waterhole, slough]}.

We would then have the landClass objects
[oid101, Ø, *landClass*,
Struct(3, "Flowing water,"
Set(oid56, oid57, oid58))] and
[oid102, Ø, *landClass*,
Struct(3, "Standing water,"
Set(oid59, oid60, oid61))].

Lastly, if the application only requires that a feature be categorized as water or nonwater for land classification, an even coarser partitioning may be used that includes all the possible water values in one equivalence class:

R^3 = {[creek, brook, stream, branch, river, lake, pond, waterhole, slough]}.

An instance of this class would be defined in a manner similar to those in R^3.

Each domain class i in the database $dom_i \in D_i$ has methods for maintaining the current level of granulation, changing the partitioning, adding new domain values to the hierarchy, and for determining equivalence based on the current indiscernibility relation imposed on the domain class.

Every domain class, then, must be able to not only store the legal values for that domain, but to maintain the grouping of these values into equivalence classes. This can be achieved through the type implementation function and class methods, and can be specified through the use of generalized constraints as in DeTré and De Caluwe (1999) for a generalized OODB.

An ordinary (nonindiscernibility) object class in this database, having one of its attributes *landClass*, may be defined as follows:

Class *RuralProperty* (
Location: geo;
Name: string;
Owner: string;
landClass: Set(string));

having particular instances of the class, for example:

[oid24, Ø, *RuralProperty*, Struct(01987345, "Sunset Farms", "Fred Farmer", Set("waterhole," "pasture"))],

[oid27, Ø, *RuralProperty*, Struct(01987355, Ø, "Dottie Farmer", Set("forest," "lake"))],

[oid31, Ø, *RuralProperty*, Struct(01987390, "Hodge Mill Runoff Lagoon", "Hodge Mill", Set("waterhole"))],

[oid32, Ø, *RuralProperty*, Struct(01987394, "Heart Lake", "Blackham County", Set("lake"))].

[oid26, Ø, *RuralProperty*, Struct(01987358, Ø, "Franklin County", Set("pond"))].

Now let us assume that we are trying to sell fish fingerlings and want to retrieve the names of all landowners that own land that contains ponds. Our query may look something like this:

SELECT Owner
FROM RuralProperty
WHERE landClass = "pond."

If our goal is selling fish for stocking small ponds, we may want our indiscernibility relation to be defined with a very fine partitioning, as discussed previously:

R^1 = {[creek, brook, stream], [branch], [river], [lake], [pond, waterhole], [slough]}.

Here "pond" and "waterhole" are considered indiscernible, and the query will match either as a certain result. Possible results will, in addition, contain those objects that have "pond" or "waterhole" as one of the values in the set for landClass.

For the partitioning R^1 and the five sample objects, our rough-set result would include the following:

$\underline{R}X$ = {Franklin County, Hodge Mill}
X = {Franklin County, Hodge Mill, Fred Farmer}

Here, oid26 (Owner is "Franklin County") provides an exact match with "pond" so it is included in the lower approximation region, which represents certain data. Because "waterhole" is in the same equivalence class as "pond", oid31 (Owner is "Hodge Mill") is also included in the lower approximation region. The upper approximation region contains those objects where at least one of the values in the set of values for that attribute matches the request. Hence, oid21 (Owner is "Fred Farmer") is returned, since "pond" and "waterhole" both belong to the class [pond], and this value is included in the set of values {waterhole, pasture}.

If we had decided that all standing water locations are likely candidates for fish stocking, then we might have coarsened the partitioning, using the equivalence relation:

R^2 = {[creek, brook, river, stream, branch], [lake, pond, waterhole, slough]}.

In this case, oid32 (Owner is Blackham County) also belongs to the lower approximation region since "lake," "pond," "waterhole," and "slough" are now indiscernible. Likewise, oid27 (Owner is Dottie Farmer) becomes part of the upper approximation since "lake" is a subset of {lake, forest}. Now the rough-set results are given as:

$\underline{R}X$ = { Franklin County, Hodge Mill, Blackham County}
X = { Franklin County, Hodge Mill, Blackham County, Fred Farmer, Dottie Farmer }.

The semantics of rough-set operations discussed for relational databases previously apply similarly for the object database paradigm. However, the implementation of these operations is done via methods associated with the individual object classes.

Fuzzy and Intuitionistic Rough Object-Oriented Database

The rough object-oriented database model can also benefit from fuzzy set uncertainty (Beaubouef & Petry, 2002). Each object has a fuzzy membership value associated with it. This is straightforward in the object paradigm. However, it is important to consider the evaluation of queries, as done for the model.

We can also extend this framework further to allow for intuitionistic rough-set indiscernibility and approximation regions for the representation of uncertainty. The intuitionistic rough object database scheme is formally defined by the type system and constraints following the definition of instance of an object type, as described with the quadruple o = [oid, N, t, v]. If we extend the rough OODB further to allow for intuitionistic types, the type specifications T can be generalized to a set $\check{T}$ so that the definitions of the domains are generalized to intuitionistic sets.

For every $ts \in T$, having domain ts being dom_{ts}, the type system $ts \in T$ is generalized to:

$$\overline{ts} \in \check{\overline{T}}$$

where domain of $\overline{ts}$ is denoted by $dom_{\overline{ts}}$ and is defined as the set $\rho(\overline{dom_{ts}})$ of intuitionistic sets on dom_{ts}, and O_{ts} is generalized to $O_{\overline{ts}}$, which contains the generalized version of the operators. This is consistent with the UFO database model of Gyseghem and De Caluwe (1997) as well.

The generalized type system then is a triple

$$GTS = [\check{\overline{T}}, P, \overline{f_{impl}^{type}}]$$

where $\check{\overline{T}}$ is the generalized type system, P is the set of all programs, and $\overline{f_{impl}^{type}}$ is a mapping that maps each $\overline{ts} \in \check{\overline{T}}$ onto that subset of P that contains the implementation for $\overline{ts}$.

An instance of this GTS is a generalized type $\overline{t} = [\overline{ts}, \overline{f_{impl}^{type}}(\overline{ts})], \overline{ts} \in \check{\overline{T}}$.

For example,

Class *Bridge* (
Location: geo;
Name: string;
Height: IntuitionisticSet(integer);
Material: Set(string)
WaterType:Set(string)
WaterFlow:Set(string));

A generalized object belonging to this class is defined by:

$$\overline{o} = [oid, N, \overline{t}, \overline{f_{impl}^{type}}(\overline{ts}), v]$$

where v draws values from the generalized domain that allows an object to contain intuitionistic membership and nonmembership values as part of the state of the object.

Both intuitionistic and rough-set uncertainty management can be used in this generalized OODB model. For example, some intuitionistic rough instances of the previously defined object type *Bridge* might include:

[oid1, Ø, *Bridge*, Struct(0289445, "Castor Creek Bridge," {(5, (.7, .2)), (7, (.9, .1))}, Set(concrete), Set(creek), Set (E,NE,N))]

where the attribute Height is shown as an intuitionistic set, and Material, WaterType, and WaterFlow are shown as ordinary sets. We assume that each of these base objects is certain, that is, each object fully exists and has a membership value of one. We further assume that we have defined the partitioning R^1 for the domain WaterType as discussed previously:

R^1 = {[creek, brook, stream], [branch], [river], [lake], [pond, waterhole], [slough]}.

It is easy to see the need for various types of uncertainty in spatial database from even this simplified example. There is indiscernibility in the labels for various types of water. Different users might use different names for the same water types, or data may have been gathered from multiple sources and currently being consolidated into a single database application. Rough sets allow us to specify this level of indiscernibility and to adjust it, when necessary, to fit the application.

There is fuzzy uncertainty in the height of the bridges. This uncertainty may be due to one of several causes. The bridge might be too high to measure accurately by a nonskilled worker, or it might be that there is uncertainty about where the top of the bridge should be marked. If there is a light on the top, do we measure to the top of the light? Uncertainty might also arise from the location at which the bridge is measured. Is it measured in height above the water? If so, the water level probably varies over time. Is it measured in height above the ground? If so, this height is likely to be different on either end of the bridge.

Direction of water flow, in this example, illustrates yet another type of uncertainty that can be modeled through the approximation regions of rough sets. Often waterways twist and turn in various directions. If a river is generally flowing toward the east, but beneath the bridge it is flowing toward the northeast, we may consider including both of these directions in the database. This decision would obviously depend on the application, and whether it is necessary to have information on the direction of water flow for the entire water body, or only at the point below the bridge.

Uncertainty in spatial databases and geographic information systems becomes an even greater issue when considering topological relationships among various objects or regions, which themselves may be uncertain (Beaubouef et.al 2007). However, fuzzy, intuitionistic, and rough-set uncertainty management, incorporated into an object-oriented spatial database, will allow for better modeling of uncertainty, extracting additional information from the underlying data.

In this section, we extended a formal framework of object-oriented databases to allow for modeling of various types of imprecision, vagueness, and uncertainty that typically occur in spatial data. The model is based on a formal type system and specified constraints, thus preserving integrity of the database, while at the same time allowing an object-oriented database to be generalized in such a way as to include both intuitionistic and rough-set uncertainty, both well-developed methods of uncertainty management.

No information is lost in using the intuitionistic rough object-oriented data model over traditional object-oriented databases. One can always examine the intuitionistic membership and nonmembership values of objects, resulting from intuitionistic rough expressions, to discover which of those are certain and which are possible. The certain results include exactly those that would be obtained in a nonintuitionistic, nonrough database. In this model, however, we are provided also with those results that are possible, along with intuitionistic membership values to aid in determining the degrees of uncertainty of each of the results. Incorporation of intuitionistic and rough-set uncertainty into the object-oriented database model is essential for representing imprecision and uncertainty in spatial data entities and in their interrelationships.

FUTURE TRENDS

Rough sets and databases will continue to be an active area of research. Since it has been well recognized that uncertainty in spatial data is an important issue in many applications, rough sets use in spatial databases and geographic information systems is being studied (Ahlqvist, Keukelaar, & Oukbir. 2000; Bittner & Stell, 2003; Beaubouef & Petry, 2005b; Beaubouef et.al 2006). Another significant area in which rough set have been applied is in the development of enhanced data-mining techniques (Beaubouef, Ladner, & Petry, 2004b, Bittner, 2000; Shi, Wang, Li, & Wang. 2003). Data systems will only get larger, and the need for information in an uncertain world will continue to be a priority.

CONCLUSION

Rough sets are an important mathematical theory, applicable to many diverse fields. They have predominantly been applied to the area of knowledge discovery in databases, offering a type of uncertainty management different from other methods such as probability, fuzzy sets, and others.

Rough-set theory can also be applied to database models and design, as discussed in this chapter. An essential feature of any database that is used for real-world applications in an "uncertain" world is its ability to manage uncertainty in data. In this chapter, we have discussed how rough sets play an important role in the management of uncertainty in databases. Furthermore, rough sets are versatile in that they can be combined with other techniques in an attempt to model various types of uncertainty that each method alone might not capture. We have shown how rough sets and fuzzy sets can be incorporated in the fuzzy rough relational database and the fuzzy rough object-oriented database models. Here the fuzziness somehow quantifies the uncertainty that rough sets alone could not capture. Another facet to the measurement of uncertainty is accomplished through the use of intuitionistic sets. In the rough intuitionistic database model, there is uncertainty represented by the indiscernibility relation and approximation regions of rough sets theory, along with membership and nonmembership values of intuitionistic set theory to provide greater information about things in the uncertain region.

REFERENCES

Ahlqvist, O., Keukelaar, J., & Oukbir, K. (2000). Rough classification and accuracy assessment. *International Journal of Geographical Information Science, 14*(5), 475-496.

Atanassov, K. (1986). Intuitionistic fuzzy sets. *Fuzzy Sets and Systems, 20*, 87-96.

Atanassov, K. (1999). *Intuitionistic fuzzy sets: Theory and applications.* Heidelberg, NY: Physica-Verlag, A Springer-Verlag Company.

Beaubouef, T., Ladner, R., & Petry, F. (2004b). Rough set spatial data modeling for data mining. *International Journal of Intelligent Systems, 19*(7), 567-584.

Beaubouef, T., & Petry, F. (1994). Rough querying of crisp data in relational databases. In *Proceedings of the Third International Workshop on Rough Sets and Soft Computing (RSSC'94)*, San Jose, California.

Beaubouef, T., & Petry, F. (2000). Fuzzy rough set techniques for uncertainty processing in a relational database. *International Journal of Intelligent System,s 15*(5), 389-424.

Beaubouef, T., & Petry F. (2002). Fuzzy set uncertainty in a rough object oriented database. In *Proceedings of North American Fuzzy Information Processing Society (NAFIPS-FLINT'02)*, New Orleans, LA (pp. 365-370).

Beaubouef, T., & Petry, F. (2004a). Rough functional dependencies. In *Proceedings of the 2004 Multiconferences: International Conference On Information and Knowledge Engineering (IKE'04)* (pp. 175-179). Las Vegas, NV.

Beaubouef, T., & Petry, F. (2005a). Normalization in a rough relational database. In *Proceedings of Tenth International Conference on Rough Sets, Fuzzy Sets, Data Mining, and Granular Computing (RSFDGrC'05)* (pp. 257-265). Regina, Canada.

Beaubouef, T., & Petry, F. (2005b). Representation of spatial data in an OODB using rough and fuzzy set modeling. *Soft Computing Journal, 9*(5), 364-373.

Beaubouef, T., Petry, F., & Buckles, B. (1995). Extension of the relational database and its algebra with rough set techniques. *Computational Intelligence, 11*(2), 233-245.

Beaubouef, T., Petry, F., & Ladner, R. (In Press). Spatial data methods and vague regions: A rough set approach. *Applied Soft Computing Journal, 7* (425-440).

Bittner, T. (2000). Rough sets in spatio-temporal data mining. In *Proceedings of International Workshop on Temporal, Spatial and Spatio-*

Temporal Data Mining, Lyon, France. Lecture Notes in Artificial Intelligence (pp. 89-104). Berlin-Heidelberg: Springer-Verlag.

Bittner, T., & Stell, J. (2003). Stratified rough sets and vagueness, in spatial information theory: Cognitive and computational foundations of geographic information science. In W. Kuhn, M. Worboys, & S. Timpf (Eds.), *International Conference (COSIT'03)* (pp. 286-303).

Buckles, B., & Petry, F. (1982). A fuzzy representation for relational data bases. *International Journal of Fuzzy Sets and Systems, 7*(3) 213-226.

Cattell, R., Barry, D., Bartels, D., Berler, M., Eastman, J., Gamerman, S., Jordan, D., Springer, A., Strickland, H., & Wade, D. (1997). *The object database standard: ODMG2.0.* San Francisco: Morgan Kaufmann.

Chanas, S., & Kuchta, D. (1992). Further remarks on the relation between rough and fuzzy sets. *Fuzzy Sets and Systems, 47,* 391-394.

De Tré, G., & De Caluwe, R. (1999). A generalized object-oriented database model with generalized constraints. In *Proceedings of North American Fuzzy Information Processing Society* (pp. 381-386). New York.

Grzymala-Busse, J. (1991). *Managing uncertainty in expert systems.* Boston: Kluwer Academic Publishers.

Gyseghem, N., & De Caluwe, R. (1997). Fuzzy and uncertain object-oriented databases: Concepts and models. *Advances in Fuzzy Systems—Applications and Theory, 13,* 123-177.

Komorowski, J., Pawlak, Z., Polkowski, L., et al, (1999). Rough sets: A tutorial. In S.K. Pal & A. Skowron (Eds.), *Rough fuzzy hybridization: A new trend in decision-making* (pp. 3-98). Singapore: Springer-Verlag.

Makinouchi, A. (1977). A consideration on normal form of not-necessarily normalized relation in the relational data model. In *Proceedings of the Third International Conference on Very Large Databases* (pp. 447-453).

Nanda, S., & Majumdar, S. (1992). Fuzzy rough sets. *Fuzzy Sets and Systems, 45,* 157-160.

Ola, A., & Ozsoyoglu, G. (1993). Incomplete relational database models based on intervals. *IEEE Transactions on Knowledge and Data Engineering, 5*(2), 293-308.

Pawlak, Z. (1984). Rough sets. *International Journal of Man-Machine Studies, 21,* 127-134.

Pawlak, Z. (1985). Rough sets and fuzzy sets. *Fuzzy Sets and Systems, 17,* 99-102.

Pawlak, Z. (1991). *Rough sets: Theoretical aspects of reasoning about data,* Norwell, MA: Kluwer Academic Publishers.

Roth, M.A., Korth, H.F., & Batory, D.S. (1987). SQL/NF: A query language for non-1NF databases. *Information Systems, 12,* 99-114.

Shi, W., Wang, S., Li, D., & Wang, X. (2003). Uncertainty-based Spatial Data Mining. In *Proceedings of Asia GIS Association,* Wuhan, China (pp. 124-35).

Wygralak, M. (1989). Rough sets and fuzzy sets—Some remarks on interrelations. *Fuzzy Sets and Systems, 29,* 241-243.

Zadeh, L. (1965). Fuzzy sets. *Information and Control, 8,* 338-353.

Zvieli, A., & Chen, P. (1986). Entity-relationship modeling and fuzzy databases. In *Proceedings of International Conference on Data Engineering* (pp. 320-327). Los Angeles, CA.

Chapter VI
Current Trends in Rough Set Flow Graphs

Cory J. Butz
University of Regina, Canada

Wen Yan
University of Regina, Canada

ABSTRACT

In this chapter, we review a graphical framework for reasoning from data, called rough set flow graphs (RSFGs), and point out issues of current interest involving RSFG inference. Our discussion begins by examining two methods for conducting inference in a RSFG. We highlight the fact that the order of variable elimination, called an elimination ordering, affects the amount of computation needed for inference. The culminating result is the incorporation of an algorithm for obtaining a good elimination ordering into our RSFG inference algorithm.

INTRODUCTION

Pawlak (2002, 2003) introduced rough set flow graphs (RSFGs) as a graphical framework for uncertainty management. RSFGs extend traditional rough set research (Pawlak, 1982, 1991) by organizing the rules obtained from decision tables as a *directed acyclic graph* (DAG). Each rule is associated with three coefficients, namely, *strength, certainty,* and *coverage,* that have been shown to satisfy Bayes' theorem (Pawlak, 2002, 2003). Pawlak stated that RSFGs are a new perspective on Bayesian inference, and provided an algorithm to answer queries in a RSFG (Pawlak, 2003).

We established in Butz, Yan, and Yang (2005) that a RSFG is, in fact, a special case of Bayesian network, and that RSFG can be carried out in polynomial time. In Butz et al. (2005), it was also shown that the traditional RSFG inference algorithm possesses exponential time complexity and a new inference algorithm with polynomial time complexity was introduced. The key difference between the two RSFG inference algorithms is that the former eliminates variables all at once, while the latter saves computation by eliminating the variables one by one. No study, however, has investigated how to obtain a good *elimination ordering*.

In this chapter, we suggest an algorithm for obtaining a good elimination ordering of the variables in a RSFG. To achieve this, we introduce a cost measure that is related to the number of incoming edges and outgoing edges for each value of a variable. The elimination process iteratively eliminates a variable with the lowest cost measure.

This algorithm is important, since the elimination ordering can dramatically affect the amount of work needed for inference.

This chapter is organized as follows. We begin by reviewing the RSFG framework and two methods for inference therein. A new RSFG inference algorithm, involving eliminating variables following a good elimination ordering, is presented. The manuscript ends with several conclusions.

ROUGH SET FLOW GRAPHS

Rough set flow graphs are built from decision tables. A *decision table* (Pawlak, 1991) represents a potential $\varphi(C, D)$, where C is a set of conditioning

Table 1. Decision tables φ (M,D), φ (D,A), φ (A,S) and φ (S,P)

M	D	$\varphi(M,D)$	D	A	$\varphi(D,A)$
Toyota	Alice	120	Alice	Old	51
Toyota	Bob	60	Alice	Middle	102
Toyota	Dave	20	Alice	Young	17
Honda	Bob	150	Bob	Old	144
Honda	Carol	150	Bob	Middle	216
Ford	Alice	50	Carol	Middle	120
Ford	Bob	150	Carol	Young	80
Ford	Carol	50	Dave	Old	27
Ford	Dave	250	Dave	Middle	81
			Dave	Young	162

A	S	$\varphi(A,S)$	S	P	$\varphi(S,P)$
Old	High	133	High	Executive	210
Old	Medium	67	High	Staff	45
Old	Low	22	High	Manager	8
Middle	High	104	Medium	Executive	13
Middle	Medium	311	Medium	Staff	387
Middle	Low	104	Medium	Manager	30
Young	High	26	Low	Executive	3
Young	Medium	52	Low	Staff	12
Young	Low	181	Low	Manager	292

attributes and D is a decision attribute. It should perhaps be emphasized immediately that all decision tables $\varphi(C, D)$ define a binary flow graph regardless of the cardinality of C. Consider a row in $\varphi(C, D)$, where c and d are the values of C and D, respectively. Then there is a directed edge from node c to node d. That is, the constructed flow graph treats the attributes of C as a whole, even when C is a nonsingleton set of attributes.

Example 1: Given five attributes: Manufacturer (M), Dealership (D), Age (A), Salary (S), Position (P). Consider the set $C = \{M\}$ of conditioning attributes and the decision attribute D. Then one decision table $\varphi(M, D)$ is shown in Table 1. Similarly, decision tables $\varphi(D, A)$, $\varphi(A, S)$ and $\varphi(S, P)$ are also depicted in Table 1.

Each decision table defines a binary RSFG. The set of nodes in the flow graph are $\{c_1, c_2, ..., c_k\} \cup \{d_1, d_2, ..., d_l\}$, where $c_1, c_2, ..., c_k$ and $d_1, d_2, ..., d_l$ are the values of C and D appearing in the decision table, respectively. For each row in the decision table, there is a directed edge (c_i, d_j) in the flow graph, where c_i is the value of C and d_j is the value of D. Clearly, the defined graphical structure is a directed acyclic graph (DAG). Each edge (c_i, d_j) is labeled with three coefficients. The strength of (c_i, d_j) is $\varphi(c_i, d_j)$ obtained from the decision table. From $\varphi(c_i, d_j)$, we can compute the certainty $\varphi(d_j| c_i)$ and the coverage $\varphi(c_i| d_j)$.

Example 2: Consider the decision tables $\varphi(M, D)$ and $\varphi(D, A)$ in Table 1. The DAGs of the binary RSFGs are illustrated in Figure 1, respectively. The strength, certainty, and coverage of the edges of the flow graphs in Figure 1 are shown in the top two tables of Table 2.

In order to combine the collection of binary flow graphs into a general flow graph, Pawlak makes the *flow conservation* assumption (Pawlak, 2003). This means that, for an attribute A appearing as a decision attribute in one decision table $\varphi_1(C_1, A)$ and also as a conditioning attribute in another decision table $\varphi_2(A, D_2)$, we have:

$$\sum_{D_1} \phi_1(C_1, A) = \sum_{D_2} \phi_2(A, D_2).$$

Figure 1. The DAGs of the binary RSFGs for the decision tables $\phi(M, D)$ and $\phi(D, A)$ in Table 1, respectively. The coefficients are given in part of Table 2.

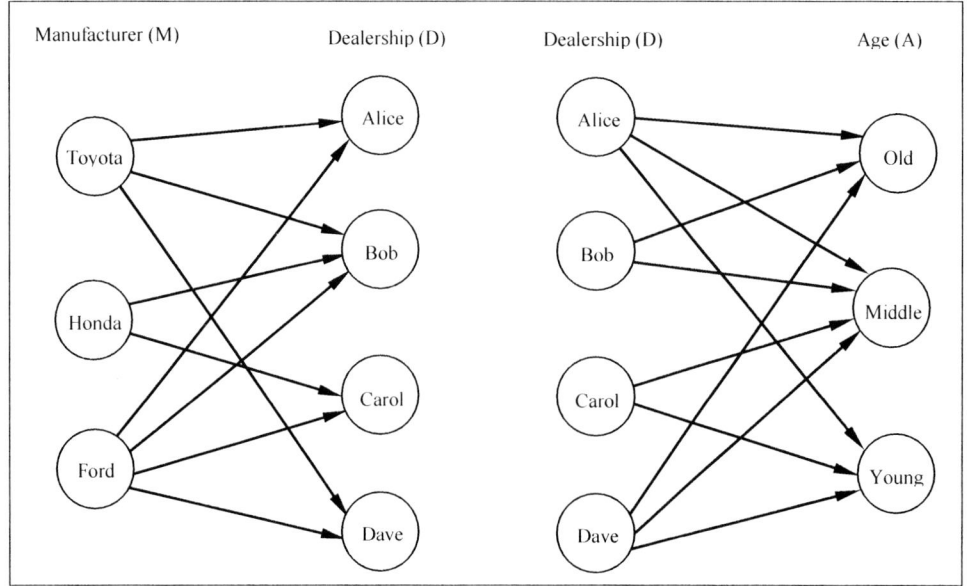

Example 3: The two binary RSFGs in Example 2 satisfy the flow conservation assumption, since in Table 2, $\varphi_1(D) = \varphi_2(D)$. For instance, $\varphi_1(D = \text{"Alice"}) = 0.170 = \varphi_2(D = \text{"Alice"})$.

Definition 1: A *rough set flow graph* (RSFG) (Pawlak, 2002, 2003) is a DAG, where each edge is associated with the strength, certainty, and coverage coefficients from a collection of decision tables satisfying the flow conservation assumption.

Example 4: The RSFG for the decision tables in Table 1 is the DAG in Figure 2, together with the strength, certainty, and coverage coefficients in Table 2.

INFERENCE IN ROUGH SET FLOW GRAPHS

RSFG inference is to compute a binary RSFG on $\{A_i, A_j\}$, namely, a DAG on $\{A_i, A_j\}$ and the coefficient table, denoted $Ans(A_i, A_j)$, which is a table containing the strength, certainty, and coverage coefficients. We use the term *query* to refer to any request involving strength, certainty, or coverage.

Table 2. The top two tables are the strength $\varphi(ai,aj)$, certainty $\varphi(aj|ai)$ and coverage $\varphi(ai|aj)$ coefficients for the edges (ai,aj) in Figure 1. These two tables together with the bottom two tables are the coefficients for the RSFG in Figure 2.

M	D	$\varphi_1(M,D)$	$\varphi_1(D|M)$	$\varphi_1(M|D)$	D	A	$\varphi_2(D,A)$	$\varphi_2(A|D)$	$\varphi_2(D|A)$
Toyota	Alice	0.120	0.600	0.710	Alice	Old	0.050	0.300	0.230
Toyota	Bob	0.060	0.300	0.160	Alice	Middle	0.100	0.600	0.190
Toyota	Dave	0.020	0.100	0.070	Alice	Young	0.020	0.100	0.080
Honda	Bob	0.150	0.500	0.420	Bob	Old	0.140	0.400	0.630
Honda	Carol	0.150	0.500	0.750	Bob	Middle	0.220	0.600	0.420
Ford	Alice	0.050	0.100	0.290	Carol	Middle	0.120	0.600	0.230
Ford	Bob	0.150	0.300	0.420	Carol	Young	0.080	0.400	0.310
Ford	Carol	0.050	0.100	0.250	Dave	Old	0.030	0.100	0.140
Ford	Dave	0.250	0.500	0.930	Dave	Middle	0.080	0.300	0.150
					Dave	Young	0.160	0.600	0.620
A	S	$\varphi_3(A,S)$	$\varphi_3(S|A)$	$\varphi_3(A|S)$	S	P	$\varphi_4(S,P)$	$\varphi_4(P|S)$	$\varphi_4(S|P)$
Old	High	0.133	0.600	0.506	High	Executive	0.210	0.800	0.929
Old	Medium	0.067	0.300	0.156	High	Staff	0.045	0.170	0.101
Old	Low	0.022	0.100	0.072	High	Manager	0.008	0.030	0.024
Middle	High	0.104	0.200	0.395	Medium	Executive	0.013	0.030	0.058
Middle	Medium	0.311	0.600	0.723	Medium	Staff	0.387	0.900	0.872
Middle	Low	0.104	0.200	0.339	Medium	Manager	0.030	0.070	0.091
Young	High	0.026	0.100	0.099	Low	Executive	0.003	0.010	0.013
Young	Medium	0.052	0.200	0.121	Low	Staff	0.012	0.040	0.027
Young	Low	0.181	0.700	0.589	Low	Manager	0.292	0.950	0.885

Figure 2. The RSFG for {M, D, A, S, P}, where the strength, certainty and coverage coefficients are given in Table 2

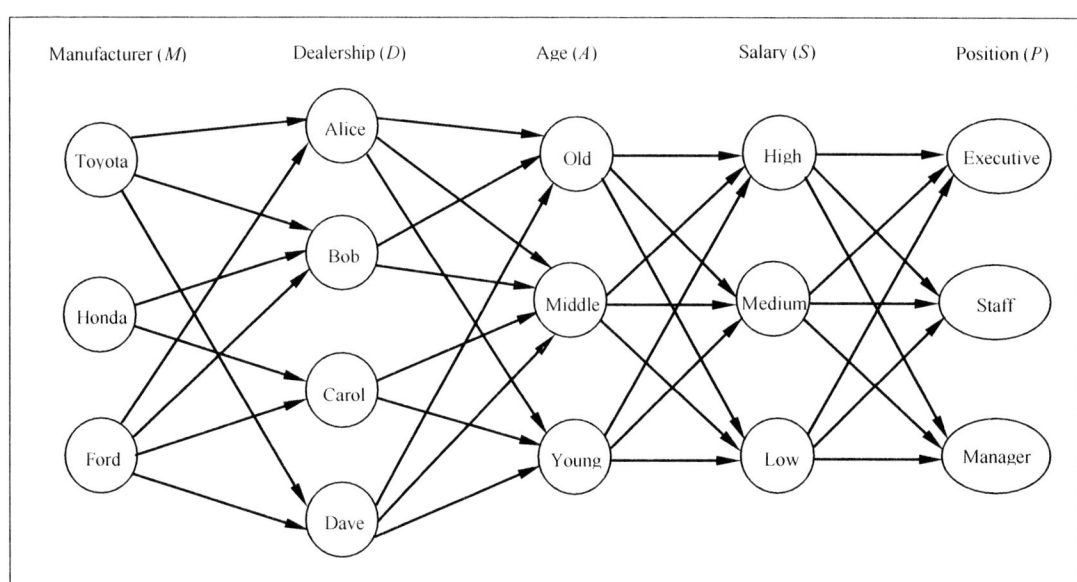

Figure 3. Answering a query on {M, P} posed to the RSFG in Figure 2 consists of the coefficient table Ans(M,P) in Table 3 and this DAG on {M, P}

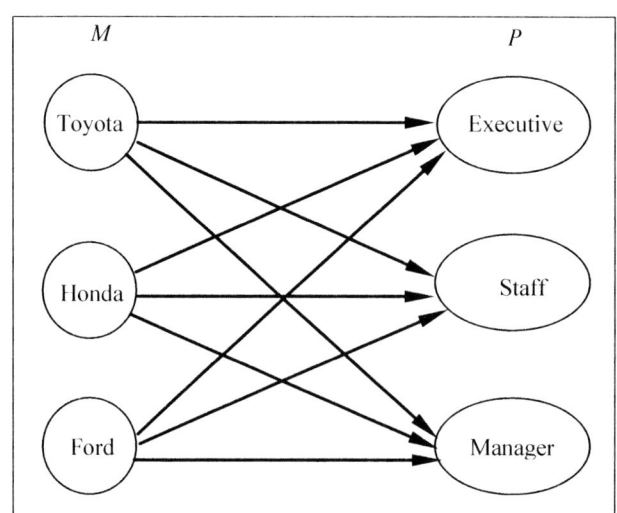

Example 5: Consider a query on {M, P} posed to the RSFG in Example 4. The answer to this query is the binary RSFG defined by Table 3 and Figure 3.

Traditional Algorithm for RSFG Inference

Pawlak (2003) proposed Algorithm 1 to answer queries in a RSFG (see Algoritm 1).

Algorithm 1 is used to compute the coefficient table of the binary RSFG on $\{A_i, A_j\}$. The DAG of this binary RSFG has an edge (a_i, a_j) provided that $\varphi(a_i, a_j) > 0$ is in $Ans(A_i, A_j)$. We illustrate Algorithm 1 with Example 6.

Example 6: Given a query on {M, P} posed to the RSFG in Figure 2. Let us focus on M = "Ford" and P = "Staff," which we succinctly write as "Ford" and "Staff," respectively. The certainty $\varphi(\text{"Staff"} | \text{"Ford"})$ is computed as:

$$\phi(\text{"Staff"} | \text{"Ford"}) = \sum_{D,A,S} \phi(D | \text{"Ford"}) \cdot \phi(A | D) \cdot \phi(S | A) \cdot \phi(\text{"Staff"} | S)$$

The coverage $\varphi(\text{"Ford"} | \text{"Staff"})$ is determined as:

$$\phi(\text{"Ford"} | \text{"Staff"})$$

Table 3. Answering a query on {M, P} posed to the RSFG in Figure 2 consists of this coefficient table Ans(M,P) and the DAG in Figure 3

M	P	$\varphi(M,P)$	$\varphi(P\|M)$	$\varphi(M\|P)$
Toyota	Executive	0.053132	0.265660	0.234799
Toyota	Staff	0.095060	0.475300	0.214193
Toyota	Manager	0.051808	0.259040	0.157038
Honda	Executive	0.067380	0.224600	0.297764
Honda	Staff	0.140820	0.469400	0.317302
Honda	Manager	0.091800	0.306000	0.278259
Ford	Executive	0.105775	0.211550	0.467437
Ford	Staff	0.207925	0.415850	0.468505
Ford	Manager	0.186300	0.372600	0.564703

Algorithm 1:

Input: A RSFG and a query on $\{A_i, A_j\}$, i < j.
Output: The coefficient table $Ans(A_i, A_j)$ of the binary RSFG on $\{A_i, A_j\}$.

$$\phi(A_j | A_i) = \sum_{A_{i+1},\ldots,A_{j-1}} \phi(A_{i+1} | A_i) \cdot \phi(A_{i+2} | A_{i+1}) \cdot \ldots \cdot \phi(A_j | A_{j-1});$$

$$\phi(A_i | A_j) = \sum_{A_{i+1},\ldots,A_{j-1}} \phi(A_i | A_{i+1}) \cdot \phi(A_{i+1} | A_{i+2}) \cdot \ldots \cdot \phi(A_{j-1} | A_j);$$

$$\phi(A_i, A_j) = \phi(A_i) \cdot \phi(A_j | A_i);$$

return ($Ans(A_i, A_j)$);

$$= \sum_{D,A,S} \phi("Ford"|D)\cdot\phi(D|A)\cdot\phi(A|S)$$
$$\cdot\phi(S|"Staff").$$

The strength φ("*Ford*", "*Staff*") is computed as:

$$\phi("Ford","Staff\ ") = \phi("Ford")$$
$$\cdot\phi("Staff\ "|"Ford").$$

The DAG of this binary RSFG on {*M, P*} is depicted in Figure 3.

In Example 6, computing the three coefficients φ("*Ford*","*Staff* "), φ("*Staff* "| "*Ford*"), and φ("*Ford*"| "*Staff* ") in *Ans*(*M*, *P*) in Table 3 required 181 multiplications and 58 additions.

AN EFFICIENT ALGORITHM FOR RSFG INFERENCE

It has been shown that Algorithm 1 has exponential time complexity (Butz, Yan, & Yang, 2006). Therefore, we (Butz et al., 2006) proposed the following algorithm for RSFG inference, and established its polynomial time complexity (see Algorithm 2).

The main idea of Algorithm 2 is to exploit the factorization to eliminate variables one by one, instead of all at once as Algorithm 1 does. In Example 7, we focus on computing the coefficient table $Ans(A_i, A_j)$ with the DAG of the output RSFG understood.

Example 7: Let us revisit Example 6 except using Algorithm 2 in place of Algorithm 1. According to the for-loop construct, variables *D*, *A*, and *S* need be eliminated in this order. When eliminating variable *D*, we compute the certainty φ(*A*| "*Ford*") as:

$$\phi(A|"Ford") = \sum_{D} \phi(D|"Ford")\cdot\phi(A|D),$$

while the coverage φ("*Ford*"| *A*) is determined as:

$$\phi("Ford"|A) = \sum_{D} \phi("Ford"|D)\cdot\phi(D|A).$$

Observe that variable *D* has been eliminated by linking variables *M* and *A* using the certainty φ(*A*| "*Ford*") and the coverage φ("*Ford*"| *A*). Similarly, variable *A* is eliminated by computing φ(*S*| "*Ford*") and φ("*Ford*"| *S*), thereby connecting variables *M* and S. Finally, we link the

---

Algorithm 2:

Input: A RSFG and a query on $\{A_i, A_j\}$, i < j.
Output: The coefficient table $Ans(A_i, A_j)$ of the binary RSFG on $\{A_i, A_j\}$.

for k = (i+1) to (j-1)

$$\phi(A_{k+1}|A_i) = \sum_{A_k} \phi(A_k|A_i)\cdot\phi(A_{k+1}|A_k);$$
$$\phi(A_i|A_{k+1}) = \sum_{A_k} \phi(A_i|A_k)\cdot\phi(A_k|A_{k+1});$$

end

$$\phi(A_i,A_j) = \phi(A_i)\cdot\phi(A_j|A_i);$$
return ($Ans(A_i, A_j)$);

query variables *M* and *P* by computing φ("Staff"| "Ford") and φ("Ford"|"Staff") when eliminating the final variable *S*:

$$\phi("Staff"|"Ford") = \sum_S \phi(S|"Ford") \cdot \phi("Staff"|S),$$

and

$$\phi("Ford"|"Staff") = \sum_S \phi("Ford"|S) \cdot \phi(S|"Staff").$$

Note that computing the strength ("Ford","Staff"), certainty φ("Staff"|"Ford"), and coverage φ("Ford"| "Staff") in *Ans*(*M*, *P*) in Table 3 only requires 45 multiplications and 30 additions.

THE ORDER OF VARIABLE ELIMINATION

Here we emphasize the importance of an elimination ordering in a RSFG, and introduce an algorithm to systemically eliminate particular variables.

The next example illustrates how changing the elimination ordering influences the amount of computation needed for RSFG inference.

Example 8: Recall computing the strength φ("Ford","Staff"), the certainty φ("Staff"|"Ford") and the coverage φ("Ford"|"Staff") in the RSFG of Figure 2 necessarily involves eliminating variables {*A*, *D*, *S*}. As shown in Example 7, 45 multiplications and 30 additions are needed, provided that these variables are eliminated in the order *D*, *A*, *S*. However, eliminating the variables in the order *A*, *D*, *S* requires more work, namely, 91 multiplications and 36 additions.

Example 8 demonstrates that finding a good elimination ordering is essential. While the problem of finding an optimal elimination ordering for Bayesian network inference is known to be NP-complete (Zhang, 1998), we note that RSFGs are a special case of Bayesian networks. Thus, even though it may be possible to determine an optimal elimination ordering for RSFG inference, we focus our efforts on determining a good elimination ordering.

The cost measure $\gamma(A_i)$ of a variable $A_i \in U$ in a RSFG is defined as:

$$\gamma(A_i) = \sum_{a_i \in dom(A_i)} no_{in}(a_i) \cdot no_{out}(a_i) \quad (1)$$

where a_i is a value of variable A_i, and $no_{in}(a_i)$ and $no_{out}(a_i)$ are the respective number of incoming edges and outgoing edges of a_i in the given RSFG. In other words, the sum of the connectedness of all values of a variable is used as a cost measure for that variable. The algorithm used to determine an elimination ordering is a greedy algorithm eliminating one variable with the lowest cost measure. After such a variable is eliminated, the RSFG is modified accordingly and new cost measures are computed only for those variables involved in the modification.

Algorithm 3 is a new RSFG inference algorithm utilizing a good elimination ordering.

Here we use the following example to illustrate Algorithm 3.

Example 9: Given a query on {*M*, *P*} posed to the RSFG in Figure 2. Let us focus on *M* = "Ford" and *P* = "Staff". By Algorithm 3, we first compute the cost measures for variables *D*, *A*, and *S* as follows:

$\gamma(D)$
$= no_{in}(Alice) \cdot no_{out}(Alice) + (no_{in} Bob)$
$\cdot no_{out}(Bob) + no_{in}(Carol) \cdot no_{out}(Carol)$
$+ no_{in}(Dave) \cdot no_{out}(Dave)$
$= 1 \times 3 + 1 \times 2 + 1 \times 2 + 1 \times 3$
$= 10,$

Algorithm 3:

Input: A RSFG and a query on $\{A_i, A_j\}$, $i < j$.
Output: The coefficient table $Ans(A_i, A_j)$ of the binary RSFG on $\{A_i, A_j\}$.

for $k = (i+1)$ to $(j-1)$
| Compute the cost measure $\gamma(A_k)$ for each variable A_k using Equation (1) ;
end
for $n = 1$ to $j-i-1$
| Select any variable A_m, with the lowest cost measure $\gamma(A_m)$;
| $\phi(A_{m+1} | A_{m-1}) = \sum_{A_m} \phi(A_m | A_{m-1}) \cdot \phi(A_{m+1} | A_m)$;
| $\phi(A_{m-1} | A_{m+1}) = \sum_{A_m} \phi(A_{m-1} | A_m) \cdot \phi(A_m | A_{m+1})$;
| if $m-1 \neq i$
| | Update the cost measure $\gamma(A_{m-1})$ for the variable A_{m-1} ;
| end
| if $m+1 \neq j$
| | Update the cost measure $\gamma(A_{m+1})$ for the variable A_{m+1} ;
| end
end
$\phi(A_i, A_j) = \phi(A_i) \cdot \phi(A_j | A_i)$;
return ($Ans(A_i, A_j)$) ;

$\gamma(A)$
$= no_{in}(Old) \cdot no_{out}(Old) + (no_{in} Middle)$
$\cdot no_{out}(Middle) + no_{in}(Young) \cdot no_{out}(Young)$
$= 3 \times 3 + 4 \times 2 + 3 \times 3$
$= 30,$

$\gamma(S)$
$= no_{in}(High) \cdot no_{out}(High) + (no_{in} Medium)$
$\cdot no_{out}(Medium) + no_{in}(Low) \cdot no_{out}(Low)$
$= 3 \times 1 + 3 \times 1 + 3 \times 1$
$= 9$

Variable S is chosen as the variable to be eliminated, since $\gamma(S) < \gamma(D) < \gamma(A)$. The coefficients $\varphi(\text{``Staff''}| A)$ and $\varphi(A| \text{``Staff''})$ are computed as follows:

$\phi(\text{"Staff"} | A)$
$= \sum_S \phi(S | A) \cdot \phi(\text{"Staff"} | S),$

$\phi(A | \text{"Staff"})$
$= \sum_S \phi(A | S) \cdot \phi(S | \text{"Staff"})$

Since variable P, one neighbour of S, is variable A_j in the query, we only need to update the cost measure of variable A as follows:

$\gamma(A)$
$= no_{in}(Old) \cdot no_{out}(Old) + (no_{in} Middle)$
$\cdot no_{out}(Middle) + no_{in}(Young) \cdot no_{out}(Young)$
$= 3 \times 1 + 4 \times 1 + 3 \times 1$
$= 10.$

As $\gamma(D) = \gamma(A)$, let us pick D for elimination. We then have:

$\phi(A | \text{"Ford"})$
$= \sum_D \phi(D | \text{"Ford"}) \cdot \phi(A | D),$

$$\phi("Ford"|A)$$
$$=\sum_D \phi(D|A)\cdot\phi("Ford"|D)$$

Since variable M, one neighbour of D, is variable A_i in the query, we only need to update the cost measure of variable A. As A is the last variable to be eliminated, we obtain:

$$\phi("Staff"|"Ford")$$
$$=\sum_A \phi("Staff"|A)\cdot\phi(A|"Ford"),$$

$$\phi("Ford"|"Staff")$$
$$=\sum_A \phi(A|"Staff")\cdot\phi("Ford"|A).$$

Calculating the strength φ("Ford";"Staff"), the certainty φ("Staff"|"Ford"), and the coverage φ("Ford"|"Staff") in $Ans(M, P)$ in Table 3 following the elimination order S, D, and A requires 45 multiplications and 30 additions.

Note that the work for determining a good elimination ordering is quite different from those used in the Bayesian network community. Our approach determines a good elimination ordering while a query is being processed in a RSFG. On the contrary, one good elimination ordering is often chosen in advance and used to process all queries in a Bayesian network (Zhang, 1998).

CONCLUSION

We have provided an overview of the brief history of *rough set flow graphs* (RSFGs) (Pawlak, 2002 & 2003). One key observation is that eliminating variables one by one, instead of all at once, can save computation during RSFG inference. At the same time, the order in which variables are eliminated will also influence the amount of work needed. A new RSFG inference algorithm utilizing a good elimination ordering was introduced. Future directions of RSFG research can be aimed at demonstrating the practical applicability of RSFGs.

REFERENCES

Butz, C. J., Yan, W., & Yang, B. (2005). The computational complexity of inference using rough set flow graphs. *The Tenth International Conference on Rough Sets, Fuzzy Sets, Data Mining, and Granular Computing, 1* (pp. 335-344).

Butz, C. J., Yan, W., & Yang, B. (2006). An efficient algorithm for inference in rough set flow graphs. *LNCS Transactions on Rough Sets, 5*, (to appear).

Pawlak, Z. (1982). Rough sets. *International Journal of Computer and Information Sciences, 11*(5), 341-356.

Pawlak, Z. (1991). *Rough sets: Theoretical aspects of reasoning about data*. Dordrecht: Kluwer Academic Publishers.

Pawlak, Z. (2002). In pursuit of patterns in data reasoning from data: The rough set way. *The Third International Conference on Rough Sets, and Current Trends in Computing* (pp. 1-9).

Pawlak, Z. (2003). Flow graphs and decision algorithms. *The Ninth International Conference on Rough Sets, Fuzzy Sets, Data Mining, and Granular Computing* (pp. 1-10).

Zhang, N.L. (1998). Computational properties of two exact algorithms for Bayesian networks. *Journal of Applied Intelligence, 9*, 173-184.

Chapter VII
Probabilistic Indices of Quality of Approximation

Annibal Parracho Sant'Anna
Universidade Federal Fluminense, Brazil

ABSTRACT

A new index of quality of approximation, called the index of mutual information, is proposed in this chapter. It measures the mutual information between the relations, respectively, determined by condition and decision attributes. Its computation is based on the comparison of two graphs, each one representing a set of attributes. Applications in the context of indiscernibility as well as in the context of dominance relations are considered. The combination of the new measurement approach with the transformation into probabilities of being the preferred option is also explored. A procedure to select the most important attributes is outlined. Illustrative examples are provided.

INTRODUCTION

The quality of the approximation is a fundamental concept of rough sets theory. It is originally measured by the index of quality of approximation (γ) of Pawlak (1982, 1991). The index γ is based on evaluations of the approximation to each decision class. Variants have been proposed for this index, always preserving this point of view. Basically, these indices measure the proportion of options indiscernible according to the condition attributes that belong to different sets in the partition determined by the decision attributes. This leads to think the quality of the approximation as a property of the options instead of a property of the relations of indiscernibility between them, and as an element to determine the usefulness of the condition attributes in foreseeing the values of the decision attributes instead of the distinctions between them.

Gediga and Düntsch (2003) notice that γ does not serve to evaluate this kind of usefulness, since, for instance, if there is only one decision attribute and only one condition attribute and both attribute distinct values for all the options, then γ assumes its maximum value, and the knowledge of the value of the condition attribute for any option nothing informs about the value of the decision attribute. In fact, asymmetry in the definition of γ makes this index assume its maximum anytime all options are discernible by the set of condition attributes, independent of the decision partition.

In this chapter, a symmetric approach is proposed to access the approximation between the partitions determined by condition and decision attributes. This approach results in evaluating the approximation according to a new index that will be called the index of mutual information. The calculus of this index is based on counting pairs of options instead of individual options.

The relation of indiscernibility generated by the decision attributes and the relation of indiscernibility generated by the condition attributes determine two graphs in the set of options in which two nodes are connected if the respective options are indiscernible and disconnected otherwise. The index of mutual information is determined by the proportion of pairs of distinct options that have the same kind of link in the two graphs. More precisely, it is calculated dividing by the total number of pairs of different options, the number of pairs of different options simultaneously discernible or indiscernible for decision attributes and for condition attributes.

The addition of a new condition attribute never reduces the value of γ, even when the new attribute presents different values for all the options. Considering all possible pairs of options, the index of mutual information may have its value increased or decreased by the inclusion of a new attribute, depending on how many differences the new attribute raises between options presenting identical values for the other attributes.

Variants of this index may be built for the case of attributes that are preference criteria, that is, not only distinguish, but also, in fact, establish relations of dominance (Greco, Matarazzo, & Slowinski, 1999) between the options. The graph that represents a dominance relation will be a directed graph. Nodes linked in the two ways are indiscernible, edges oriented in one only direction denote dominance, and the lack of a link between two nodes denote contradictory preference according to different elements of the set of criteria.

In this situation of dominance, an index of mutual information may be built, excluding from counting the pairs of options with contradictory evaluations either when we consider the decision attributes or when we consider the condition attributes. As for the numerator, we would then count, with equal weights, the number of pairs of options presenting, according to the two sets of attributes, the same relation between the options, either of dominance (oriented edge) or of indiscernibility (double edge).

The pairs of options with contradictory evaluations, according to each set of attributes, constitute sets of different size in different contexts. Their size can bring information on the ability of approximating the evaluation, according to the decision attributes, by the evaluations, according to the condition attributes. The problem with excluding the pairs with contradictory evaluations is that the number of such pairs may grow with the number of attributes to the point of risking an excessive reduction of the number of cases employable in the calculation of the index. An exit to reduce the number of contradictions, eliminating those involving small distances, may be found in raising the granularity, that is, using rougher approaches in the measurement of the attributes. This may be coupled with the transformation of the vectors of original measures into vectors of probabilities of being the preferred option (Sant'Anna & Sant'Anna, 2001), which reduces the small differences.

A simpler alternative consists of, following Greco et al. (1999), counting as indiscernible the options with contradictory evaluations. Another alternative that allows consideration of every pair of options consists of counting as indiscernible the options that have the same number of preferences for one and another element of the pair, and to count as dominant that option that is preferable according to a larger number of attributes. All these alternatives fit into the measurement of mutual information framework.

After formally defining the indices described, this chapter shows how to employ them to coherently evaluate the importance of each attribute to the approximation, taking into account the interaction between attributes. Finally, it will be shown how a previous probabilistic transformation of data may turn these indices more informative and, conversely, how the measures obtained for the importance of the attributes may be used to reduce the number of variables in a probabilistic evaluation.

QUALITY OF APPROXIMATION

The basic concept of rough sets theory is that of indiscernibility. Indiscernibility is an equivalence relation. The elements of the partition that it determines are called elementary sets. Any set in the universe of interest is described in terms of the elementary sets, through two approximations, possibly coincident, the lower approximation and the upper approximation. The set of these two approximations constitutes the rough set. The lower approximation of X is the union of the elementary sets contained in it. The upper approximation is the union of the elementary sets with a nonempty intersection with it. Thus, if a given option belongs to X, all options indiscernible of it belong to its upper approximation, and only those options indiscernible of no option not in it belong to the lower approximation.

The elements of the upper approximation that do not belong to the lower approximation constitute the boundary of the rough set. The boundaries are the regions of uncertainty. If the boundary is empty, the rough set is a common, crisp set. The lower approximation of a set is the complement of the upper approximation of the complement of such set, in such a way that the boundary of any set, and of its complement, coincide.

The evaluations of the options, with respect to the different attributes, are usually presented in a spreadsheet, each row corresponding to an option and each column to an attribute. Each cell of this spreadsheet presents the evaluation of the option represented in its row with respect to the attribute of its column. Formally, any classification is characterized by a quadruple (U, Q, V, f), where U is a nonempty set, the universe of the options evaluated, Q is a set of attributes, each attribute $q \in Q$ being able to receive evaluations in a set V_q, V is the union of the V_q, and f, the information function, an application of the Cartesian product of U and Q in V such that, for each $u \in U$ and $q \in Q$, $f(u, q) \in V_q$ is the evaluation of u by q.

The Classical Index

To each $P \subseteq Q$, nonempty, is associated a relation of indiscernibility in U, denoted I_P. The I_P equivalence classes are formed by the options with identical evaluations by all the attributes in P. If $(u_1, u_2) \in I_P$, we say that u_1 and u_2 are P-indiscernible. The equivalence class containing option u is denoted $I_P(u)$ and is called a P-elementary set. The P-lower approximation and the P-upper approximation of a subset X of U are defined, respectively, as $P_{inf}(X) = \{x \in X: I_P(x) \subseteq X\}$, the set of the options of X that are only P-indiscernible of options in X, and $P^{sup}(X) = \cup_{x \in X} I_P(x)$, the set of all options P-indiscernible of some option in X. It is easy to see that $P_{inf}(X) \subseteq X \subseteq P^{sup}(X)$. The P-boundary of X is the set $B_P(X) = P^{sup}(X) - P_{inf}(X)$.

For each particular set, the cardinality of its P-boundary is an indication of its identifiability by the attributes in P, that is, of the accuracy of its determination by means of such attributes. An increasing measure, varying between 0 and 1 of such accuracy, is then given by the reason between the cardinality of $P_{inf}(X)$ and the cardinality of $P^{sup}(X)$. This measure is denoted by $\alpha_P(X)$, and can be thought as the probability of an option indiscernible of some option in X relatively to the attributes in P not being also indiscernible of options not in X. If $\alpha_P(X) = 1$, then X is a crisp set.

The idea behind the Pawlak index of quality of approximation is that, if decreasing the size of the boundary of a set increases its identifiability, then decreasing the size of the union of the boundaries of the sets of a partition increases the identifiability of such partition. Thus, it measures the quality of the approximation of a classification C by means of a set of attributes P by the reason $\gamma_P(C)$ between the sum of the cardinalities of the P-lower approximations of the elements of C and the cardinality of U.

The loss of quality of the approximation when some attribute is withdrawn from P is an indication of the importance of that attribute to explain the classification. Comparing the values of $\gamma_P(C)$ for different values of P, we have a mechanism to identify which attributes are relevant for the classification.

Index of Mutual Information

The classification C being approached is, in practice, the set of the classes of indiscernibility according to some set of attributes D. Thus, we have, in the spreadsheet, the set Q of attributes divided into two parts, D and P. The first part, D, is the set of decision attributes responsible for the classification C that must be explained using the attributes in the set of condition attributes, $P = Q - D$. Looking only at the boundaries of the partition determined by D, the index γ counts only the cases of indiscernible options according to P being in distinct classes according to D. It does not take into account the possibility of options discernible according to P being in the same class according to D. The index of mutual information between the sets of attributes D and P will count, besides the number of cases of simultaneous indiscernibility according to D and P, the number of cases of simultaneous discernibility. This index will be computed dividing by the number of pairs of options of the universe U, the sum of the number of pairs of indiscernible options according to P that belong to the same class according to D, and the number of pairs discernible according to P that belong to distinct classes according to D.

This definition may be set in a graphs theory framework. Consider two graphs with the same set of nodes representing the options in the universe U. In the first, two different nodes are linked if and only if the respective options belong to the same class in the partition generated by D. This graph is determined by the indicator function G_D, defined in the Cartesian product $U \times U$ by $G_D(u_1, u_2) = 1$ if $f_D(u_1, qi) = f_D(u_2, qi)$ for every attribute q_i in D and $G_D(u_1, u_2) = 0$, otherwise, f_D denoting the information function of the classification determined by the set of decision attributes D. In the second graph, analogously, there is a link between two nodes if and only if the respective options present the same values for all the attributes in the set of condition attributes P. This second graph is represented by the indicator function G_P, defined in the Cartesian product $U \times U$ by $G_P(u_1, u_2) = 1$ if $f_P(u_1, pi) = f_P(u_2, pi)$ for every attribute p_i in P and $G_P(u_1, u_2) = 0$, otherwise, for f_P denoting the information function of the classification determined by the set of condition attributes P. The index of mutual information is then defined by the ratio $G(D, P)$ between the number of pairs of different options $\{u_1, u_2\}$ with $G_P(u_1, u_2) = G_D(u_1, u_2)$ divided by the total number of possible such pairs, given by $\#(U)_* [\#(U)-1]/2$.

This approach allows for the production of variants of the index that may be employed to measure different kinds of information. For instance, we may wish a measure conceptually more similar to that provided by the index γ, in the sense of not taking into account the possibility that the condition attributes discriminate within classes of indiscernibility according to the decision attributes. An index keeping this property would be obtained subtracting, from the numerator and from the denominator of the quotient described, the number of pairs of options indiscernible with respect to the condition attributes. Complementarily, another index would take out of consideration those pairs of options indiscernible according to the decision attributes.

To complete the frame, we should then compute two other indices. The first would take out of consideration those pairs of options indiscernible according to both sets of attributes, of decision and of condition. It would count the proportion of pairs with identical classification as indiscernible by decision and condition attributes among the pairs indiscernible by at least one of such sets of attributes. This index might be called the index of concordance on indiscernibility. The second would take out of consideration those pairs discernible according to both sets of attributes. It would measure the proportion of concordances on classifying the pairs of options as distinct, among the pairs considered as distinct by at least one of the sets of attributes. This would be called the index of concordance on discernibility.

Dominance Relations

Greco et al. (1999) adapted rough sets theory to take into account the importance of the order in the case where the attributes are preference criteria. Their basic idea is to substitute for the concept of indiscernibility that of dominance. They explore the fact that, dealing with preference relations, we may restrict analysis to partitions $(C1, \ldots, C_t)$ such that, if $r < s$, then, for all $x \in C_r$ and $y \in C_s$, the preference for x is higher or equal to the preference for y.

In this context, the partition $(C1, \ldots, C_t)$ may be replaced by the series of nested unions $(C1^{\geq}, \ldots, C_t^{\geq})$ and $(C1^{\leq}, \ldots, C_t^{\leq})$, where, for any r, $C_r^{\geq}$ is the union of the C_s with $s > r$ and $C_r^{\leq}$ is the union of the C_s with $s < r$. The lower approximation to $C_r^{\geq}$ with respect to the set of criteria P is the set $P^-(C_r^{\geq})$ of the options u such that all options dominating u with respect to P belong to $C_r^{\geq}$. And the upper approximation is the set $P^+(C_r^{\geq})$ of the options dominating some option u in $C_r^{\geq}$; conversely for the $C_r^{\leq}$.

Combined Dominance

The dominance, according to a set of criteria, may be defined in different ways. Greco et al.'s (1999) approach corresponds to assume that, for a set P of preference criteria, an option u_1 dominates an option u_2 with respect to P if and only if the preference for u_1 is higher or equal to the preference for u_2 according to all the criteria in P. Otherwise, that means, if there are contradictory preferences according to the criteria in P, in the sense that, for at least one of them, u_1 is preferable to u_2 and, for at least one other, u_2 is preferable to u_1, they are taken as indiscernible, as much as if they had the same evaluation according to every criteria in P.

As the number of attributes increases, the chance of finding pairs of attributes with contradictory evaluations increases, and such contradictions may be due to a relatively small set of criteria. Another form of classifying all pairs of options is assuming dominance of one option u_1 over another option u_2 according to a set of criteria, if the number of criteria in this set for which u_1 is preferable for u_2 is larger or equal to the number of criteria in this set for which u_2 is preferable to u_1. If these numbers are equal the options are considered indiscernible; otherwise, there is strict dominance.

Still another alternative could be taken. It would consist of considering not indiscernible, but incomparable, those pairs of options with contradictory evaluations. They would be taken out of account, and we would evaluate the agreement between the two classifications only among the remaining pairs.

Index of Mutual Information for Preference Relations

Once established the rule to determine joint dominance, and established for each pair of options the relation of indiscernibility, dominance, or incomparability, measures of quality of approximation may be built, as in the case of absence of order. The index of mutual information between an ordinal classification corresponding to a set of decision criteria, D, and the classification generated by a set of condition attributes that are preference criteria, P, will be given by the proportion of pairs for which there is simultaneously indiscernibility according to P and D, or dominance in the same direction according to P and D.

This definition may be also formulated in terms of graphs theory. The sets of criteria D and P now determine two oriented graphs with directed edges if there is strict dominance, and simple edges if there is indiscernibility. For the last variant of the global dominance concept, which excludes that part of the pairs considered incomparable, there is also the case of unconnected nodes. The index of mutual information is again defined by the ratio between the number of pairs of different options with the same kind of link in the two graphs, divided by the total number of pairs presenting some kind of link in each graph. In this case, it will be denoted $G^>(D,P)$.

IMPORTANCE OF ATTRIBUTES FOR THE APPROXIMATION

The importance of any attribute isolated, or set of attributes together, for the approximation, may be evaluated by comparing the values of the indices discussed. Importance values for each attribute A_i may be derived, composing with proper weights, the variations produced in the indices of different sets of attributes by the addition of A_i. Greco, Matarazzo, and Slowinski (1998) propose the use of the weights implicit in Banzhaf's (1965) and Shapley's (1953) approaches. Based on such values and on global values analogously derived for the joint importance of two or more sets of attributes, we may decide which attributes are relevant for the approximation and which may be withdrawn from the set of condition attributes.

Banzhaf's (1965) and Shapley's (1953) values take into account all the parts of the set of condition attributes. This makes their computation complex. Besides, the final result may be distorted by the effect of considering the addition of the attribute evaluated to sets of attributes finally found to be irrelevant for the approximation. If the probability of concordance does not increase as a given condition attribute is added to any set of attributes that will be kept as explanatory attributes, this addition is not necessary and such attribute need not be taken into account. Its influence on the evaluation of the importance of the other attributes is then restricted to revealing the relevance of those attributes to make it dischargeable. This suggests employing a quick selection procedure based on simultaneous exclusion of attributes made irrelevant by other attributes, and definitive inclusion of those attributes that make some other irrelevant.

A procedure to select condition attributes approaching a set of decision attributes D according to these lines will have the following steps:

1. Start ranking the unitary sets of condition attributes according to the most suitable index.
2. Select the condition attribute A_1 with the highest value of the index, and rank the pairs of attributes (A_1, A_j) with $j \neq 1$ according to the value of that index. Exclude from the set of attributes those A_j whose addition to A_1 does not increase the value of the index.
3. Select, among the attributes not withdrawn in Step 2, that A_j with the pair (A_1, A_j) presenting the highest value of the index in the ranking of the previous step, and rank the triples formed by adding to that pair (A_1, A_j)

one attribute still not excluded. Exclude from the set of attributes any attribute forming a triple with a value of the index smaller or equal to that of the starting pair.
4. Proceed to quadruples, including the triple with the highest values of the index among those ranked in the previous step.
5. Proceed, increasing the size of the set in the same way until there are no more attributes to be added.

EXAMPLES

To see how this works, let us consider a few examples with small sets. In the first example, three options, O_1, O_2 and O_3, are evaluated according to three attributes, the condition attributes A_1 and A_2 and the decision attribute D, as shown in Table 1. The three attributes agree in the value 0 to O_1. D gives the value 1 to O_2 and O_3. A_1 gives the value 0 to O_2 and 1 to O_3, while A_2 gives the value 1 to O_2 and 0 to O_3.

Since the three pairs of values of A_1 and A_2 are different, $\gamma(A_1,A_2) = 1$. But, isolated, A_1 and A_2 have a γ of 1/3, because they give the same value, zero, to two options with different values by D. These options will then be in the boundaries.

Isolated, A_1 and A_2 discriminate O_1 from one of the other options, in agreement with D, and discriminate between O_2 and O_3, contradicting D.

Then $G(D,A_1) = G(D,A_2) = 1/3$. But $G(D,\{A_1,A_2\}) = 2/3$, taking into account the different evaluations that A_1 and A_2 give to the pair of options that D does not discern.

If the attributes are thought as preference criteria, the value of the index γ for A_1 and A_2 isolated does not change, because A_1 and A_2 are nondecreasing with D. The same happens for A_1 and A_2 taken together.

The index of mutual information for the case of preference relations will not be affected by the choice of considering options with contradictory evaluations as indiscernible or incomparable. In the first case, we ought to take into account all the three pairs of options; in the second case, only two. But, in any of these instances, $G^>(D,\{A_1,A_2\}) = 1$. And if we consider each condition attribute isolated, we still have for $G^>(D,A_1)$ and $G^>(D,A_2)$ the same value, 1/3.

Summarizing, there is only one difference in this case between the index γ and the index of mutual information. This happens in the case of joint evaluation of A_1 and A_2 in absence of dominance. And, assuming the relations as of dominance or not, all indices increase as the two attributes are taken together, confirming that the set of condition attributes cannot be reduced without loosing the quality of the approximation to D.

In the second example, there are six options, one decision attribute, D, and two condition attributes, A_1 and A_2. Their values are given in Table 2.

Table 1. Three attributes for three options

Options	Condition Attributes		Decision Attribute
	A_1	A_2	D
O_1	0	0	0
O_2	0	1	1
O_3	1	0	1

Table 2. Three attributes for six options

Options	Condition Attributes		Decision Attribute
	A_1	A_2	D
O_1	0	0	0
O_2	0	1	0
O_3	1	1	0
O_4	0	0	1
O_5	1	0	1
O_6	1	1	1

Each attribute divides the set of options into two subsets of equal size. A_1 agrees with D in two of the three elements of each of the subsets determined by D and A_2 in only one. Notwithstanding, $\gamma(A_1) = \gamma(A_2) = 0$. We will have $\gamma(A_1,A_2) = 1/3$, because, although O_1 and O_4, as well as O_3 and O_6, have the same evaluation according to both condition attributes and different evaluations according to the decision attribute, O_2 and O_5 distinguish the classes of equivalences of D. It is interesting to notice that evaluating O_2 and O_5, A_1 and A_2 fully contradict each other.

Finally, if the attributes are thought as preference criteria, then $\gamma(A_1) = \gamma(A_2) = \gamma(A_1,A_2) = 0$, because, in this case, O_2 dominates O_4 and O_3 dominates O_5 according to the condition attributes, the contrary happening according to the decision attribute.

Summarizing, taking the classical approach, we would make no difference between A_1 and A_2 and, in the case of preference relations, would be advised to disregard both.

To compute the index of mutual information, we ought to examine a set of 15 pairs of options. In six of these pairs, there is agreement according to D. Isolated, A_1 and A_2 agree in two of such pairs (A_1 gives the same value to O_1 and O_2 and to O_5 and O_6 and A_2 to O_2 and O_3 and to O_4 and O_5). If these attributes are taken together, this number falls to zero. For the nine pairs with different evaluations by the decision attribute, A_1 and A_2 isolated distinguish the elements of five pairs. If they are applied together, from these nine pairs, only (O_1,O_4) and (O_3,O_6) are pairs of indiscernible options. Thus, $G(D,A_1) = G(D,A_2) = G(D,\{A_1,A_2\}) = 7/15$.

If order is taken into account, slightly different values result for the index of mutual information for A_1 and A_2 together, according as we consider O_2 and O_5 indiscernible or take them out of account. $G^>(D,\{A_1,A_2\})$ will be equal to 3/14 in this last case, and to 3/15 = 1/5 in the first. The important point here is that the index captures the difference between A_1 and A_2. In fact, $G^>(D,A_1) = 2/5$, while $G^>(D,A_2) = 1/5$. The pairs (O_1,O_5), (O_2,O_5), and (O_2,O_6) answer for this difference.

Example 2 clarifies the usefulness of taking into account order in the dominance criteria instances. If order is not considered, A_1 and A_2 are of equal importance, the equal values of the index would advise to choose one of them indifferently. If order is taken into account, we find a clear recommendation to discard A_2 and employ A_1 alone.

In Example 3, we face the problem brought by larger variability. The attributes are kept two valued, but we have now sets of four condition and three decision attributes. The number of options increases to eight, with data in Table 3.

The three first condition attributes are enough to make discernible the eight options. Thus, any set containing them, that means $\{A_1,A_2,A_3\}$ and $\{A_1,A_2,A_3,A_4\}$, will necessarily present a value of 1 for the index γ.

The decision partition determined by the set of decision attributes $D = \{D_1,D_2,D_3\}$ is $\{\{O_1,O_2\},\{O_3,O_4\},\{O_5,O_6\},\{O_7,O_8\}\}$, a set of four pairs of indiscernible options. And these four decision equivalence classes are in an increasing order if the decision attributes are taken as preference criteria. A_1 just introduces distinction inside the decision equivalence classes. A_2 makes the same, agreeing with A_1 in the classes (O_1,O_2) and (O_5,O_6) and inverting A_1 in (O_3,O_4) and (O_7,O_8). A_3 distinguishes the two first classes from the two last ones and A_4 distinguishes the first class from the three last ones. Thus, isolated, all attributes, except A_4 with a γ of 0,25, have γ equal to zero.

The pairs of attributes not including A_4 also have γ equal to zero, while those including A_4 have γ equal to 0,25. This is due again to the first decision class, (O_1,O_2) presenting an empty boundary anytime A_4 is considered, and all the options in the other classes, according to D, always including options in more than one class according to any pair of condition attributes.

Finally, the sets of three attributes including A_4 have a γ of 05, due to the presence of options presenting the same evaluations according to the three condition attributes in two distinct decision classes, the two last ones.

If the attributes are thought as preference criteria, 1 meaning higher preference than 0, the index γ for the attributes isolated or for the sets of two of attributes, except $\{A_1,A_4\}$ and $\{A_2,A_4\}$, will remain unchanged, with the value 0,5 for $\{A_3,A_4\}$, 0,25 for A_4 alone, and null in the other cases. But for any other combination of attri-

Table 3. Seven attributes for eight options

Options	Condition Attributes				Decision Attributes		
	A_1	A_2	A_3	A_4	D_1	D_2	D_3
O_1	0	0	0	0	0	0	0
O_2	1	1	0	0	0	0	0
O_3	0	1	0	1	1	0	0
O_4	1	0	0	1	1	0	0
O_5	0	0	1	1	1	1	0
O_6	1	1	1	1	1	1	0
O_7	0	1	1	1	1	1	1
O_8	1	0	1	1	1	1	1

butes, the value of γ will be reduced to 0,125. In fact, for these combinations only O_1 will not be a boundary option. This fall reflects the distinctions established by A_1 and A_2 inside the classes of indiscernibility of D.

These evaluations show that to inform about the decision classes determined by D, the attributes A_1 and A_2 may be discharged, and the set of four condition attributes can be reduced to $\{A_3, A_4\}$ whether order is taken into account or not.

To obtain the indices of mutual information we must look for concordances and discordances in the 28 pairs of options. Table 4 presents the values of this index for all nonempty subsets of the set of attributes, first assuming simple relations, that is, absence of order considerations, and then assuming dominance. In this last case, contradictory evaluations are counted as indiscernibility to make easier the comparison to values of γ.

We can see in Table 4 that now the value of the index related to A_3 alone is not null. In fact, $G(D,A_3) > G(D,A_4)$ and $G^{\succ}(D,A_3) > G^{\succ}(D,A_4)$, taking into account that A_3 discriminates a larger number of distinct classes in the decision classification. We can see also that the fall associated to the addition of A_1 and A_2 is not as high as in the values of γ.

Applying the selection procedure to reduce the set of attributes, again A_3 and A_4 will be found to be enough to inform about the decision classification, and A_1 and A_2 will be found to be dischargeable.

PROBABILISTIC TRANSFORMATION

As the number of options increase, different sets of condition attributes may approach, better than others, the decision attributes in different sections of the universe of options. This may reduce the efficiency of the selection procedure designed, which pays more attention to isolated attributes and to combinations of a small number of attributes. Nevertheless, some of such sections may involve small differences or differences unimportant for the main decisions to be taken.

Specifically, in the case of preference relations, the ranks according to the different criteria, as well as, in many cases, the measurements in the scale in which the attributes of interest appear naturally, provide many contradictory evaluations that may be disregarded. For instance, if the goal is the choice of the best option, a transformation of variables, placing in the same class options of

Table 4. Indices of mutual information for subsets of condition attributes

Attribute	Simple	Dominance	Attributes	Simple	Dominance	Attributes	Simple	Dominance
A_1	0.43	0.21	A_1A_2	0.71	0.39	$A_1A_2A_3$	0.86	0.54
A_2	0.43	0.21	A_1A_3	0.71	0.50	$A_1A_2A_4$	0.71	0.43
A_3	0.71	0.71	A_2A_3	0.71	0.50	$A_1A_3A_4$	0.79	0.57
A_4	0.57	0.57	A_1A_4	0.64	0.43	$A_2A_3A_4$	0.79	0.57
			A_2A_4	0.64	0.43	$A_1A_2A_3A_4$	0.86	0.54
			A_3A_4	0.86	0.86			

lower preference will simplify the task of rough set analysis, while better representing the problem. One such transformation is that on probabilities of reaching the position of highest preference (Sant'Anna & Sant'Anna, 2001).

Taking into account the possibility of disturbances affecting the evaluations, this approach treats the measurements or the ranks, according to each attribute, as estimates for location parameters of random distributions. Then, assuming classical hypothesis of independent disturbances, each vector of observations of an attribute is translated into a vector of probabilities of being chosen according to such attribute.

The calculus of the probability of reaching the position of highest preference involves measuring distances between all the production units, not only distances of each one to the boundary. This, besides making this procedure resistant to the influence of random disturbances, results in very small probabilities for most options, in such a way that, with suitable granularity, they become indiscernible.

For instance, in Example 3, the decision attributes rank the eight options in four classes of two options each. The vector of probabilities of presenting the highest preference evaluation derived from a vector of preferences (1, 1, 2, 2, 3, 3, 4, 4), assuming uniform disturbances with range 3 (the lowest value for the range parameter that allows for an observed value of the least preferred options tying an observed value of the most preferred), is (0, 0, 0, 0, 0.08, 0.08, 0.42, 0.42). Assuming normal distributions with the standard deviation corresponding to an expected range of 3 in samples of size 8, this vector of probabilities would be (0, 0, 0.02, 0.02, 0.11, 0.11, 0.37, 0.37). Similar results will be obtained, assuming other forms for the distributions, in such a way that, in any case, taking a granulation by multiples of 5%, options O_1, O_2, O_3 and O_4 will form a decision class of four least preferred options.

Computing the index of mutual information between the decision classification with three classes in increasing order $D = \{\{O_1, O_2, O_3, O_4\}, \{O_5, O_6\}, \{O_7, O_8\}\}$, and the two-valued condition attributes of Example 3 taken as preference criteria with 1 denoting higher preference, we obtain $G^2(D, A_3) = 24/28$ against $G^2(D, A_4) = G^2(D, \{A_3, A_4\}) = 16/28$. Thus, any reduction procedure leads now to exclude A_1, A_2 and A_4, leaving A_3 as the only relevant variable. The index γ, by its turn, does not take into account the distinction that A_4 determines inside the class $\{O_1, O_2, O_3, O_4\}$, and will give the same value for A_3 and A_4.

Probabilistic Scoring

The transformation into probabilities of being the preferred option, according to each criterion, provides direct ways to combine, into a global score, evaluations derived from different attributes. A decision strategy may consist in simply choosing the option with the highest probability of reaching maximal preference according to at least one of the criteria, or, alternatively, with the largest probability of being the preferred according to all of them.

In computing these joint probabilities, to avoid the need of taking into the computation the possible correlations, it is interesting to exclude criteria that repeat information brought by other criteria. Sant'Anna (2004) proposes to remake the calculus of the joint probability after a rough-set analysis, with the joint probability as the decision criterion, has reduced the set of criteria initially combined. An application of such strategy, employing Banzhaf and Shapley values, was made in Sant'Anna and Sant'Anna (2006) to formulate a methodology for efficiency evaluation.

Before applying the rough-sets analysis, some rounding of the probabilities is needed to reduce the number of classes to a manageable size. This granulation process will efface differences in the factors that, nevertheless, when cumulated, may become important in differentiating according to the decision variable. A consequence of this is null values for γ. On the other side, in this context it is important to value variations in particular factors that may be effaced when combining them to the others. The symmetric treatment of differences in decision and condition attributes makes the index of mutual information sensitive also to these last kinds of variations, which are not considered in the computation of γ.

Applying directly the index of mutual information, a simple strategy to the selection of factors in a combination of probabilistic criteria would be the following:

1. Compute the global probabilistic scores according to whatever composition algorithm is considered the best for the kind of application under consideration.
2. Round the values of such global scores and of the probabilities of being the preferred option according to each attribute isolated to assure suitable granulation.
3. Obtain the values of index of mutual information between the classification according to the global scores, taken as the decision attribute, and the classification according to the probabilities related to individual attributes, pairs of attributes and so on, as needed to apply the procedure designed to select relevant attributes, and apply it to reduce the number of factors.
4. Remake the computation of global scores, now employing only the variables found relevant in the previous stage.
5. Repeat the process until the reduction procedure finds all attributes relevant to the global score entering as decision attribute in the iteration.

CONCLUSION

The index of mutual information proposed in this chapter takes, symmetrically, into account, variations in the classifications according to the decision and condition attributes. Counting all kinds of concordances and discordances, it processes a larger volume of information than the indices based on the size of the boundaries of the decision classes. This results not only in a more informative evaluation but also in more stable measures.

The index of mutual information is also suitable to the case of attributes that are preference criteria so that order should be taken into account. In that case it can be applied after a previous transformation into probabilities of reaching the position of preferred option improves the granularity. Conversely, the procedure to select the most important attributes can be applied to simplify the computation of probabilistic scores of global preference. The new feature of this application is that the decision classification to be explained changes at each iteration.

Strong resistance to accidental discordances makes the index of mutual information reliable when applied in strategies to reduce the number of attributes. This enables us to employ it in a quick selection procedure to that end. The examples presented in this chapter compare the index of mutual information with the index of quality of information of Pawlak in that context.

Such a selection procedure may employ different indices. But the approach of counting concordances and discordances in the comparison of pairs of options allows improving it in ways not conceivable if it is based on comparing the size of boundaries. For instance, the choice of the new attribute to be added in successive steps, as well as the choice of the attributes to be excluded, may take into account only the pairs of options for which the attributes already included contradict the decision attributes. Advantages of such strategy are that it would make clear the gain in each step, and easier to follow the forward orientation in the explanation generated.

REFERENCES

Banzhaf, J. F. (1965). Weighted voting doesn't work: A mathematical analysis. *Rutgers Law Review, 19*, 317-343.

Gediga, G., & Düntsch, I. (2003). On model evaluation, indices of importance and interaction values in rough set analysis. In S.K. Pal, L. Polkowski, & A. Skowron (Eds.), *Rough-neuro computing: A way for computing with words* (pp. 251-276). Heidelberg: Physica Verlag.

Greco, S., Matarazzo, B., & Slowinski, R. (1998). Fuzzy measures as a technique for rough set analysis. In *Proceedings of the 6th European Congress on Intelligent Techniques & Soft Computing, Aachen, 1* (pp. 99-103).

Greco, S., Matarazzo, B., & Slowinski, R. (1999). Rough approximation of a preference relation by dominance relations. *European Journal of Operational Research, 117*, 63-83.

Pawlak, Z. (1982). Rough sets. *International Journal of Computer and Information Sciences, 11*(5), 341-356.

Pawlak, Z. (1991). *Rough sets: Theoretical aspects of reasoning about data*. Dordrecht: Kluwer.

Sant'Anna, A. P. (2004). Rough sets in the probabilistic composition of preferences. In B. de Baets, R. de Caluwe, G. de Tré, J. Fodor, J. Kacprzyk, & S. Sadrosny (Eds.), *Current issues in data and knowledge engineering* (pp. 407-414). Warszawa: EXIT.

Sant'Anna, L. A. F. P., & Sant'Anna, A. P. (2001). Randomization as a stage in criteria combining. In F. S. Fogliato, J. L. D. Ribeiro, & L. B. M. Guimarães (Eds.), *Production and distribution challenges for the 21st century*, (pp. 248-256). Salvador: ABEPRO.

Sant'Anna, L. A. F. P., & Sant'Anna, A. P. (in press). A probabilistic approach to evaluate the exploitation of the geographic situation of hydroelectric plants. *Annals of ORMMES 2006*.

Shapley, L. (1953). A value for n-person games. In H. Kuhn, & A. Tucker, *Contributions to the theory of games II* (pp. 307-317). Princeton: Princeton University Press.

Chapter VIII
Extended Action Rule Discovery Based on Single Classification Rules and Reducts

Zbigniew W. Ras
University of North Carolina at Charlotte, USA

Elzbieta M. Wyrzykowska
University of Information Technology and Management, Poland

ABSTRACT

Action rules can be seen as logical terms describing knowledge about possible actions associated with objects that are hidden in a decision system. Classical strategy for discovering them from a database requires prior extraction of classification rules that next are evaluated, pair by pair, with a goal to build a strategy of action based on condition features, in order to get a desired effect on a decision feature. An actionable strategy is represented as a term $r = [(\omega) \wedge (\alpha \rightarrow \beta)] \Rightarrow [\phi \rightarrow \psi]$, where ω, α, β, ϕ, and ψ are descriptions of events. The term r states that when the fixed condition ω is satisfied and the changeable behavior $(\alpha \rightarrow \beta)$ occurs in objects represented as tuples from a database, so does the expectation $(\phi \rightarrow \psi)$. With each object, a number of actionable strategies can be associated, and each one of them may lead to different expectations and the same to different reclassifications of objects. This chapter will focus on a new strategy of construction of action rules directly from single classification rules instead of pairs of classification rules. This way we do not only gain on the simplicity of the method of action rules construction, but also on its time complexity. The chapter will present a modified tree-based strategy for constructing action rules, followed by a new simplified strategy of constructing them. Finally, these two strategies will be compared.

INTRODUCTION

There are two aspects of interestingness of rules that have been studied in data mining literature, objective and subjective measures (Adomavicius & Tuzhilin, 1997; Liu, Hsu, Chen, 1997; Silberschatz & Tuzhilin, 1995, 1996). Objective measures are data driven and domain independent.

Generally, they evaluate the rules based on their quality and similarity between them. Subjective measures, including unexpectedness, novelty, and actionability, are user driven and domain dependent.

The notion of an action rule, constructed from certain pairs of association rules, has been proposed in Ras and Wieczorkowska (2000). Its different definition was given earlier in Geffner and Wainer (1998). Also, interventions introduced in Greco, Matarazzo, Pappalardo, and Slowinski (2005) are conceptually very similar to action rules. Action rules have been investigated further in Tsay and Ras (2005, 2006), Tzacheva and Ras (2005), and Ras and Dardzinska (2006). To give an example justifying the need of action rules, let us assume that a number of customers have closed their accounts at one of the banks. We construct possibly the simplest description of that group of people and next search for a new description, similar to the one we have, with a goal to identify a new group of customers from which no one left that bank. If these descriptions have a form of rules, then they can be seen as actionable rules. Now, by comparing these two descriptions, we may find the cause why these accounts have been closed, and formulate an action that, if undertaken by the bank, may prevent other customers from closing their accounts. For example, an action rule may say that by inviting people from a certain group of customers for a glass of wine by the bank, it is almost guaranteed that these customers will not close their accounts and they do not move to another bank. Sending invitations by regular mail to all these customers, or inviting them personally by giving them a call, are examples of an action associated with that action rule.

In Tzacheva and Ras (2005), the notion of a cost and feasibility of an action rule was introduced. The cost is a subjective measure and feasibility is an objective measure. Usually, a number of action rules or chains of action rules can be applied to reclassify a certain set of objects. The cost associated with changes of values within one attribute is usually different than the cost associated with changes of values within another attribute. The strategy for replacing the initially extracted action rule by a composition of new action rules, dynamically built and leading to the same reclassification goal, was proposed in Tzacheva and Ras (2005). This composition of rules uniquely defines a new action rule. Objects supporting the new action rule also support the initial action rule, but the cost of reclassifying them is lower or even much lower for the new rule. In Ras and Dardzinska (2006), authors propose a new simplified strategy for constructing action rules. In this chapter, we present an algebraic extension of this method, and show the close correspondence between the rules generated by tree-based strategy (Tsay & Ras, 2005) and rules constructed by this newest method.

BACKGROUND

In the paper by Ras and Wieczorkowska (2000), the notion of an action rule was introduced. The main idea was to generate, from a database, special types of rules that basically form a hint to users showing a way to reclassify objects with respect to some distinguished attribute (called a decision attribute). Values of some attributes used to describe objects stored in a database can be changed, and this change can be influenced and controlled by the user. However, some of these changes (for instance "profit") cannot be done directly to a decision attribute. In such a case, definitions of this decision attribute in terms of other attributes (called classification attributes) have to be learned. These new definitions are used to construct action rules, showing what changes in values of some attributes for a given class of objects are needed to reclassify objects the way users want. But users may still be either unable or unwilling to proceed with actions leading to such changes. In all such cases, we may search for definitions of values of any classification attribute listed in an action rule. By replacing a value of such an attribute by its definition, extracted either locally or at remote sites (if system is distributed), we construct new action rules that might be of more interest to users than the initial rule (Tzacheva & Ras, 2005).

We start with a definition of an information system given in Pawlak (1991).

By an information system, we mean a pair S = (U, A), where:

1. U is a nonempty, finite set of objects (object identifiers),
2. A is a nonempty, finite set of attributes, that is, $a:U \to V_a$ for $a \in A$, where V_a is called the domain of a.

Information systems can be seen as decision tables. In any decision table, together with the set of attributes, a partition of that set into conditions and decisions is given. Additionally, we assume that the set of conditions is partitioned into stable and flexible (Ras & Wieczorkowska, 2000).

Attribute $a \in A$ is called stable for the set U if its values, assigned to objects from U, cannot be changed in time; otherwise, it is called flexible. "Place of birth" is an example of a stable attribute. "Interest rate" on any customer account is an example of a flexible attribute. For simplicity reasons, we consider decision tables with only one decision. We adopt the following definition of a decision table:

By a decision table, we mean an information system $S = (U, A_{St} \cup A_{Fl} \cup \{d\})$, where $d \notin A_{St} \cup A_{Fl}$ is a distinguished attribute called the decision.

The elements of A_{St} are called stable conditions, whereas the elements of $A_{Fl} \cup \{d\}$ are called flexible. Our goal is to change values of attributes in A_{Fl} for some objects in U so the values of the attribute d for these objects may change as well. Certain relationships between attributes from $A_{St} \cup A_{Fl}$ and the attribute d will have to be discovered first.

By Dom(r) we mean all attributes listed in the IF part of a rule r extracted from S. For example, if r = [(a1,3)*(a2,4) → (d,3)] is a rule, then Dom(r) = {a1,a2}. By d(r) we denote the decision value of rule r. In our example d(r) = 3.

If r1, r2 are rules and $B \subseteq A_{Fl} \cup A_{St}$ is a set of attributes, then r1/B = r2/B means that the conditional parts of rules r1, r2 restricted to attributes B are the same.

For example if r1 = [(a1,3) → (d,3)], then r1/{a1} = r/{a1}.

Assume also that (a, v → w) denotes the fact that the value of attribute a has been changed from v to w. Similarly, the term (a, v → w)(x) means that a(x)=v has been changed to a(x)=w; in other words, the property (a,v) of an object x has been changed to property (a,w). Assume now that rules r1, r2 have been extracted from S and r1/[Dom(r1)∩Dom(r2)∩A_{St}]= r2/[Dom(r1)∩Dom(r2)∩A_{St}], d(r1)=k1, d(r2)=k2. Also, assume that (b1, b2,..., bp) is a list of all attributes in Dom(r1) ∩ Dom(r2) ∩ A_{Fl} on which r1, r2 differ and r1(b1)= v1, r1(b2)= v2,..., r1(bp)= vp, r2(b1)= w1, r2(b2)= w2,..., r2(bp)= wp.

By (r1,r2)-action rule on $x \in U$ we mean a statement r:

Table 1.

a (St)	b (Fl)	c (St)	e (Fl)	g (St)	h (Fl)	d (Decision)
a1	b1	c1	e1			d1
a1	b2			g2	h2	d2

$[r2/A_{St} \wedge (b1, v1 \to w1) \wedge (b2, v2 \to w2) \wedge ... \wedge (bp, vp \to wp)](x) \Rightarrow [(d, k1 \to k2)](x)$.

Object $x \in U$ supports action rule r, if x supports r1, $(\forall a \in Dom(r2) \cap A_{St}) [a(x) = r2(a)]$, $(\forall i \le p)[bi(x)=vi]$, and $d(x)=k1$. The set of all objects in U supporting r is denoted by $U^{<r>}$.

To define an extended action rule (Ras & Tsay, 2003), let us assume that two classification rules are considered. We present them in Table 1 to better clarify the process of constructing extended action rules. Here, "St" means stable classification attribute and "Fl" means flexible one.

In a classical representation, these two rules have a form:

r1 = [a1 ∧ b1 ∧ c1 ∧ e1 → d1] , r2 = [a1 ∧ b2 ∧ g2 ∧ h2 → d2].

It is easy to check that [[a1 ∧ g2 ∧ (b, b1 → b2)] ⇒ (d, d1 → d2)] is the (r1,r2)-action rule.

Assume now that object x supports rule r1, which means that x is classified as d1. In order to reclassify x to class d2, we need to change its value b from b1 to b2, but also we have to require that g(x)=g2 and that the value h for object x has to be changed to h2. This is the meaning of the extended (r1,r2)-action rule r given next:

[[a1 ∧ g2 ∧ (b, b1 → b2) ∧ (h, → h2)] ⇒ (d, d1→ d2)].

Let us observe that this extended action rule can be replaced by a class of new action rules. First, we need to define a new relation $\approx_h$ on the set $U^{<r>}$ as:

$x \approx_h y$ iff $h(x)=h(y)$, for any $x, y \in U^{<r>}$.

Now, let us assume that $V_h = \{h1,h2,h3\}$ and $U^{<r,hi>} = U^{<r>}/\approx_h = \{y \in U^{<r>}: h(y)=hi\}$, for any i=1,2,3. The extended action rule r can be replaced by two action rules:

[[a1 ∧ g2 ∧ (b, b1 → b2) ∧ (h, h1→ h2)] ⇒ (d, d1→ d2)] with supporting set $U^{<r,h1>}$

and

[[a1 ∧ g2 ∧ (b, b1 → b2) ∧ (h, h3→ h2)] ⇒ (d, d1→ d2)] with supporting set $U^{<r,h3>}$.

This example shows that extended action rules can be seen as generalizations of action rules. Also, it gives us a hint of how to look for the most compact representations of action rules.

Main Thrust of the Chapter

1. Issues, Controversies, Problems

In this section, we present a modification of the action-tree algorithm (Tsay & Ras, 2005) for discovering action rules. Namely, we partition the set of classification rules R discovered from a decision system $S = (U, A_{St} \cup A_{Fl} \cup \{d\})$, where A_{St} is the set of stable attributes, A_{Fl} is the set of flexible attributes, and, $V_d = \{d_1, d_2,..., d_k\}$ is the set of decision values, into subsets of rules having the same values of stable attributes in their classification parts and defining the same value of the decision attribute. Classification rules can be extracted from S using, for instance, discovery system LERS (Grzymala-Busse, 1997).

Action-tree algorithm for discovering extended action rules from a decision system S is as follows:

Build Action Tree:

Step 1.
- Partition the set of classification rules R in a way that two rules are in the same class if their stable attributes are the same.
- Find the cardinality of the domain V_{vi} for each stable attribute v_i in S.

- Take v_i which card(V_{vi}) is the smallest, as the splitting attribute, and divide R into subsets, each of which contains rules having the same value of the stable attribute v_i.
- For each subset obtained in step 2, determine if it contains rules of different decision values and different values of flexible attributes. If it does, go to step 2. If it does not, there is no need to split the subset further, and we place a mark.

Step 2.
- Partition each resulting subset into new subsets, each of which contains only rules having the same decision value.

Step 3.
- Each leaf of the resulting tree represents a set of rules that do not contradict on stable attributes, and also it uniquely defines decision value d_i. The path from the root to that leaf gives the description of objects supported by these rules.

Generate Extended Action Rules

- Form extended action rules by comparing all unmarked leaf nodes of the same parent.

The algorithm starts at the root node of the tree, called action tree, representing all classification rules extracted from S. A stable attribute is selected to partition these rules. For each value of that attribute, an outgoing edge from the root node is created, and the corresponding subset of rules that have the attribute value assigned to that edge is moved to the newly created child node. This process is repeated recursively for each child node. When we are done with stable attributes, the last split is based on a decision attribute for each current leaf of action tree. If at any time all classification rules representing a node have the same decision value, then we stop constructing that part of the tree. We still have to explain which stable attributes are chosen to split classification rules representing a node of action tree. The algorithm selects any stable attribute that has the smallest number of possible values among all the remaining stable attributes. This step is justified by the need to apply a heuristic strategy (Ras, 1999) that will minimize the number of edges in the resulting tree and at the same time, make the complexity of the algorithm lower.

We have two types of nodes: a leaf node and a nonleaf node. At a nonleaf node, the set of rules is partitioned along the branches, and each child node gets its corresponding subset of rules. Every path to the decision attribute node, one level above the leaf node, represents a subset of the extracted classification rules when the stable attributes have the same value. Each leaf node represents a set of rules that do not contradict on stable attributes, and also define decision value d_i. The path from the root to that leaf gives the description that objects supporting these rules have to satisfy.

Instead of splitting the set of rules R by stable attributes and next by the decision attribute, we can also start the partitioning algorithm from a decision attribute. For instance, if a decision attribute has three values, we get three initial

Table 2.

	a	b	c	d
x_1	2	1	2	L
x_2	2	1	2	L
x_3	1	1	0	H
x_4	1	1	0	H
x_5	2	3	2	H
x_6	2	3	2	H
x_7	2	1	1	L
x_8	2	1	1	L
x_9	2	2	1	L
x_{10}	2	3	0	L
x_{11}	1	1	2	H
x_{12}	1	1	1	H

subtrees. In the next step of the algorithm, we start splitting these subtrees by stable attributes, following the same strategy as the one presented for action trees. This new algorithm is called action-forest algorithm.

Now, let us take Table 2 as an example of a decision system S. Attributes a, c are stable, and b, d are flexible. Assume now that our plan is to reclassify some objects from the class (d,L) into the class (d,H).

Table 3 shows the set of classification rules extracted from Table 2 by LERS algorithm (Grzymala-Busse, 1997). The first column presents sets of objects supporting these rules.

First, we represent the set R of certain rules extracted from S as a table (see Table 3). The first column of this table shows objects in S supporting the rules from R (each row represents a rule). For instance, the second row represents the rule $[[(a,2) \wedge (c, 1)] \rightarrow (d, L)]$. The construction of an action tree starts with the set R as a table (see Table 3) representing the root of the tree T_1 in Fig. 1. The root node selection is based on a stable attribute, with the smallest number of values among all stable attributes. The same strategy is used for a child node selection. After labeling the nodes of the tree by all stable attributes, the tree is split, based on the value of the decision attribute. Referring back to the example in Table 3, we use stable attribute a to split that table into two subtables defined by values {1, 2} of attribute a. The domain V_a of attribute a is {1, 2}. Since $card[V_a] < card[V_c]$, then we partition the table into two: one table with rules containing the term (a,1) and another with rules containing the term (a,2). Corresponding edges are labeled by values of attribute a. All rules in the subtable T_2 have the same decision value. So, action rules cannot be constructed from subtable T_2, which means it is not divided any further. Because rules in the subtable T_3 contain different decision values and a stable attribute c, T_3 is partitioned into three subtables, one with rules containing the term (c, 0), one with rules containing (c, 1), and one with rules containing (c, 2). Now, rules in each of the subtables do not contain any stable attributes. Subtable T_6 is not split any further for the same reason as subtable T_2. All objects in subtable T_4 have the same value of flexible attribute b. So, there is no way to construct a workable strategy from this subtable, which means it is not partitioned any further. Subtable T_5 is divided into two new subtables. Each leaf represents a set of rules that do not contradict on stable attributes, and also define decision value d_1.

The path from the root of the tree to that leaf gives the description of objects supported by these rules. Following the path described by the term $(a, 2) \wedge (c, 2) \wedge (d, L)$, we get table T_7. Following the path described by the term $(a, 2) \wedge (c, 2) \wedge (d, H)$, we get table T_8. Because T_7 and T_8 are sibling nodes, we can directly compare pairs of rules belonging to these two tables and construct an action rule $[[(a,2) \wedge (b, 1 \rightarrow 3)] \Rightarrow (d, L \rightarrow H)]$.

The action-tree algorithm proposed in this section requires the extraction of all classification rules from the decision system before any action rule is constructed. Additionally, the strategy of action rules extraction presented has $O(k^2)$ complexity in the worth case, where k is the number of classification rules. The question is, if any action rule can be constructed from a single classification rule by a strategy that guarantees the same time complexity of action rules construction as the time complexity of classification rules discovery?

Table 3.

Objects	a	b	c	d
$\{x_3, x_4, x_{11}, x_{12}\}$	1			H
$\{x_1, x_2, x_7, x_8\}$	2		1	L
$\{x_7, x_8, x_9\}$	2		0	L
$\{x_3, x_4\}$		1	0	H
$\{x_5, x_6\}$		3	2	H

Extended Action Rule Discovery Based on Single Classification Rules and Reducts

2. **Solutions and recommendations**. Let us assume again that $S = (U, A_{St} \cup A_{Fl} \cup \{d\})$ is a decision system, where $d \notin A_{St} \cup A_{Fl}$ is a distinguished attribute called the decision. The elements of A_{St} are called stable conditions, whereas the elements of $A_{Fl} \cup \{d\}$ are called flexible. Assume that $d_1 \in V_d$ and $x \in U$. We say that x is a d_1-object if $d(x)=d_1$. We also assume that $\{a_1, a_2, ..., a_p\} \subseteq A_{Fl}$, $\{b_1, b_2, ..., b_q\} \subseteq A_{St}$, $a_{[i,j]}$ denotes a value of attribute a_i, $b_{[i,j]}$ denotes a value of attribute b_i, for any i, j and that:

$$r = [[a_{[1,1]} \wedge a_{[2,1]} \wedge ... \wedge a_{[p,1]}] \wedge [b_{[1,1]} \wedge b_{[2,1]} \wedge ..... \wedge b_{[q,1]}] \to d_1]$$

is a classification rule extracted from S supporting some d_1-objects in S. By sup(r) and conf(r), we mean the support and the confidence of r, respectively. Class d_1 is a preferable class, and our goal is to reclassify d_2-objects into d_1 class, where $d_2 \in V_d$.

By an extended action rule $r_{[d2 \to d1]}$ associated with r and the reclassification task $(d, d_2 \to d_1)$ we mean the following expression:

$$r_{[d2 \to d1]} = [[a_{[1,1]} \wedge a_{[2,1]} \wedge ... \wedge a_{[p,1]}] \wedge [(b_1, \to b_{[1,1]}) \wedge (b_2, \to b_{[2,1]}) \wedge ..... \wedge (b_q, \to b_{[q,1]})]]$$

In a similar way, by an extended action rule $r[\to d_1]$ associated with r and the reclassification task $(d, \to d_1)$ we mean the following expression:

$$r_{[\to d1]} = [[a_{[1,1]} \wedge a_{[2,1]} \wedge ... \wedge a_{[p,1]}] \wedge [(b_1, \to b_{[1,1]}) \wedge (b_2, \to b_{[2,1]}) \wedge ..... \wedge (b_q, \to b_{[q,1]})] \, d_1)].$$

The term $[a_{[1,1]} \wedge a_{[2,1]} \wedge ... \wedge a_{[p,1]}]$, built from values of stable attributes, is called the header of the action rule $r[d_2 \to d_1]$, and its values cannot be changed.

The support set of the action rule $r[d_2 \to d_1]$ is defined as:

Figure 1.

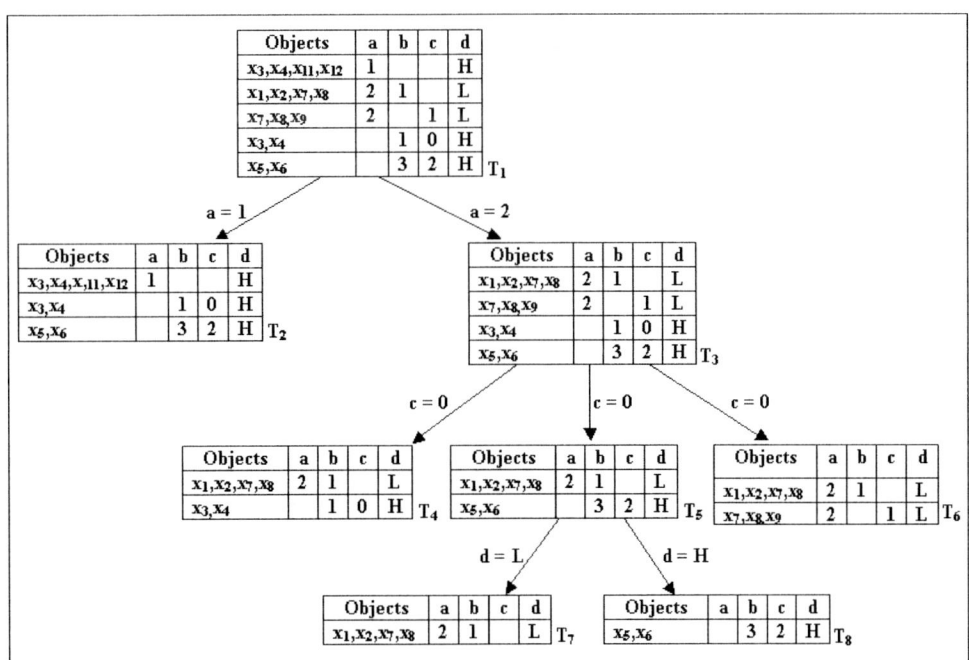

$Sup(r_{[d2 \to d1]}) = \{x \in U: (a_1(x)=a_{[1,1]} \land a_2(x)=a_{[2,1]} \land \ldots \land a_p(x)=a_{[p,1]}) \land (d(x)=d_2)\}$.

Clearly, if $conf(r) \neq 1$, then some objects in S satisfying the description

$[a_{[1,1]} \land a_{[2,1]} \land \ldots \land a_{[p,1]} \land b_{[1,1]} \land b_{[2,1]} \land \ldots \land b_{[q,1]}]$

are classified as d_2. According to the rule $r_{[d2 \to d1]}$ they should be classified as d_1, which means that the confidence of $r_{[d2 \to d1]}$ will get decreased.

If $Sup(r_{[d2 \to d1]}) = \emptyset$, then $r_{[d2 \to d1]}$ cannot be used for reclassification of objects. Similarly, $r[\to d_1]$ cannot be used for reclassification if $Sup(r_{[d2 \to d1]})= \emptyset$ for each d_2 where $d_2 \neq d_1$. From the point of view of actionability, such rules are not interesting (Silberschatz & Tuzhilin, 1995, 1996).

In the following paragraph, we show how to calculate the confidence of action rules and extended action rules. Let $r_{[d2 \to d1]}$, $r'_{[d2 \to d3]}$ be two action rules extracted from S. We say that these rules are p-equivalent ($\approx$), if the following condition holds for every $b_i \in A_{Fl} \cup A_{St}$:

if r/b_i, r'/b_i are both defined, then $r/b_i = r'/b_i$.

Now, we explain how to calculate the confidence of $r_{[d2 \to d1]}$.

Let us take d_2-object $x \in Sup(r_{[d2 \to d1]})$. We say that x positively supports $r_{[d2 \to d1]}$ if there is no classification rule r' extracted from S and describing $d_3 \in V_d$, $d_3 \neq d_1$, which is p-equivalent to r, such that $x \in Sup(r'_{[d2 \to d3]})$. The corresponding subset of $Sup(r_{[d2 \to d1]})$ is denoted by $Sup^+(r_{[d2 \to d1]})$. Otherwise, we say that x negatively supports $r_{[d2 \to d1]}$. The corresponding subset of $Sup(r_{[d2 \to d1]})$ is denoted by $Sup^-(r_{[d2 \to d1]})$.

By the confidence of $r_{[d2 \to d1]}$ in S we mean:

$Conf(r_{[d2 \to d1]}) = [card[Sup^+(r_{[d2 \to d1]})]/card[Sup(r_{[d2 \to d1]})]] \cdot conf(r)$.

Now, let us go back to the definition of an extended action rule $r_{[d2 \to d1]}$ associated with r:

$r_{[d2 \to d1]} = [[a_{[1,1]} \land a_{[2,1]} \land \ldots \land a_{[p,1]}] \land [(b_1, \to b_{[1,1]}) \land (b_2, \to b_{[2,1]}) \land \ldots \land (b_q, \to b_{[q,1]})] \Rightarrow (d, d_2 \to d_1)]$.

In the previous section, we introduced the relation $\approx_{bi}$ defined on the set $U^{<r>}$ as:

$x \approx_{bi} y$ iff $b_i(x)=b_i(y)$, for any $x, y \in U^{<r>}$.

Now, assume that $B = \{b_1, b_2, \ldots, b_q\}$ and $\approx_B = \cap\{\approx_{bi}: b_i \in B\}$. We say that B is b_i-reducible with respect to an extended action rule r in S, if $U^{<r>}/\approx_B = U^{<r>}/\approx_{B-\{bi\}}$, for any $b_i \in B$.

We say that $C \subseteq B$ is a reduct with respect to an extended action rule r if $U^{<r>}/\approx_B = U^{<r>}/\approx_C$ and for any $b_i \notin C$, the set C is not b_i-reducible with respect to an extended action rule r in S.

Theorem

Each reduct with respect to an extended action rule r in S defines uniquely a new extended action rule in S.

Proof (sketch). Assume that:

$r_{[d2 \to d1]} = [[a_{[1,1]} \land a_{[2,1]} \land \ldots \land a_{[p,1]}] \land [(b_1, \to b_{[1,1]}) \land (b_2, \to b_{[2,1]}) \land \ldots \land (b_q, \to b_{[q,1]})] \Rightarrow (d, d_2 \to d_1)]$

is an extended action rule in S and for instance $[B-\{b_1,b_2\}]$ is a reduct with respect to $r_{[d2 \to d1]}$.

It basically means that all objects in $U^{<r>}$ have the same value in a case of attributes b_1, b_2. Let us assume that $b_{[1,i]}$, $b_{[2,j]}$ are these two values. Because $[B-\{b_1,b_2\}]$ is a reduct, then:

$\{x: [a_1(x)=a_{[1,1]} \land a_2(x)=a_{[2,1]} \land \ldots \land a_p(x)=a_{[p,1]}] \land d(x)=d_2 \land [b_1(x)=b_{[1,1]} \land b_2(x)=b_{[2,j]}]\}$ is not an empty set. The rule:

$[[a_{[1,1]} \land a_{[2,1]} \land \ldots \land a_{[p,1]}] \land [(b_1, b_{[1,1]} \to b_{[1,1]}) \land (b_2, b_{[2,j]} \to b_{[2,1]}) \land \ldots \land (b_q, \to b_{[q,1]})] \Rightarrow (d, d_2 \to d_1)]$

is a new extended action rule.

Extended action rules, whose construction is based on reducts, are called optimal.

Future Trends

Assume that d is a decision attribute in S, $d_1, d_2 \in V_d$, and the user would like to reclassify objects in S from the group d_1 to the group d_2. Assuming that the cost of reclassification with respect to each attribute is given, he/she may look for an appropriate action rule, possibly of the lowest cost value, to get a hint which attribute values need to be changed (Tzacheva & Ras, 2005). To be more precise, let us assume that $R_S[(d, d_1 \rightarrow d_2)]$ denotes the set of all action rules in S having the term $(d, d_1 \rightarrow d_2)$ on their decision site. Additionally, we assume that the cost representing the left-hand site of the rule in $R_S[(d, d_1 \rightarrow d_2)]$ is always lower than the cost associated with its right hand site.

Now, among all action rules in $R_S[(d, d_1 \rightarrow d_2)]$, we may identify a rule that has the lowest cost value. But the rule we get may still have the cost value that is much too high to be of any help. Let us notice that the cost of the action rule:

$r = [(b_1, v_1 \rightarrow w_1) \wedge (b_2, v_2 \rightarrow w_2) \wedge ... \wedge (b_p, v_p \rightarrow w_p)] \Rightarrow (d, d_1 \rightarrow d_2)$

might be high only because of the high cost value of one of its subterms in the conditional part of the rule.

Let us assume that $(b_j, v_j \rightarrow w_j)$ is that term. In such a case, we may look for an action rule in $R_S[(b_j, v_j \rightarrow w_j)]$ that has the smallest cost value.

Assume that:

$r_1 = [(b_{j1}, v_{j1} \rightarrow w_{j1}) \wedge (b_{j2}, v_{j2} \rightarrow w_{j2}) \wedge ... \wedge (b_{jq}, v_{jq} \rightarrow w_{jq})] \Rightarrow (b_j, v_j \rightarrow w_j)$

is such a rule. Now, we can compose r with r_1, getting a new action rule given as follows:

$r = [(b_1, v_1 \rightarrow w_1) \wedge (b_2, v_2 \rightarrow w_2) \wedge ... \wedge [(b_{j1}, v_{j1} \rightarrow w_{j1}) \wedge (b_{j2}, v_{j2} \rightarrow w_{j2}) \wedge ... \wedge (b_{jq}, v_{jq} \rightarrow w_{jq})]$
$\wedge ... \wedge (b_p, v_p \rightarrow w_p)] \Rightarrow (d, d_1 \rightarrow d_2)$

Clearly, the cost of this new rule is lower than the cost of r. However, if its support in S gets too low, then such a rule has no value to us. Otherwise, we may recursively follow this strategy, trying to lower the cost needed to reclassify objects from the group d_1 into the group d_2. Each successful step will produce a new action rule whose cost is lower than the cost of the current rule. Obviously, this heuristic strategy always ends. Interestingness of rules is closely linked with their cost. It means that new algorithms showing how too look for rules of the lowest cost are needed. An example of such an algorithm can be found in Tzacheva and Ras (2005).

CONCLUSION

Attributes are divided into two groups: stable and flexible. By stable attributes we mean attributes whose values cannot be changed (for instance, age or maiden name). On the other hand, attributes (like percentage rate or loan approval to buy a house) whose values can be changed are called flexible. Classification rules are extracted from a decision table, using standard KD methods, with preference given to flexible attributes, so mainly they are listed in a classification part of rules. Most of these rules can be seen as actionable rules and the same used to construct action rules. Two methods for discovering extended action rules are presented. The first one, based on action trees, requires comparing pairs of classification rules and, depending on the result of this comparison, an action rule is either built or not. The second strategy shows how to construct extended action rules from a single classification rule. In its first step, the most general extended action rule is built, and next, it is partitioned into a number of atomic expressions representing a new class of extended action rules that jointly represent the initial extended action rule. The first strategy can be seen as bottom-up method, whereas the second strategy is a classical example of a top-down method.

REFERENCES

Adomavicius, G., & Tuzhilin, A. (1997). Discovery of actionable patterns in databases: The action hierarchy approach. In *Proceedings of KDD97 Conference*. Newport Beach, CA: AAAI Press

Geffner, H., & Wainer, J. (1998). Modeling action, knowledge and control. In H. Prade (Ed.), *ECAI 98, 13th European Conference on AI* (pp. 532-536). John Wiley & Sons.

Greco, S., Matarazzo, B., Pappalardo, N., & Slowinski, R. (2005). Measuring expected effects of interventions based on decision rules. *Journal of Experimental and Theoretical Artificial Intelligence, Taylor Francis, 17*(1-2).

Grzymala-Busse, J. (1997). A new version of the rule induction system LERS. *Fundamenta Informaticae, 31*(1), 27-39.

Liu, B., Hsu, W., & Chen, S. (1997). Using general impressions to analyze discovered classification rules. In *Proceedings of KDD97 Conference*. Newport Beach, CA: AAAI Press.

Pawlak, Z. (1991). Information systems—theoretical foundations. *Information Systems Journal, 6*, 205-218.

Ras, Z. (1999). *Discovering rules in information trees* In J. Zytkow & J. Rauch (Eds.), *Principles of Data Mining and Knowledge Discovery, Proceedings of PKDD'99*, Prague, Czech Republic, LNAI, No. 1704 (pp. 518-523). Springer.

Ras, Z.W., & Dardzinska, A. (2006). Action rules discovery, a new simplified strategy. In F. Esposito et al., *Foundations of Intelligent Systems, Proceedings of ISMIS'06*, Bari, Italy, LNAI, No. 4203 (pp. 445-453). Springer.

Ras Z., & Tsay, L.-S. (2003). Discovering extended actions-rules, System DEAR. In Intelligent Information Systems 2003, Advances in software computing. In *Proceedings of IIS2003 Symposium*, Zakopane, Poland (pp. 293-300).

Ras, Z.W., Tzacheva, A., & Tsay, L.-S. (2005). Action rules. In J. Wang (Ed.), *Encyclopedia of data warehousing and mining* (pp. 1-5). Idea Group Inc.

Ras, Z., & Wieczorkowska, A. (2000). Action rules: How to increase profit of a company. In D.A. Zighed, J. Komorowski, & J. Zytkow (Eds.), *Principles of data mining and knowledge discovery, Proceedings of PKDD'00*, Lyon, France, LNAI, No. 1910 (pp. 587-592). Springer.

Silberschatz, A., & Tuzhilin, A., (1995). On subjective measures of interestingness in knowledge discovery. In *Proceedings of KDD'95 Conference*. AAAI Press

Silberschatz, A., & Tuzhilin, A. (1996). What makes patterns interesting in knowledge discovery systems. *IEEE Transactions on Knowledge and Data Engineering, 5*(6).

Tsay, L.-S., & Ras, Z.W. (2005). Action rules discovery system DEAR, method and experiments. *Journal of Experimental and Theoretical Artificial Intelligence, 17*(1-2), 119-128.

Tsay, L.-S., & Ras, Z.W. (2006). Action rules discovery system DEAR3. In F. Exposito et al. (Eds.), *Foundations of Intelligent Systems, Proceedings of ISMIS'06*, Bari, Italy, LNAI, No. 4203 (pp. 483-492). Springer.

Tzacheva, A., & Ras, Z.W. (2005). Action rules mining. *International Journal of Intelligent Systems, 20*(7), 719-736.

Section III
Rough Sets and Hybrid Systems

Chapter IX
Monocular Vision System that Learns with Approximation Spaces

James F. Peters
University of Manitoba, Canada

Maciej Borkowski
University of Manitoba, Canada

Christopher Henry
University of Manitoba, Canada

Dan Lockery
University of Manitoba, Canada

ABSTRACT

This chapter introduces a monocular vision system that learns, with approximation spaces, to control the pan and tilt operations of a digital camera that is tracking a moving target. This monocular vision system has been designed to facilitate inspection by a line-crawling robot that moves along an electric power transmission line. The principal problem considered in this chapter is how to use various forms of reinforcement learning to control movements of a digital camera. Prior work on the solution to this problem was done by Chris Gaskett, using neural Q-learning starting in 1998, and more recently by Gaskett in 2002. However, recent experiments have revealed that both classical targets tracking as well as other forms of reinforcement learning control outperform Q-learning. This chapter considers various forms of the actor critic (AC) method to solve the camera movement control problem. Both the conventional AC method, as well as a modified AC method that has a built-in run-and-twiddle (RT) control strategy mechanism, is considered in this chapter. The RT mechanism, introduced by Oliver Selfridge in 1981, is an action control strategy, where an organism continues what it has been doing while things are improving (increasing action reward), and twiddles (changes its action strategy) when past actions yield diminishing rewards. In this work, RT is governed by measurements (by a critic) of the degree of overlap between past behaviour patterns and a behavior pattern template representing a standard, carried out within the framework provided by approximation spaces, introduced by Zdzisław Pawlak during the early 1980s. This chapter considers how to guide reinforcement learning based on knowledge of acceptable behavior patterns. The contribution of this chapter is an introduction to actor critic learning methods that benefit from approximation spaces in controlling camera movements during target tracking.

INTRODUCTION

The problem considered in this chapter is how to guide action choices by an actor that is influenced by a critic governed by the evaluation of past actions. Specifically, one might ask how to measure the value of an action relative to what has been learned from experience (i.e., from previous patterns of behavior), and how to learn good policies for choosing rewarding actions. The solution to this problem stems from a rough-set approach to reinforcement learning by cooperating agents. It is an age-old adage that experience is a good teacher, and one learns from experience. This is at the heart of reinforcement learning, where estimates of the value of an action are based on past experience.

In reinforcement learning, the choice of an action is based on estimates of the value of a state and/or the value of an action in the current state. A swarm learns the best action to take in each state by maximizing a reward signal obtained from the environment. Two different forms of actor critic (AC) method are investigated in this chapter, namely, a conventional AC method and a form of AC method that includes an adaptive learning strategy, called run—and—twiddle (RT), played out in the context of remembered behavior patterns that accumulate in what are known as ethograms. An *ethogram* is a table of stored behavior patterns (i.e., vectors of measurements associated with behavior features) borrowed from ethology by Tinbergen (1963). Quantitative comparisons of past behavior patterns, with a template representing "normal" or desirable behavior, are carried out within the framework of an approximation space. Approximation spaces were introduced by Zdzislaw Pawlak (1981) during the early 1980s, elaborated by Orlowska (1982), Pawlak (1982), and generalized by Skowron and Stepaniuk (1995) and Stepaniuk (1998). The motivation for considering approximation spaces, as an aid to reinforcement learning, stems from the fact that it becomes possible to derive pattern-based action preferences (see, *e.g.*, Peters & Henry 2005a, 2005b).

Prior work on the reinforcement learning control was done by Chris Gaskett (2002), using neural Q-learning, starting in 1998. However, recent experiments have revealed that both classical target tracking, as well as other forms of reinforcement learning control, outperform Q-learning. Consideration of Gaskett's particular version of Q-learning and neural networks as means of camera movement control is outside the scope of this chapter. This chapter considers various forms of the actor critic (AC) method to solve the camera movement control problem. AC methods have been studied extensively (see, *e.g.*, Barto, **Sutton, Anderson**, 1983; Berenji, 2003; Bertsekas & **Tsitsiklis**, 1996; Konda & **Tsitsiklis**, 2000; Rosenstein, 2003; Sutton & Barto, 1998; Watkins & Dayan, 1992; Wawrzynski, 2005). The conventional actor critic method evaluates whether things have gotten better or worse than expected, as a result of an action selection in the previous state. A temporal difference (TD) error term, δ, is computed by the critic to evaluate an action previously selected. An estimated action preference in the current state is then determined by an actor using δ. Swarm actions are generated by a policy that is influenced by action preferences. In the study of swarm behavior of multiagent systems, such as systems of cooperating bots, it is helpful to consider ethological methods (see, e.g., Tinbergen, 1963), where each proximate cause (stimulus) usually has more than one possible response. Swarm actions with lower TD error tend to be favored. A second form of actor critic method is defined in the context of an approximation space (see, e.g., Peters, 2004a, 2004b, 2005; Peters & Henry, 2006; Peters & Ramanna, 2004; Peters, Skowron, Synak, & Ramanna, 2003; Skowron & Stepaniuk, 1995; Stepaniuk, 2002), which is an extension of recent work with reinforcement comparison (Peters, 2005; Peters & Henry, 2005a, 2005b, 2006). This form of actor critic method utilizes what is known as a reference reward, which is pattern based and action specific. Each action has its own reference reward that is computed within an approximation space that makes it possible to measure the closeness of action-based blocks of equivalent behaviors to a standard. The contribution of this chapter is an introduction to a form of monocular vision system that learns with approximation spaces used as frameworks for pattern-based evaluation of behavior during reinforcement learning.

This chapter is organized as follows. Rough-set theory is briefly introduced in Section 2. The basic idea of an approximation space is presented in Section 3. An approach to commanding a monocular vision system is described in Section 4. A description of a test bed for learning experiments with a monocular vision systems is given in Section 5. A model for the design of a monocular vision system that learns with approximation spaces is given in Section 6. A rough coverage form of actor critic method is presented in Section 7. Finally, the results of learning by a monocular vision system in a nonnoisy and in a noisy environment are presented in Sections 8 and 9.

ROUGH SETS: BASIC CONCEPTS

This section briefly presents some fundamental concepts in rough-set theory resulting from the seminal work by Zdzisław Pawlak (for an overview, see, e.g., Peters & Skowron, 2006, 2007), and that provides a foundation for a new approach to reinforcement learning by collections of cooperating agents. The rough-set approach introduced by Zdzisław Pawlak (1981a, 1981b, 1982) and Pawlak and Skowron (2007a, 2007b, 2007c) provides, for example, a ground for concluding to what degree a set of equivalent objects is covered by a set of objects representing a standard. The term "coverage" is used relative to the extent that a given set is contained in a standard set. An overview of rough-set theory and applications is given by Polkowski (2002). For computational reasons, a syntactic representation of knowledge is provided by rough sets in the form of data tables. A data (information) table IS is represented by a pair (U, A), where U is a nonempty, finite set of elements and A is a nonempty, finite set of attributes (features), where $a: U \rightarrow V_a$ for every $a \in A$ and V_a is the value set of a. For each $B \subseteq A$, there is associated an equivalence relation $Ind_{IS}(B)$ such that $Ind_{IS}(B) = \{(x, x') \in U2 \mid \forall a \in B, a(x)=a(x')\}$. Let $U/Ind_{IS}(B)$ denote a partition of U determined by B (i.e., $U/Ind_{IS}(B)$ denotes the family of all equivalence classes of relation $Ind_{IS}(B)$), and let $B(x)$ denote a set of B-indiscernible elements containing x. $B(x)$ is called a block, which is in the partition $U/Ind_{IS}(B)$. For $X \subseteq U$, the sample X can be approximated from information contained in B by constructing a B-lower and B-upper approximation denoted by B_*X and B^*X, respectively, where $B_*X = \cup\{B(x) \mid B(x) \subseteq X\}$ and $B^*X = \cup\{B(x) \mid B(x) \cap X \neq \varnothing\}$. The B-lower approximation B_*X is a collection of blocks of sample elements that can be classified with full certainty as members of X using the knowledge represented by attributes in B. By contrast, the B-upper approximation B^*X is a collection of blocks of sample elements representing both certain and possibly uncertain knowledge about X. Whenever B_*X is a proper subset of B^*X, that is, $B_*X \subset B^*X$, the sample X has been classified imperfectly, and is considered a rough set.

Approximation Spaces

This section gives a brief introduction to approximation spaces. The basic model for an approximation space was introduced by Pawlak (1981a), elaborated by Orlowska (1982) and Pawlak (1981b), generalized by Skowron and Stepaniuk (1995) and Stepaniuk (1998), and applied in a number of ways (see, e.g., Peters, 2005; Peters & Henry, 2006; Skowron, Swiniarski, & Synak, 2005). An approximation space serves as a formal counterpart of perception or observation (Orlowska, 1982), and provides a framework for approximate reasoning about vague concepts. A *generalized approximation space* is a system $GAS = (U, N, \nu)$ where:

- U is a nonempty set of objects, $P(U)$ is the powerset of U
- $N : U \rightarrow P(U)$ is a neighborhood function
- $\nu : P(U) \times P(U) \rightarrow [0, 1]$ is an overlap function

A set $X \subseteq U$ is definable in a GAS if, and only if, X is the union of some values of the neighborhood function. In effect, the uncertainty function N defines, for every object x, a set of similarly defined objects. That is, N defines a neighborhood of every sample element x belonging to the universe U (see, e.g., Peters et al., 2003). Specifically, any information system $IS = (U, A)$ defines for any feature set $B \subseteq A$ a parameterized approximation space $AS_B = (U, N_B, \nu)$, where $N_B = B(x)$, a B-indiscernibility class in the partition of U. The overlap function ν computes the degree of overlap between two subsets of U. Let $P(U)$ denote the powerset of U. We are interested in the larger of the two sets (assume that the $card(Y) \geq card(X)$) because we want to see how well Y "covers" X, where Y represents a standard for evaluating sets of similar objects. Standard rough coverage (SRC) ν_{SRC} can be defined as in equation (1).

$$\nu_{SRC}(X,Y) = \begin{cases} \dfrac{|X \cap Y|}{|Y|}, & \text{if } Y \neq \varnothing, \\ 1, & \text{if } Y = \varnothing. \end{cases} \quad (1)$$

Figure 1. ALiCE II camera

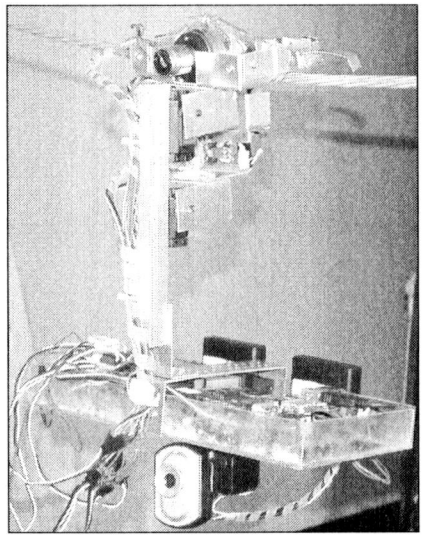

Figure 2. ALiCE II wireless card

In other words, $v_{SRC}(X, Y)$ returns the degree that Y covers X. In the case where $X = Y$, then $v_{SRC}(X, Y) = 1$. The minimum coverage value $v_{SRC}(X, Y) = 0$ is obtained when $X \cap Y = \varnothing$ (i.e., X and Y have no elements in common).

ENVIRONMENT FOR LINE-CRAWLING ROBOT

This section briefly describes the environment to be inspected by a second-generation autonomous line-crawling robot named ALiCE II, which is a member of a family of line-crawling bots designed to inspect electric power transmission towers (see Figures 1 and 2).

ALiCE II has been designed to crawl along a wire stretched between a particular class of steel towers in the Manitoba Hydro power transmission system. Sample steel towers in the Manitoba hydro system are shown in Figures 3 and 5. ALiCE II has been designed to crawl along the top, lightning guard wire (called a sky wire) of a tower like the one shown in Figure 5. A bot suspended from a sky wire stretched between electric power transmission towers usually sways from side to side as it moves along a wire, due to buffeting by the wind. For example, in Manitoba, towers usually range from 20 to 50 metres in height (see, e.g., Figure 4). The tallest transmission towers in the Manitoba Hydro system are more than 100 metres high to support the long crossing of the Nelson River (see **Berger, 1995**). A broad range of target objects (e.g., insulators, pins, bolts, cross braces) are part of the repetoire of objects that ALiCE II inspects. A sample insulator group is shown in Figure 6 (notice that the top insulator in Figure 6 is damaged, and needs to be replaced).

Figure 3. Sample steel towers

Figure 5. Sample steel tower

Figure 4. Sample tower measurements

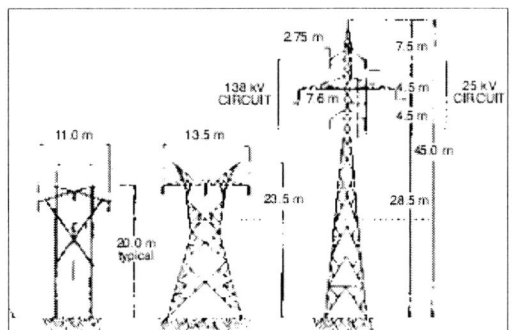

Figure 6. Sample insulator group

Hence, once a bot identifies a target (e.g., insulator or tower cross-brace), target tracking is necessary to move a camera to compensate for wind action so that it continues to inspect a target object.

COMMANDING

Commanding of ALiCE II is made possible by a wireless network, as shown in Figure 2. Specifically, ALiCE II uses the Sierra AirCard 580 wireless WAN modem (see, e.g., Figure 2) to establish the connection to the point-to-point protocol (PPP) server over the Telus cellular network. PPP has three main components corresponding to RFC 1332 from the IETF (2006), a method for sending data over the connection between two nodes, a link control protocol (LCP) for maintaining the data link, and a suite of network control protocols (NCPs) used to establish and conFigureure different network-layer protocols. Specifically, the NCP protocol used to establish the IP address is the PPP Internet protocol control protocol (IPCP), which negotiates the IP address only after the data link has been established and tested by the IETF (2006). What follows is a comparison of several types of tracking systems, using classical as well as reinforcement learning methods.

Only the tilt servo is visible in Figure 2. A bot suspended from a sky wire stretched between electric power transmission towers usually sways from side to side as it moves along a wire. Hence, once a bot identifies a target (e.g., insulator or tower cross—brace), target tracking is necessary to move a camera so that it continues to inspect a target object. What follows is a comparison of several types of tracking systems using classical as well as reinforcement learning methods.

MONOCULAR VISION SYSTEM TEST BED

This section gives a brief overview of a test bed for the ALiCE II vision system (see Figures 1, 2, and 9). The main processing unit of ALiCE II is the TS-5500 compact full-featured PC-compatible single board computer based on the AMD Elan520 processor from Elan (2006). The TS-5500 is running the Linux operating system (see Torvalds, 2006) on the Elan 0x586 class processor from Technologic (2006) using the Elan microcontroller. The TS-5500 is conFigureured to use an Orinoco wireless ethernet card that can be programmed as a wireless access point (AP), or to connect to an existing AP, such as the DLink wireless router, to send camera images to a base

Figure 7. ALiCE II wireless network

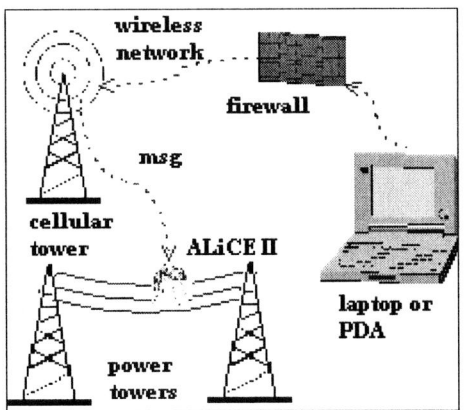

Figure 8. UML comm. dialogue

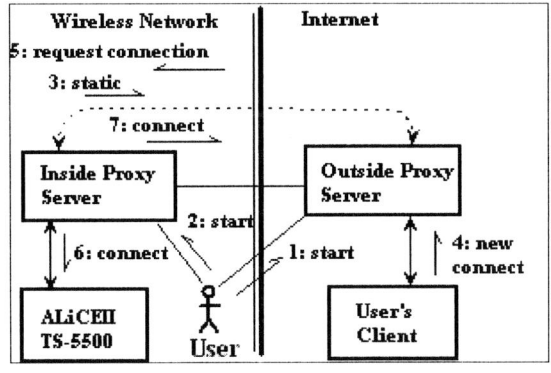

station. Furthermore, the TS-5500 is also con-Figureured to use a Sierra 555 wireless network card that can connect to the Internet through a cellular CDMA 1X network (see Figure 7). The main constraint limiting the possible algorithms that can be used for target tracking is the relatively low computational power of the CPU, which is run by a 133 MHz clock. As a result, the TS-5500 has computational power comparable to a Pentium III processor with an approximately 70 MHz clock. The TS-5500 is also equipped with a Creative NX Ultra Web cam that is mounted on two Hobbico mini servos. The servos provide the camera with two degrees of freedom (DOF) since one servo swings the camera in a horizontal plane (panning), and a second servo swings the camera up and down (tilting). Each servos movement can be controlled separately. Two additional computers (IBM Thinkpads) are included in the hardware test setup (see Figure 9). One IBM ThinkPad displays a randomly moving target. A second Thinkpad serves as an observation PC, which is connected to a server running on the TS-5500. This second Thinkpad provides visual feedback and records data for comparison purposes.

MONOCULAR VISION SYSTEM

The goal of the monocular vision system is to point a camera at a moving target, continuously. To achieve this goal, a fast and efficient image-processing technique is required to detect the position of a target in an input frame obtained from the Web cam. The approach utilized in our system is called template matching. However, before template matching can occur, preprocessing of the input image from the camera is necessary, due to the computational restraints of the TS-5500 (comparable to a Pentium III processor with a clock speed $\approx$ 70 MHz). Preprocessing consists of both spatially decimating the input image, as well as transforming the RGB colours into grey levels. Spatial decimation is performed in each dimension (both x and y) by taking every n^{th} pixel. In our case, we selected every 4th pixel to transform the smallest available output of the Web cam from 160 × 120 pixels to 40 × 30 pixels. Furthermore, conversion into grey levels is preformed using equation (2).

$$I = \frac{(R+G+B)}{3}. \quad (2)$$

The result of preprocessing is an image that is over 40 times smaller (160 × 120 × 3 = 57600 bytes compared with 40 × 30 × 1 = 1320 bytes) and still contains sufficient information to perform template matching. Template matching is implemented in this system using the sum of squared differences (SSD), which is similar to the approach used by Gaskett (2002; Gaskett, Ude, Cheng, 2005). The idea is to find a match of the template in the input image, where it is assumed that the template is the smaller of the two images. The SSD model is given in equation (3), and is calculated for every position of the template in the input image.

$$SSD(x,y) = \sum_j \sum_k [input(j,k) - template(j-k, k-y)]^2 \quad (3)$$

Figure 9. Test bed for monocular vision system

The result is that equation (3) is minimal at the point in which the template occurs in the input image. Next, the coordinates of the minimum are passed on to the target-tracking algorithms. Finally, it is important to note that the coordinates of the target refer to the centre of the template, and will only span a subset of size 28 × 21 of the 40 × 30 image. For example, consider Figure 10, which is a simple example showing, in grey, all the possible coordinates of the centre of a template of size 5 × 5 within an image of size 8 × 8.

Reinforcement Learning by Vision System

At the heart of reinforcement learning is the principle of state and state transitions. The learning agent determines the state from observed changes in its environment. Furthermore, for each perceived state, there is one or more desirable actions that will cause a change in the environment and produce a transition into a beneficial state. Rewards can then be defined by the perceived desirability of the new state. States in the monocular vision system are based on the coordinates of the target obtained from the template matching process and are shown in Figure 11. These states are applied to the 28 × 21 subset described previously. The selection of states in a learning environment is more of an art than a science. The states given in Figure 11 were arrived at through trial and error. The goal was to select areas small enough to allow a few ideal servo movements in each state, but not so small as to severely limit the set of possible movements. Similarly, the actions available in each state are based on the maximum distance from the centre of the 28 × 21 area to the outside edge in any single dimension, which is approximately 14 pixels. Consequently, the actions available in each state range from 1 - 12 and represent increments to the servo's position. Each step increment to the servo provides a rotation of 0.9 degrees with an accuracy of ±0.25%. Furthermore, the two numbers located below each state identifier, in Figure 11, represent the direction of the pan and tilt servos, respectively. Finally, reward is calculated using equation (4).

$$reward = 1 - \frac{distance}{maxdistance}, \quad (4)$$

where distance is calculated using equation (5)

$$distance = \sqrt{x^2 + y^2}. \quad (5)$$

Note that x and y represent the coordinates of the target, and the origin is selected as the centre

Figure 10. Image template

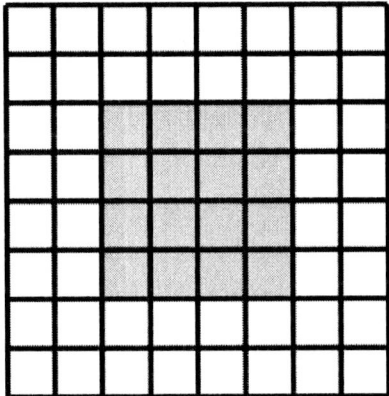

Figure 11. System states

s20 -1 1	s21 -1 1	s22 0 1	s23 1 1	s24 1 1
s15 -1 1	s16 -1 1	s17 0 1	s18 1 1	s19 1 1
s10 -1 0	s11 -1 0	s12 0 0	s13 1 0	s14 1 0
s5 -1 -1	s6 -1 -1	s7 0 -1	s8 1 -1	s9 1 -1
s0 -1 -1	s1 -1 -1	s2 0 -1	s3 1 -1	s4 1 -1

of the camera's field of view; thus, there is no need to include the coordinates of the centre in the distance calculation. A result computed using equation (4) is a normalized reward that equals 1 when the target is located at the centre, and 0 when the target is at the outside edge of the field of view.

Classical Target Tracking

This section presents the classical tracking system implemented in the monocular vision system. The classical target tracking method is a deterministic algorithm and does not perform any learning. Consequently, this algorithm defines its state space separately from the reinforcement learning algorithms. In effect, every possible target coordinate can be considered a state, and the only action available in each state is to move the servo's the calculated distance (using equation (5)) to the target. In a sense, it selects the right action to take in every state. As a result, this algorithm is important because it demonstrates the desired behaviour of the reinforcement learning algorithms, and provides a baseline for comparison when plotting the results. The classical target tracking method is provided in Algorithm 1.

Algorithm 1: Classical target tracking
Input: States $s \in S$; //1 state for each possible set of target coordinates
Output: Deterministic policy $\pi(s)$; //selects the same action in every state.
    while (true) **do**
        get current state; //i.e., coordinates of target
        pan ← target's horizontal distance from center of camera view;
        tilt ← target's vertical distance from center of camera view;
        move servos by (pan, tilt);
    end

Actor Critic Learning Method

Actor critic (AC) learning methods are temporal difference (TD) learning methods with a separate memory structure to represent policy independent of the value function used. The AC method considered in this section is an extension of reinforcement comparison found in Sutton and Barto (1998). The estimated value function $V(s)$ is an average of the rewards received while in state s. After each action selection, the critic evaluates the quality of the selected action using δ in equation (6), which represents the error (labeled the TD error) between successive estimates of the expected value of a state.

$$\delta_t = r_t + \gamma V(s_{t+1}) - V(s_t), \quad (6)$$

where γ is the discount rate, and the value of a state implemented by the critic is given in equation (7).

$$V(s_t) = V(s_t) + \alpha_t \delta_t, \quad (7)$$

where α_t is the critic's learning rate at time t. For simplicity, the subscript t is omitted in what follows. If $\delta > 0$, then it can be said that the expected return received from taking action a_t is larger than the expected return in state s_t resulting in an increase to action preference $p(s, a)$. Conversely, if $\delta < 0$, the action a_t produced a return that is worse than expected and $p(s, a)$ is decreased (see, e.g., Wawrzynski, 2005). The run-and-twiddle (RT) method, introduced by Selfridge (1984) and elaborated by Watkins (1989), is a control strategy inspired by behaviour that has been observed in biological organisms such as E. coli and silk moths, where an organism continues its current action until the strength of a signal obtained from the environment falls below an acceptable level, and then it "twiddles" (i.e., works out a new action strategy). This idea can be applied to the value of δ in Algorithm 2. Whenever $\delta < 0$ occurs too often, it can be said that an agent is performing below expectations, and that a "twiddle" is necessary to improve the current situation. The preferred action a in state s is calculated using equation (8).

$$p(s,a) = p(s,a) + \beta\delta, \quad (8)$$

where β is the actor's learning rate. The preferred action $p(s, a)$ is employed by an actor to choose actions stochastically using the Gibbs softmax method as shown in equation (9).

$$\pi(s,a) = \frac{e^{p(s,a)}}{\sum_{b=1}^{|A(s)|} e^{p(s,b)}}. \quad (9)$$

Algorithm 2 gives the actor critic method, which is an extension of the reinforcement comparison method given in Sutton and Barto (1998). It is assumed that the behaviour represented by Algorithm 2 is episodic, and is executed continuously.

Algorithm 2: Actor critic method
Input: States $s \in S$, actions $a \in A$, initialized α, γ, β.
Output: Policy $\pi(s, a)$ //controls selection of action a in state s.
for (all $s \in S, a \in A$) **do**
    p(s, a) ← 0;
    $\pi(s,a) = \dfrac{e^{p(s,a)}}{\sum_{b=1}^{|A(s)|} e^{p(s,b)}}$;
end
while (true) **do**
   initializes;
   for (t = 0; t < T_m; t = t + 1) **do**
   choose a from s using π(s, a);
   take action a, observe r, s'; //s' = s_{t+1}
   $\delta = r + \gamma V(s') - V(s)$;
   $V(s) = V(s) + \alpha\delta$;
   $p(s,a) = p(s,a) + \beta\delta$;
   $\pi(s,a) = \dfrac{e^{p(s,a)}}{\sum_{b=1}^{|A(s)|} e^{p(s,b)}}$;
   s ← s';
  end
end

ACTOR CRITIC METHOD USING ROUGH COVERAGE

This section presents a modified actor critic method using rough coverage derived within the context of an approximation space that is constructed relative to a decision table known as an ethogram. An ethogram is a decision table where each object in the table represents an observed behavior, which has features such as state, action, and proximate cause, inspired by Tinbergen (1963). This is explained in detail in Peters, Henry, and Ramanna (2005a), and not repeated, here.

Average Rough Coverage

This section illustrates how to derive average rough coverage using an ethogram. During a swarm episode, an ethogram is constructed, which provides the basis for an approximation space and the derivation of the degree that a block of equivalent behaviours is covered by a set of behaviours representing a standard (see, e.g., Peters, 2005b; Peters et al., 2005a; Tinbergen, 1963). Let $x_i, s, PC, a, p(s, a), r, d$ denote ith observed behaviour, current state, proximate cause (Tinbergen, 1963), possible action in current state, action-preference, reward for an action in previous state, and decision (1 = choose action, 0 = reject action), respectively.

Assume, for example, $B_a(x) = \{y \in U_{beh} \mid xIND(B \cup \{a\})y\}$. Let $B = \{B_a(x) \mid x \in \Omega\}$ denote a set of blocks representing actions in a set of sample behaviours Ω. Let $\bar{r}$ denote average rough coverage as shown in equation (10).

$$\bar{r} = \frac{1}{card(\mathrm{B})} \sum_{i=1}^{card(\mathrm{B})} \nu_\mathrm{B}(B_a(x), B_*D), \qquad (10)$$

where $B_a(x) \in \mathrm{B}$. Computing the average lower rough-coverage value for action blocks extracted from an ethogram implicitly measures the extent that past actions have been rewarded. What follows is a simple example of how to set up a lower approximation space relative to an ethogram. The calculations are performed on the feature values shown in Table 1.

Rough Coverage Actor Critic Method

The rough coverage actor critic (RCAC) method is one among many forms of the actor critic method (see, *e.g.*, Barto et al., 1983; Berenji, 2003; Bertsekas & Tsitsiklis, 1996; Konda & **Tsitsiklis,** 1995; Peters & Henry, 2005; Sutton & Barto, 1998;, Watkins & Dayan, 1992; Wawrzyński, 2005; Wawrzyński & Pacut, 2004). Common variations include additional factors that vary the amount of credit assigned to selected actions. This is most commonly seen in calculating preference, $p(s, a)$. The rough coverage form of the actor critic method calculates preference values as shown in equation (11).

$$p(s,a) \leftarrow p(s,a) + \beta \left[\delta - \bar{r}\right], \qquad (11)$$

where $\bar{r}$ denotes average rough coverage computed using equation (10). This is reminiscent of the idea of a *reference reward* used during reinforcement comparison. Recall that incremental reinforcement comparison uses an incremental average

Table 1. Sample ethogram

x_i	s	PC	a	$p(s,a)$	r	d
x0	1	3	4	0.010	0.010	0
x1	1	3	5	0.010	0.010	1
x2	1	3	4	0.010	0.010	0
x3	1	3	5	0.020	0.011	1
x4	1	3	4	0.010	0.010	0
x5	1	3	5	0.031	0.012	1
x6	1	3	4	0.010	0.010	0
x7	1	3	5	0.043	0.013	1
x8	1	3	4	0.010	0.010	1
x9	1	3	5	0.056	0.014	0

$B = \{s_i, PC_i, a_i, p(s,a)_i, r_i\}$,

$D = \{x \in U \mid d(x) = 1\} = \{x1, x3, x5, x7, x8\}$

$B_a(x) = \{y \in U_{beh} \mid xIND(B \cup \{a\})y\}$, hence

$B_{a=4}(x0) = \{x0, x2, x4, x6, x8\}, B_{a=5}(x1) = \{x1\}$,

$B_{a=5}(x3) = \{x3\}, B_{a=5}(x5) = \{x5\}, B_{a=5}(x7) = \{x7\}, B_{a=5}(x9) = \{x9\}$,

$B_*D = \cup\{B_a(x) \mid B_a(x) \subseteq D\} = \{x1, x3, x5, x7\}$

$\nu_B(B_{a=4}(x0), B_*D) = 0, \nu_B(B_{a=5}(x1), B_*D) = 0.25$,

$\nu_B(B_{a=5}(x3), B_*D) = 0.25, \nu_B(B_{a=5}(x5), B_*D) = 0.25$,

$\nu_B(B_{a=5}(x7), B_*D) = 0.25, \nu_B(B_{a=5}(x9), B_*D) = 0$,

$\overline{r = \nu_B(B_a(x), B_*D)} = 0.1667$

of all recently received rewards, as suggested by Sutton and Barto (1998). Intuitively, this means action probabilities are now governed by the coverage of an action by a set of equivalent actions that represent a standard. Rough-coverage values are defined within a lower approximation space. Algorithm 3 is the RCAC learning algorithm used in the monocular vision system. Notice that the only difference between Algorithms 2 and 3 is the addition of the reference reward $\bar{r}$, which is calculated using rough coverage.

Algorithm 3: Rough coverage actor critic method
Input: States $s \in S$, Actions $a \in A$, initialized α, γ, β.
Output: Policy $\pi(s, a)$ //controls selection of action a in state s.
for (all $s \in S, a \in A$) **do**
    $p(s, a) \leftarrow 0$;
    $$\pi(s,a) = \frac{e^{p(s,a)}}{\sum_{b=1}^{|A(s)|} e^{p(s,b)}};$$
end
while (true) **do**
    initialize s;
    for ($t = 0; t < T_m; t = t + 1$) **do**
    choose a from s using $\pi(s, a)$;
    take action a, observe r, s'; //$s' = s_{t+1}$
    $\delta = r + \gamma V(s') - V(s)$;
    $V(s) = V(s) + \alpha \delta$;
    $p(s,a) = p(s,a) + \beta [\delta - \bar{r}]$;
    $$\pi(s,a) = \frac{e^{p(s,a)}}{\sum_{b=1}^{|A(s)|} e^{p(s,b)}};$$
    $s \leftarrow s'$;
    end
    Extract ethogram table $IS_{swarm} = (U_{beh}, A, d)$;
    Discretize feature values in IS_{swarm};
    Compute $\bar{r}$ as in equation (10) using IS_{swarm};
end

Run-and-Twiddle Actor Critic Method

This section briefly presents a run-and-twiddle (RT) form of AC method in Algorithm 4. Both AC methods use preference values to compute action selection probabilities. However, the RT AC method uses $\bar{v}_a$ (average rough coverage of all the blocks containing action a) to control the rate of learning instead of β in the actor. A "twiddle" entails advancing the window of the ethogram (recorded behavior patterns of ecosystem organisms, or, in the monocular vision system, behavior patterns of a moving camera) and recalibrating $\bar{v}_a \forall a \in A$. This form of twiddling mimics the behaviour of E. coli bacteria (diminishing food supply results in change in movement) or a male silk moth following the perfume emitted by a female silk moth (diminishing perfume signal results in a change of the search path), which is described by Selfridge (1984). This idea can be applied to the value of δ in Algorithm 3. When $\delta < 0$ occurs too often, then it can be said that the agent is performing below expectations, and that a "twiddle" is necessary to improve the current situation. This is the basic approach underlying Algorithm 4.

Algorithm 4: Rough coverage actor critic method
Input: States $s \in S$, Actions $a \in A$, initialized $\alpha, \gamma, \beta, th$. //$th$ = threshold
Output: Policy $\pi(s, a)$ //controls selection of action a in state s.
for (all $s \in S, a \in A$) **do**
    $p(s, a) \leftarrow 0$;
    $$\pi(s,a) = \frac{e^{p(s,a)}}{\sum_{b=1}^{|A(s)|} e^{p(s,b)}};$$
end
    while (true) **do**
    initialize s; $v \leftarrow 0$;
    for ($t = 0; t < T_m; t = t + 1$) **do**
    choose a from s using $\pi(s, a)$;
    take action a, observe r, s'; //$s' = s_{t+1}$

$$\delta = r + \gamma V(s') - V(s);$$
$$V(s) = V(s) + \alpha \delta;$$
$$p(s,a) = p(s,a) + \beta \left[\delta - \overline{r}\right];$$
$$\pi(s,a) = \frac{e^{p(s,a)}}{\sum_{b=1}^{|A(s)|} e^{p(s,b)}};$$
if ($\delta < 0$) **then**
    $\nu \leftarrow \nu + 1$;
    if ($\nu <$ th) **then**
        Extract ethogram table $IS_{swarm} = (U_{beh}, A, d)$;
        Discretize feature values in IS_{swarm};
        Compute $\overline{\nu}_a \forall a \in A$;

        $\nu \leftarrow 0$;
    end
end
$s \leftarrow s'$;
  end

end

RESULTS

The values shown in the plots in Figures 12(a)-12(d) represent the RMS error between the distance

Figure 12(a). RC AC vs. RT AC

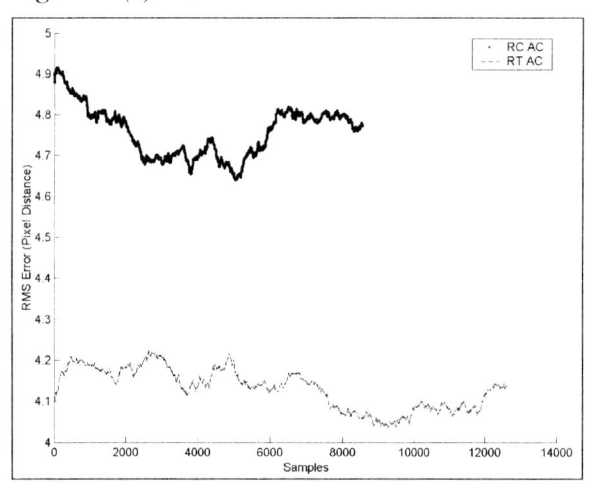

Figure 12(b). AC, RC AC, classical

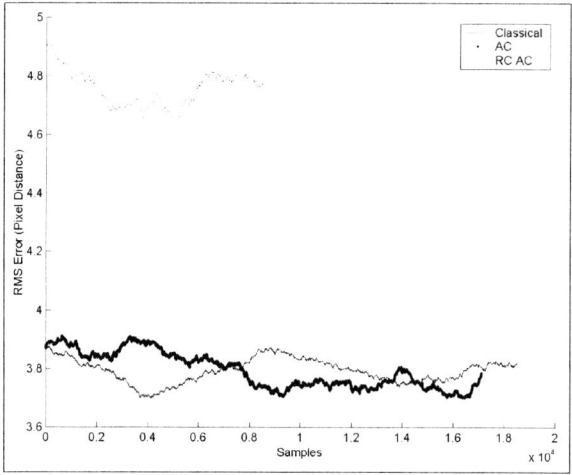

Figure 12(c). RMS for 4 AC methods

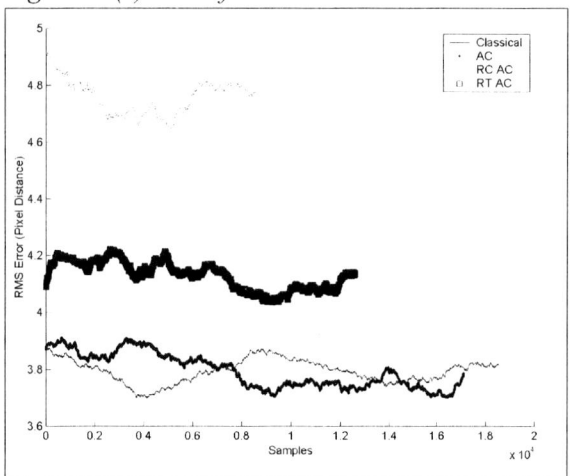

Figure 12(d). AC vs. RT AC

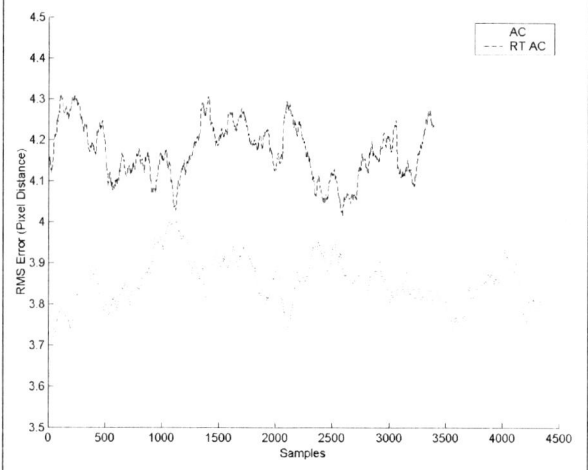

of the target from the centre of the field of view of each camera image and a running average of the distance values. Preliminary experiments suggest that the RC AC method has the poorest performance among the methods (see sample RMS plots in Figures 12(a), 12(b), 12(c)). This is due largely to the fact that the RC AC method must construct a lower approximation space at the end of each episode (in contrast to the run and twiddle method, which only constructs a lower approximation space whenever things appear to be going badly). As a result, the RC AC algorithm hangs, due to the limited processing power of the TS-5500 and, as a result, the target moves a significant distance from the centre of the image. However, note that the RC AC method performs quite similar to the AC and RT AC when observing the normalized totals of the state value function V (see Figure 13). This suggests that the RC AC method is comparable to the other two AC methods in terms of converging to an optimal policy. The RC AC method just does poorly in this real-time learning environment because its "calculate hang" causes the algorithm to lose the target. This can be corrected by implementing the RC AC method on a system with more processing power.

Figure 13. Comparison of normalized total state values

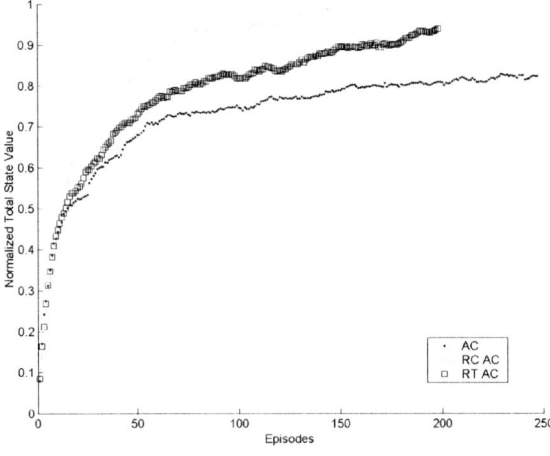

In preliminary tests with camera control in the monocular vision system, the performance of run-and-twiddle (RT AC) method compares favorably to the actor critic (AC) method. Evidence of this can be seen in the plots in Figure 12(a) and Figure 12(d).

Finally, it is important to note that there are two reasons why the classical tracking method outperforms the reinforcement learning algorithms in the sample RMS plots in Figures 12(b), 12(c). At the beginning of the tests, the RL algorithms are performing more exploration than greedy selection to determine the best possible action to select in each state. Toward the end of the tests (in later episodes), the RL algorithms consistently converge to an optimum policy. However, even during later episodes, the AC methods still perform exploration some of the time. As a result, the classical method should outperform the RL methods every time. The RL algorithms are ideally suited to deal with noise, which has not been considered in the form of SSD given in equation (3). In this context, the term *noise* denotes irregular fluctuations in either movements of a target or in the environment that influences the quality of camera images. Noise has a number of sources (e.g., camera vibration due to buffeting of the wind, electromagnetic field, and weather conditions such has rain or glare from the sun). This is very important because of the noisy environment in which the monocular vision system must operate. In the presence of noise, the classical algorithm will not do as well as RL because it has no provision for exploration whenever it is not clear what action should be performed next to find a moving target.

LEARNING EXPERIMENTS IN A NOISY ENVIRONMENT

To induce noise, the platform that the camera was sitting on (see Figure 9) was purposely placed on a slope rather than a flat surface. The result of this placement caused the entire platform to vibrate when the camera servos were activated. This provided a degree of realism analogous to what would be experienced in a suspended system, where any type of motion onboard ALiCE II will disturb the system somewhat. The extra motion was intended to provide a more challenging movement of the target as it varied with the step sizes; greater motion resulted from larger step sizes. The resulting motion of the platform was a rocking action from back to front only.

The illumination for the experiments was kept as uniform as possible, with the addition of papers surrounding the monitor in an attempt to restrict the available ambient light. The experiments were all conducted for 5 minutes each, providing a reasonable time period to get an idea of their individual performance. Overall, this resulted in a controlled environment with some extra fluctuations in movement, and a slightly more difficult target-tracking problem than the previous set of experiments.

One thing that can readily be seen in Figures 14-16 is that there is a periodicity in the experimental procedure. This is due to the trajectory that the target takes. The peaks in the cycle correspond to the highest error, and this coincides to part of the trajectory where the target travels down the right-hand side, where the field of view can only see part of the target and often the camera would lose the target, depending on the method in question. Note that the performance of the actor critic method (both the conventional AC with sample RMS values in Figure 14, and rough coverage AC with RMS values in Figures 15-16) is excellent, and it has managed to outperform the classical tracking method.

The classical tracking system is being outperformed by the AC and RC AC learning methods.

Figure 15. Rough coverage actor critic in noisy environment

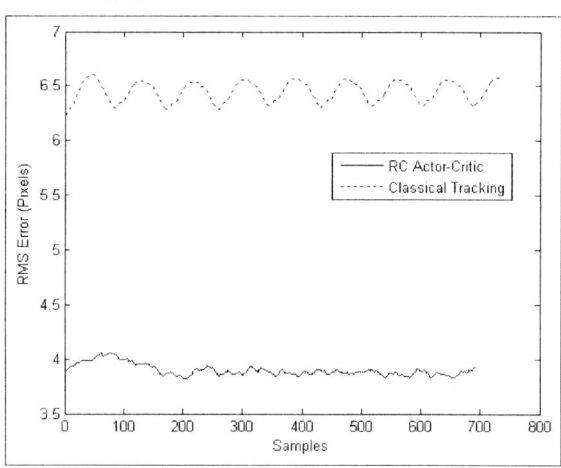

Figure 14. Actor critic in noisy environment

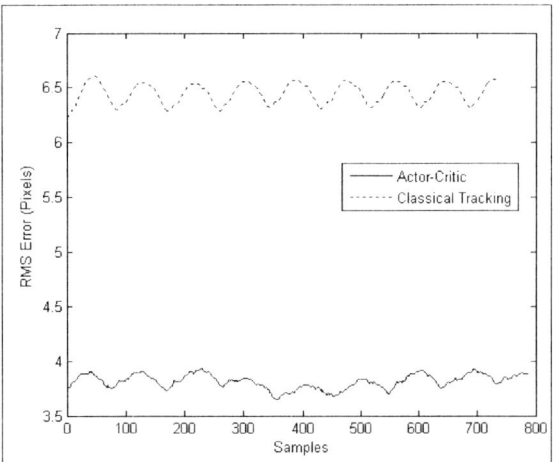

Figure 16. RC AC, AC, classical tracking in noisy environment

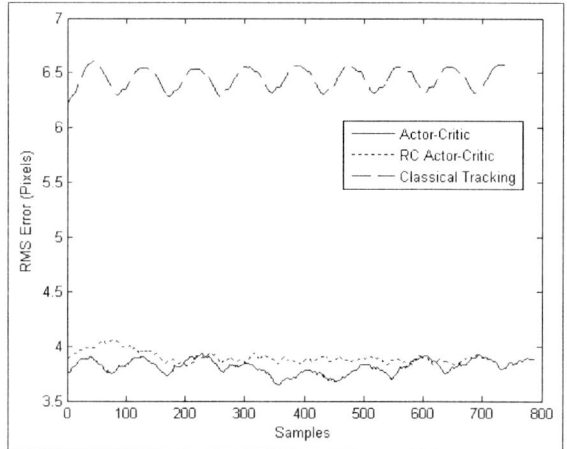

The likely explanation for the worsened performance of the classical tracking algorithm is the extra motion from the camera platform, and a tendency for the target to slip down so that the field of view of the camera includes only part or none of the target. With these new problems, the AC and RC AC learning methods led to improved performance, since these reinforcement learning methods were able to explore, in a somewhat noisy environment, to discover a better tracking profile. Notice that the rough coverage form of actor critic method tends to have slightly better performance than the conventional AC method (see, for example, Figure 16).

CONCLUSION

This chapter introduces a monocular vision system that learns with the help of approximations. This results in a vision system with actions that are influenced by patterns of behavior discovered in the context of approximation spaces that have been constructed periodically from ethograms. In a noisy environment, the rough coverage form of actor critic method tends to do better than the classical tracking method. This is in keeping with the basic approach introduced in 1981 by Zdzislaw Pawlak, who suggested classifying objects by means of their features. In the context of target tracking by a monocular vision system, an object can observe behavior of the system. Classification of vision system behavior is made possible by comparing feature values of observed system behaviors in different states that have been captured in ethogram tables. The basic approach in the design of the vision system described in this chapter is derived from ethology introduced by Tinbergen in 1963. The current work on vision systems is part of a swarm intelligence approach, where pairs of cameras on separate, cooperating bots synchronize to form binocular vision systems that learn. Future work will consider n-ocular vision systems that learn with approximation spaces, as well as various reinforcement learning methods not considered here.

ACKNOWLEDGMENT

Many thanks to Wes Mueller, who has been a great help in obtaining images showing hydro transmission equipment. This research has been supported by Natural Sciences and Engineering Research Council of Canada (NSERC) grant 185986 and grant T247 from Manitoba Hydro.

REFERENCES

Barto, A. G., Sutton, R. S., & Anderson, C. W. (1983). Neuronlike elements that can solve difficult problems. *IEEE Trans. on Systems, Man, and Cybernetics, 13*, 834-846.

Berenji, H. R. (2003). A convergent actor-critic-based FRL algorithm with application to power management of wireless transmitters. *IEEE Trans. on Fuzzy Systems, 11*(4), 478-485.

Berger, R. P. (1995). *Fur, feathers & transmission lines*. Retrieved from http://www.hydro.mb.ca/environment/publications/fur_feathers.pdf

Bertsekas, D. P., & Tsitsiklis, J. N. (1996). *Neuro-dynamic programming*. Belmont, MA: Athena Scientific.

Elan. (2006). *Elan SC520 Microcontroller User's Manual*. Retrieved from http://www.embeddedarm.com/

Gaskett, C. (2002). *Q-learning for robot control*. PhD thesis, Supervisor: A. Zelinsky, Department of Systems Engineering, The Australian National University.

Gaskett, C., Ude, A., & Cheng, G. (2005). Hand-eye coordination through endpoint closed-loop and learned endpoint open-loop visual servo control. In *Proceedings of the International Journal of Humanoid Robotics, 2*(2), 203-224.

Internet Engineering Task Force (IEFT). (2006). *RFC 1332 The PPP Internet Protocol Control Protocol (IPCP)*. Retrieved from http://www.ietf.org/rfc/rfc1332.txt

Konda, V. R., & Tsitsiklis, J. N. (1995). Actor—critic algorithms. *Adv. Neural Inform. Processing Sys.*, 345-352.

Orlowska, E. (1982). *Semantics of vague concepts. Applications of rough sets*. Institute for Computer Science, Polish Academy of Science Report 469.

Pawlak, Z. (1981a). Classification of objects by means of attributes. *Institute for Computer Science, Polish Academy of Sciences Report 429*.

Pawlak, Z. (1981b). Rough sets. *Institute for Computer Science, Polish Academy of Sciences Report 431*.

Pawlak, Z. (1982). Rough sets. *International J. Comp. Inform. Science, 11*, 341-356.

Pawlak, Z., & Skowron, A. (2007a). Rudiments of rough sets. *Sciences, 177*, 3-27.

Pawlak, Z., & Skowron, A. (2007b). Rough sets: Some extensions. *Sciences, 177*, 28-40.

Pawlak, Z., & Skowron, A. (2007c). Rough sets and Boolean reasoning. *Sciences, 177*, 41-73.

Peters, J.F. (2004a). Approximation space for intelligent system design patterns. *Engineering Applications of Artificial Intelligence, 17*(4), 1-8.

Peters, J.F. (2004b). Approximation spaces for hierarchical intelligent behavioral system models. In B.D. Kepli/c{c}z, A. Jankowski, A. Skowron, M. Szczuka (Eds.), *Monitoring, security and rescue techniques in Multiagent systems. Advances in Computing* (pp. 13-30). Physica-Verlag, Heidelberg.

Peters, J.F. (2005). Rough ethology: Towards a biologically-inspired study of collective behavior in intelligent systems with approximation spaces. *Transactions on Rough Sets III, LNCS 3400*, 153-174.

Peters, J.F. (2005b). Rough ethology: Towards a biologically-inspired study of collective behavior in intelligent systems with approximation spaces. *Transactions on Rough Sets, III, LNCS 3400*, 153-174.

Peters, J.F., & Henry, C. (2005). Reinforcement learning in swarms that learn. In *Proceedings of the IEEE/WIC/ACM International Conference on Intelligent Agent Technology (IAT 2005)*, Compiègne Univ. of Tech., France (pp. 400-406).

Peters, J.F., & Henry, C. (2006). Reinforcement learning with approximation spaces. *Fundamenta Informaticae, 71*(2-3), 323-349.

Peters, J.F., Henry, C., & Ramanna, S. (2005a). Rough ethograms: Study of intelligent system behavior. In: M. A. Kłopotek, S. Wierzchoń, & K. Trojanowski (Eds.), *New trends in intelligent information processing and Web mining (IIS05)*, Gdańsk, Poland (pp. 117-126).

Peters, J.F., Henry, C., & Ramanna, S. (2005b). Reinforcement learning in swarms that learn. In *Proceedings of the 2005 IEEE/WIC/ACM International Conference on Intelligent Agent Technology (IAT, 2005)* (pp. 400-406). Compiegne University of Technology, France.

Peters, J.F., & Pawlak, Z. (2007). Zdzisław Pawlak life and work (1906-2006). *Information Sciences, 177*, 1-2.

Peters, J.F., & Ramanna, S. (2004). Measuring acceptance of intelligent system models. In, M. Gh. Negoita et al. (Eds.), *Knowledge-based intelligent information and engineering systems, Lecture Notes in Artificial Intelligence 3213*(1), 764-771.

Peters, J.F., & Skowron, A. (2006). Zdzislaw Pawlak: Life and work. *Transactions on Rough Sets V*, 1-24.

Peters, J.F., Skowron, A., Synak, P., & Ramanna, S. (2003). Rough sets and information granulation. In T. Bilgic, D. Baets, & O. Kaynak (Eds.), Tenth Int. Fuzzy Systems Assoc. World Congress IFSA, Instanbul, Turkey. *Lecture Notes in Artificial Intelligence 2715*, 370-377. Heidelberg: Physica-Verlag.

Polkowski, L. (2002). *Rough sets. Mathematical foundations*. Heidelberg: Springer-Verlag.

Rosenstein, M.T. (2003). *Learning to exploit dynamics for robot motor coordination*. PhD Thesis, Supervisor: A.G. Barto, University of Massachusetts Amherst.

Selfridge, O. G. (1984). Some themes and primitives in ill-defined systems. In O. G. Selfridge, E. L. Rissland, & M. A. Arbib (Eds.), *Adaptive control of ill-defined systems*. London: Plenum Press.

Skowron, A., & Stepaniuk, J. (1995). Generalized approximation spaces. In T. Y. Lin & A. M. Wildberger (Eds.), *Soft computing, simulation councils* (pp. 18-21). San Diego.

Skowron, A., Swiniarski, R., & Synak, P. (2005). Approximation spaces and information granulation. *Transactions on Rough Sets III*, 175-189.

Stepaniuk, J. (1998). Approximation spaces, reducts and representatives. In L. Polkowski, A. Skowron (Eds.), Rough sets in knowledge discovery 2. *Studies in Fuzziness and Soft Computing, 19*, 109-126. Heidelberg: Springer-Verlag.

Sutton, R.S., & Barto, A.G. (1998). *Reinforcement learning: An introduction*. Cambridge, MA: The MIT Press.

Technologic. (2006). *TSUM user's manual*. Retrieved from http://www.embeddedarm.com/

Tinbergen, N. (1963). On aims and methods of ethology. *Zeitschrift für Tierpsychologie, 20*, 410-433.

Torvalds, L. (2006). *Linux operating system*. Retrieved from http://www.linux.org/

Watkins, C.J.C.H. (1989). *Learning from delayed rewards*. PhD thesis, supervisor: Richard Young, King's College, University of Cambridge, UK.

Watkins, C. J. C. H., & Dayan, P. (1992). Technical note: Q-learning. *Machine Learning, 8*, 279-292.

Wawrzyński. P. (2005). *Intensive reinforcement learning*. PhD dissertation, supervisor: Andrzej Pacut, Institute of Control and Computational Engineering, Warsaw University of Technology.

Wawrzyński. P., & Pacut, A. (2004). Intensive versus nonintensive actor-critic algorithms of reinforcement learning. In *Proceedings of the 7th International Conference on Artificial Intelligence and Soft Computing* (pp. 934-941). Springer 3070.

Chapter X
Hybridization of Rough Sets and Multi-Objective Evolutionary Algorithms for Classificatory Signal Decomposition

Tomasz G. Smolinski
Emory University, USA

Astrid A. Prinz
Emory University, USA

Jacek M. Zurada
University of Louisville, USA

ABSTRACT

Classification of sampled continuous signals into one of a finite number of predefined classes is possible when some distance measure between the signals in the dataset is introduced. However, it is often difficult to come up with a "temporal" distance measure that is both accurate and efficient computationally. Thus, in the problem of signal classification, extracting particular features that distinguish one process from another is crucial. Extraction of such features can take the form of a decomposition technique, such as principal component analysis (PCA) or independent component analysis (ICA). Both these algorithms have proven to be useful in signal classification. However, their main flaw lies in the fact that nowhere during the process of decomposition is the classificatory aptitude of the components taken into consideration. Thus, the ability to differentiate between classes, based on the decomposition, is not assured. Classificatory decomposition (CD) is a general term that describes attempts to improve the effectiveness of signal decomposition techniques by providing them with "classification-awareness." We propose a hybridization of multiobjective evolutionary algorithms (MOEA) and rough sets (RS) to perform the task of decomposition in the light of the underlying classification problem itself.

INTRODUCTION

Many "real-life" classification problems are concerned with the ability to classify a digitized continuous signal (*i.e.*, time series) into one of a finite number of predefined classes (Fayyad, Piatetsky-Shapiro, Smyth, & Uthurusamy, 1996; Rajan, 1994). For example, neurobiological evoked potentials can be used to characterize the functioning of a patient's central nervous (*e.g.*, normal vs. abnormal), speech waveforms can be used to recognize a speaker, *and so forth*.

As an illustration, Figure 1 shows a sample taken from one of the most commonly studied benchmarking datasets within the field of time series classification, the Cylinder-Bell-Funnel (CBF) task that was originally proposed by Saito (1994).

The CBF task is to solve an artificially formed three-category classification problem, with the three classes generated by the equations shown in (1).

$$c(t) = (6+\eta) \cdot X_{[a,b]}(t) + \varepsilon(t),$$
$$b(t) = (6+\eta) \cdot X_{[a,b]}(t) \cdot \frac{(t-a)}{(b-a)} + \varepsilon(t),$$
$$f(t) = (6+\eta) \cdot X_{[a,b]}(t) \cdot \frac{(b-a)}{(b-t)} + \varepsilon(t),$$
(1)

where $X_{[a,b]}=\{1, \text{ if } a \leq t \leq b \text{ else } 0\}$, η, and $\varepsilon(t)$ are drawn from a standard normal distribution $N(0,1)$, a is an integer drawn uniformly from the range [16,32], and $b-a$ is an integer drawn uniformly from the range [32, 96] (Chu, Keogh, Hart, & Pazzani, 2002).

Classification of such signals is possible when some measure D, which describes distances (*e.g.*, Euclidean, L2, *etc.*) between particular signals in the dataset, is introduced. However, it is often very difficult to come up with a distance measure that is not only accurate, but also efficient computationally. Thus, in the problem of signal classification, extracting particular features that distinguish one process from another (*i.e.*, preserve the original distances) is crucial.

Figure 1. Example of a time series classification problem – Cylinder-Bell-Funnel (CBF) task

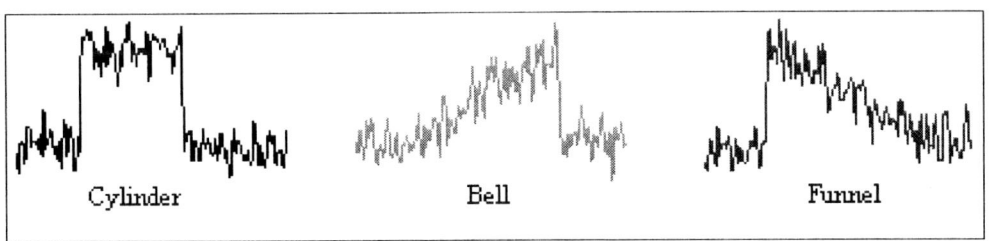

There are many ways to extract such features, but the general idea is always the same: the original signal of dimensionality n should be replaced with a new so-called feature vector of dimensionality m, where m ≤ n (desirably m << n), and for any signal **x** and **y** in the database, $D(\mathbf{x}, \mathbf{y})$ is preserved by some new distance measure $D_F(\mathbf{x}_F, \mathbf{y}_F)$ in the new feature space **F** (i.e., if two signals were close to each other in the original feature space, they will also be "similar" in the new feature space).

Extraction of such feature vectors can be based upon a determination of a set of simple characteristics of a given signal (e.g., minimum, maximum, average, etc.) or can be in the form of a more complex transformation or signal decomposition technique, such as principal component analysis (PCA) (Flury, 1988) or independent component analysis (ICA) (Hyvarinen, & Oja, 2000). Statistical criteria utilized in those methodologies, however, are often insufficient to build a reliable classifier.

Classificatory decomposition (CD) is a general term that describes attempts to improve the effectiveness of signal decomposition techniques by providing them with "classification-awareness" (Smolinski, 2004). The main idea here is to look for basis functions whose coefficients allow for an accurate classification while preserving the reconstruction. To achieve that goal, we propose a hybridization of multiobjective evolutionary algorithms (MOEA) and rough sets (RS) to perform the task of decomposition in the light of the underlying classification problem itself.

DECOMPOSITION TECHNIQUES IN SIGNAL CLASSIFICATION

As already indicated, signal classification very often makes use of some descriptive features of the signal, rather than the sequence itself. Signal decomposition techniques prove to be useful for extraction of such feature vectors.

The general idea of signal decomposition is to represent the original signal **x** in terms of some basis functions **M**, which are fixed for all signals in the database, and a set of coefficients **a**, which

Figure 2. Example of signal decomposition for classification – raw signals

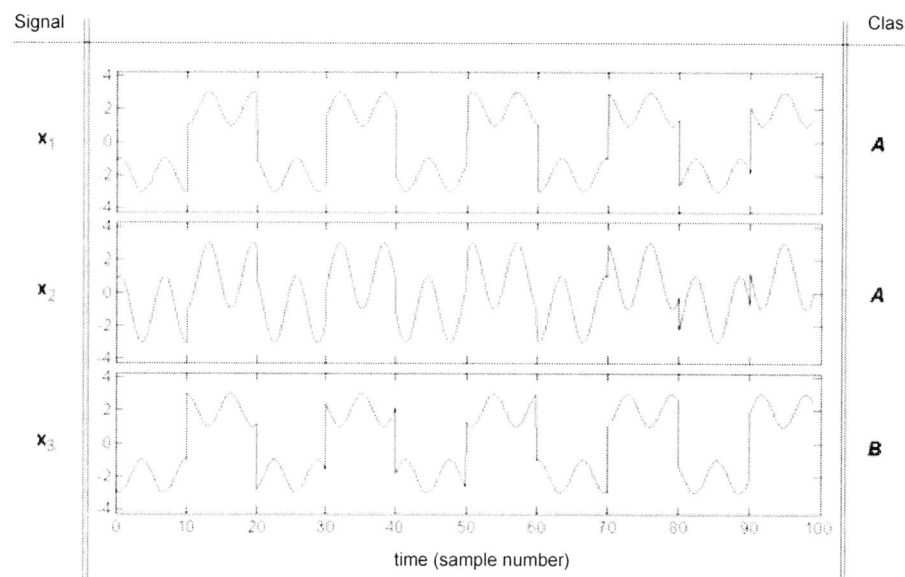

can be (and most likely will be) different for each signal. This representation also allows for some noise, or reconstruction error, **e**:

$$\mathbf{x} = \mathbf{Ma} + \mathbf{e} \qquad (2)$$

The temporal properties of the system are preserved by the basis functions, while the original sequences are replaced by a set of scalar coefficients that represent the original data in the space spanned by the basis functions. The process of reconstruction into the original input space is based upon a linear combination of the basis functions (*i.e.*, a sum of the basis functions weighted by the coefficients). The process of classification takes place within the database of coefficients.

For example, the following artificially generated dataset consisting of three sequences $\mathbf{x}_1$, $\mathbf{x}_2$, $\mathbf{x}_3$, each belonging to one of the two categories *A* and *B* (Figure 2), can be replaced by two basis functions $\mathbf{m}_1$, $\mathbf{m}_2$ (Figure 3), and a new dataset consisting of the coefficients $\mathbf{a}_1$, $\mathbf{a}_2$ (Figure 4), for the basis functions $\mathbf{m}_1$, $\mathbf{m}_2$ respectively, that will represent the original vectors $\mathbf{x}_1$, $\mathbf{x}_2$, $\mathbf{x}_3$ in the new attribute space.

After such transformation, not only has the feature space been tremendously reduced (*i.e.*, instead of operating on vectors with 100 values each, just two numbers are being used), but also the classification problem becomes straightforward.

In the example in Figure 4, the time series of class *A* are only those that have positive values of the coefficients for the sinusoidal component. This fact is not easy to deduce directly from the analysis of the shapes of the examples in the database, as visually $\mathbf{x}_1$ and $\mathbf{x}_3$ seem to be more closely related than $\mathbf{x}_1$ and $\mathbf{x}_2$. However, after the above decomposition, a single decision rule will be sufficient to classify the signals in the database without error (*e.g.*, IF $a_1 > 0$, THEN Class is *A*,

Figure 3. Example of signal decomposition for classification – generated basis functions

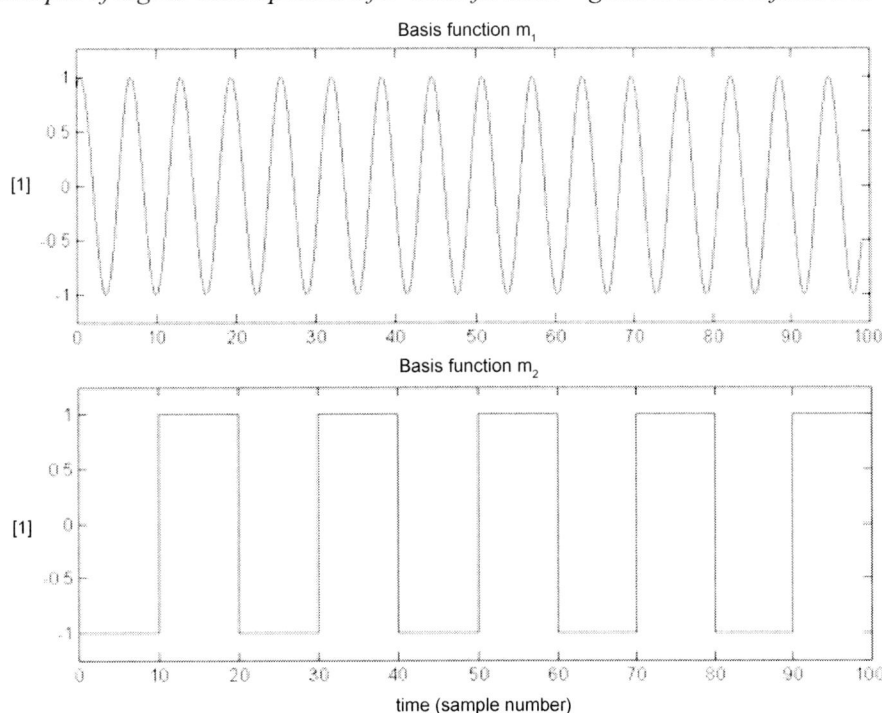

Figure 4. Example of time series decomposition for classification – coefficients for representation of the original time series in new attribute space

Signal	a_1	a_2	Class
x_1	1	2	**A**
x_2	2	1	**A**
x_3	-1	2	**B**

ELSE Class is **B**). This is a very simple, synthetic example, as the decomposition is known *a priori*; however, one can easily imagine a "real-life" problem where such a solution would be very desirable; not only the appropriate decomposition (*i.e.*, underlying phenomena) of the input signals is found (*i.e.*, two basis functions: sinusoid and square pulse), but also the most important one among the components is determined (*i.e.*, the sinusoidal component, or more precisely its phase, determines the correct classification).

Over the years, various transformation/decomposition techniques have been successfully applied to signal classification. The two most commonly used methods for signal classification are principal component analysis (PCA) and independent component analysis (ICA). Both of these techniques assume that some statistical criterion related to the discovered components (orthogonality in PCA and independence in ICA) will be sufficient for a successful differentiation between signals that belong to different classes. Obviously, this is not always true. Thus, the idea of combining the premises of a reliable and accurate decomposition of a signal (verifiable via the reconstruction process) with the determination of the components that really matter in terms of segregation of the input sequences into separate categories (validated via the classification accuracy measure) seems plausible.

In order to improve the classification abilities of decomposition techniques, some research has been done in the area of hybridization of those methods with rough-set (RS) theory-based schemes. For instance, Swiniarski (1999) combined the traditional PCA with rough sets to improve the categorization of texture images, and Smolinski et al. (Smolinski, Boratyn, Milanova, Zurada, & Wrobel, 2002) amalgamated the concept of reducts with the sparse coding with overcomplete bases (SCOB) methodology in the problem of analysis and classification of cortical evoked potentials. Although both of those studies were quite successful, their main limitation was their dependence on a two-stage procedure. Thus, the idea for the incorporation of classification aptitude measures into the process of decomposition itself was born.

In classificatory decomposition (CD), rather than utilize the traditional statistical criteria to determine the usefulness of discovered components, we propose to exploit rough-sets-derived classificatory accuracy measures, to ensure classification usefulness of the decomposition. We propose to employ multiobjective evolutionary algorithms (MOEA) with fitness functions designed to deal with two considerations simultaneously: the first constituent of the function evaluation is based upon the reconstruction error, while the second utilizes rough-sets-based classification accuracy measures.

Based on this idea, we have already performed a number of empirical studies on "real-life" data and have obtained very promising results (Smolinski, Milanova, Boratyn, Buchanan, & Prinz, 2006a; Smolinski, Boratyn, Milanova, Buchanan, & Prinz, 2006b; Smolinski, Buchanan, Boratyn, Milanova, Prinz, 2006c). In those studies, we investigated several different variants for the implementation of the methodology, but a systematic analysis and comparison of those is still missing. In this chapter, we present results of such a study.

THEORETICAL BACKGROUND

Theory of Rough Sets

Information Systems and Decision Tables

The theory of rough sets (RS) deals with the classificatory analysis of data tables (Pawlak, 1982). The main idea behind it is the so-called indiscernibility relation that describes objects indistinguishable from one another (Marek, & Pawlak, 1984; Pawlak, 1991). The main goal of rough-set analysis is to synthesize approximation of concepts from acquired data (Komorowski, Pawlak, Polkowski, & Skowron, 1999). The concepts are represented by lower and upper approximations.

In the theory of rough sets, a dataset is in the form of a table in which each row represents an object or a case (*e.g.*, a signal), and every column represents an attribute that can be measured for that object (*e.g.*, the signal's value at time *t*). Such a table is called an **information system**, which can be represented formally as a pair: ***IS*** = (*U*, *A*), where *U* is a nonempty finite set of objects (universe), and *A* is a nonempty finite set of attributes such that a : $U \Rightarrow V_a$ for every a $\in$ *A*. The V_a is called the value set of a.

In many applications, the outcome of classification is given *a priori*. This knowledge is expressed by one or more distinguished attributes called the **decision attributes**. Information systems of this kind are called **decision tables**. Such a system can be formally represented by ***DT*** = (*U*, *A* $\cup$ {d}), where d $\in$ *A* is the decision attribute. The elements of A are called **conditional attributes** or, simply, **conditions**.

Indiscernibility

A decision table expresses all the knowledge about a given model. This table may be unnecessarily large because it is redundant in at least two ways: "*vertical*" (*i.e.*, objectwise) and "*horizontal*" (*i.e.*, attribute wise). The same or indistinguishable-from-one-another objects may be represented several times, or some of the attributes may be overabundant.

In order to take a closer look at those issues, let us recall the notion of equivalence first. A binary relation $R \subseteq X \times X$ that is reflexive (*i.e.*, an object is in relation with itself x*R*x), symmetric (if x*R*y, then y*R*x), and transitive (if x*R*y and y*R*z, then x*R*z) is called an **equivalence relation**. This relation divides a given set of elements (objects) into a certain number of disjoint equivalence classes. The equivalence class of an element x $\in$ *X* consists of all objects y $\in$ *X* such that x*R*y.

Let ***IS*** = (*U*, *A*) be an information system, then with any $B \subseteq A$ there is associated an equivalence relation $IND_{IS}(B)$:

$$IND_{IS}(B) = \{(x, x') \subset U^2 \mid \forall a \in B, a(x) = a(x')\}$$

(3)

$IND_{IS}(B)$ is called a *B*-**indiscernibility relation**. If (x,x') $\in IND_{IS}(B)$, then objects x and x' are indistinguishable from each other, in the light of the attributes in *B*. The equivalence classes of the *B*-indiscernibility relation are denoted $[x]_B$. The subscript ***IS*** in the indiscernibility relation is usually omitted if it is clear which information system is being considered.

Set Approximation

An equivalence relation induces a partitioning of the universe. These partitions can be used to build new subsets of the universe. Subsets that are most often of interest are those that contain objects with the same value of the outcome attribute (*i.e.*, belong to the same decision class). It may happen, however, that a concept cannot be defined in a crisp manner. That is where the notion of rough sets emerges. Although it may be impossible to precisely define some concept (*i.e.*, in a "crisp set" manner), it may be viable to determine which objects certainly belong to the concept, which objects certainly do not belong to the concept, and, finally, which belong to the boundary region between those two categories. Formally, the notion of a **rough set** can be defined as follows:

- Let $IS = (U, A)$ be an information system and let $B \subseteq A$ and $X \subseteq U$. We can approximate X using only the information contained in B by constructing the B-lower and B-upper approximations of X, denoted by $\underline{B}X$ and $\overline{B}X$ respectively, where $\underline{B}X = \{x \mid [x]_B \subseteq X\}$ and $\overline{B}X = \{x \mid [x]_B \cap X \neq \emptyset\}$.
- The objects in $\underline{B}X$ can be, with certainty, classified as members of X, based on the knowledge conveyed by B, while the objects in $\overline{B}X$ can only be classified as possible members of X, based on the knowledge in B. The set $BN_B(X) = \overline{B}X - \underline{B}X$ is called the B-**boundary region** of X, and thus consists of those objects that we cannot decisively classify into X on the basis of knowledge in B. The set $U - \overline{B}X$ is called B-**outside region** of X and consists of those objects that can be classified with certainty as not belonging to X. A set is said to be *crisp* if the boundary region is empty. Consequently, a set is said to be *rough* if the boundary region is nonempty.

A rough set can also be characterized numerically by the following coefficient:

$$\alpha_B(X) = \frac{card \, \underline{B}X}{card \, \overline{B}X} \quad (4)$$

This coefficient is called the **accuracy of approximation**.

Obviously, $0 \leq \alpha_B(X) \leq 1$. If $\alpha_B(X) = 1$, then X is *crisp* (precise) in terms of B, and if $\alpha_B(X) < 1$, then X is *rough* (vague) with respect to B.

Another coefficient that can be very useful for characterization of rough sets is called the **quality of classification** and is given by the following expression:

$$\gamma_B(X) = \frac{card \, \underline{B}X \cup \underline{B}\neg X}{card \, U} \quad (5)$$

where $\underline{B}X$ is the lower approximation of X, $\underline{B}\neg X$ is the lower approximation of the set of objects that do not belong to X, U is the set of all objects, and *card* stands for cardinality of a given set.

Dependency and Significance of Attributes

Discovering dependencies between attributes is another very important issue in data analysis. Intuitively, a set of attributes D depends totally on a set of attributes C, denoted as $C \Rightarrow D$, if all attributes from D are uniquely determined by the values of attributes from C.

Formally, dependency between the sets of attributes C and D can be defined as follows (Komorowski et al., 1999):

Let D and C be subsets of A. We will say that D depends on C in a degree k ($0 \leq k \leq 1$), denoted $C \Rightarrow_k D$, if:

$$k = \varphi(C, D) = \frac{card POS_C(D)}{card U} \quad (6)$$

where

$$POS_C(D) = \bigcup_{X \in U/D} \underline{C}(X) \quad (7)$$

is called a **positive region** of the partition U/D with respect to C. This is a set of all elements of

U that can be uniquely classified to blocks of the partition U/D by means of C.

Consequently, a **significance coefficient** of an attribute a, in the light of all attributes A, and the decision attribute(s) D can be computed as follows:

$$\sigma_{(A,D)}(a) = \frac{card(POS_A(D) - POS_{A-a}(D))}{cardU} \quad (8)$$

Reducts

All the considerations stated previously represent one natural way of reducing data that is to identify equivalence classes, that is, objects that are indistinguishable using the available attributes. That would be a much more efficient representation, since only one element of the equivalence class is needed to characterize the entire class. The other consideration in terms of data reduction is to keep only those attributes that preserve the indiscernibility relation and, consequently, the set approximation. The rejected attributes are redundant, since their removal cannot worsen the classification.

There are usually several such subsets of attributes, and those that are minimal are called **reducts**. The following presents a formal definition of a reduct:

Reduct is a "minimal" $R \subseteq A$, such that:

$$\forall k, n \forall a_i \in R : ((a_i(o_k) = a_i(o_n)) \Rightarrow (d(o_k) = d(o_n))) \quad (9)$$

where $k,n = 1..N$ (N is the number of objects in the decision table), $o_{k|n}$ is the $k^{th}|n^{th}$ object, $i = 1..R$ (R is the number of conditional attributes in the reduct), and d is the decision attribute.

The "minimal" criterion stands for:

$$\exists P \subset R : (9) \text{ is satisfied for } P \quad (10)$$

Computing equivalence classes is straightforward. Finding a global minimal reduct, on the other hand (*i.e.*, reduct with a minimal cardinality of attributes among all reducts), is NP-hard. For more details and examples of reduct-finding algorithms and heuristics, including utilization of genetic algorithms, see for example, Øhrn (2001) and Wroblewski (1995).

Motivation for the Use of Rough Sets

The choice of the theory of rough sets for the classificatory part of this project is motivated by several of its characteristics. First of all, rough sets provide a number of straightforward and easy to use classification performance measures, including the coefficient of *accuracy of approximation* (4) and the coefficient of *quality of classification* (5) introduced in this section. Those measures can be interchangeably applied and investigated in order to come up with the best possible solutions. Furthermore, since the concept of reducts is inherently embedded in the theory, and there are numerous algorithms and heuristics for computing them, this approach seems to be very well suited for searching for the minimal set of basis functions in the problem of time-series decomposition.

Theory of Multiobjective Evolutionary Algorithms

Multiobjective Evolutionary Algorithms

Many decision-making or design problems involve optimization of multiple, rather than single, objectives, simultaneously. In the case of a single objective, the goal is to obtain the best global minimum or maximum (depending on the nature of the given optimization problem), while with multiobjective optimization, there usually does not exist a single solution that is optimal with respect to all objectives. Since evolutionary algorithms (EA) work with a population of individuals, a number of solutions can be found in a single run. Therefore, an application of EAs to multiobjective optimization seems natural. For a detailed introduction to multiobjective optimization using evolutionary algorithms see Deb (2001).

Pareto-Optimality

In a set of trade-off solutions yielded by some optimization algorithm for a multiobjective, let us say bi-objective, optimization problem, also referred to as multiobjective programming or vector optimization (Cohon, 1978), one can observe that one solution can be better than another solution in the first objective, but, at the same time, worse in the other. This simply suggests that one cannot conclude that the first solution dominates the second, nor can one say that the second dominates the first. In that case, it is customary to say that the two solutions are nondominated with respect to each other (Deb, 2001; Gen, & Cheng, 2000).

For a given set of solutions *P*, yielded by some optimization algorithm, one can perform all possible pair-wise comparisons and disclose its nondominated subset of solutions *P'*. If *P* is the entire search space, the nondominated subset *P'* is called a globally **Pareto-optimal** or **nondominated set**. If *P* is just a subspace of the entire search space (*e.g.*, one population of candidate solutions in a multiobjective evolutionary algorithm), the nondominated subset *P'* is called a locally Pareto-optimal or nondominated set (Deb, 2001).

Goals of Multiobjective Optimization

Based on the concept of a Pareto-optimal set, the first main goal of the multiobjective optimization task can be formulated as convergence towards the, so-called, Pareto-optimal front (which basically means that the globally Pareto-optimal set is being targeted). Since evolutionary algorithms (EA) work with a population of individuals, a number of Pareto-optimal solutions can be found in a single run. Therefore, an application of EAs to multiobjective optimization seems natural. However, since the process of evolutionary optimization results in a collection of solutions rather than just one optimal answer, maintaining a diverse set of the nondominated solutions is also essential (which basically means that the outcome of the algorithm should be flexible in terms of the trade-off options). These are the main two goals of multiobjective optimization and another fundamental difference with single-objective optimization, in which just one goal, namely reaching the globally optimal solution, is being considered (Coello, Van Veldhuizen, & Lamont, 2002; Cohon, 1978; Deb, 2001).

Motivation for Use of Multiobjective Evolutionary Algorithms

Since the methodology of classificatory decomposition for signal classification obviously encompasses a bi-objective optimization problem (*i.e.*, minimization of the reconstruction error vs. maximization of the classificatory relevance of components), an application of a multiobjective evolutionary algorithm for that purpose seems natural. Utilization of single-objective evolutionary algorithms for both signal decomposition (Milanova, Smolinski, Boratyn, Zurada, & Wrobel, 2002; Smolinski et al., 2002) as well as rough-sets-driven reduct search (Wroblewski, 1995) has proven to be an efficient and useful approach. Thus hybridization of these two for solving the combined problem of classification and/or clustering-competent decomposition, by drawing the best from both techniques, seems highly plausible.

UTILIZED MULTIOBJECTIVE EVOLUTIONARY ALGORITHMS

Vector Evaluated Genetic Algorithm (VEGA)

This algorithm was the first implementation of a multiobjective genetic algorithm to yield a set of nondominated solutions (Schaffer, 1984). With this approach, the genetic algorithm evaluates an objective vector, instead of a scalar objective function, with each element of the vector representing each objective function. Since there are several objectives to be considered, $e.g.$, $i = 1..O$, the idea is to randomly divide the population, at every generation, into O equal subpopulations P_i. Each subpopulation is assigned fitness based on a different objective function f_i. Thus, each of the O fitness functions is used to evaluate some individuals in the population. A schematic of VEGA is shown in Figure 5.

Advantages of VEGA

The main advantage of VEGA is that it is very easy to implement. Only slight changes are required to convert a simple genetic algorithm to a multiobjective EA. Also, it does not incur any additional computational complexity.

Disadvantages of VEGA

Each solution in VEGA is evaluated only with one objective function. Thus, every solution is not tested for other (O–1) objectives, all of which are also important in the light of multiobjective optimization. During a simulation run of a VEGA, it is likely that the selection operator in a subpopulation would prefer solutions near the optimum of an individual objective function. Obviously, such preferences also take place in parallel with other objective functions in different subpopulations. It is assumed that the crossover operator would combine these individual "champion" solutions to find a trade-off solution in the Pareto-optimal region. However, it is possible that VEGA will not be able to find diverse enough solutions in the population, and will eventually converge to some individual "champion" solutions. This problem is referred to as speciation (*i.e.*, emergence of species, each fit to its own objective).

Modified Vector Evaluated Genetic Algorithm (M-VEGA)

This algorithm is based on the same scheme as the original VEGA technique; however, in order to deal with the aforementioned disadvantages of the former, several modifications were introduced (Smolinski, 2004).

Figure 5. Schematic of VEGA

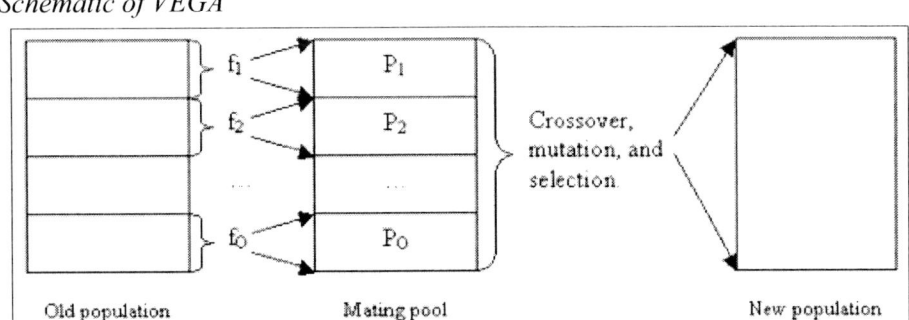

The first modification in M-VEGA is that all solutions can be evaluated in the light of all objectives. This slightly increases computational complexity of the algorithm but, as will be shown, has a significant advantage. In the process of selection, the standard "roulette wheel" approach is used. However, the wheel is divided into as many sections as there are objectives. The population is iteratively evaluated for each objective, and the "champion" representatives in a given objective are placed in the section of the wheel assigned to that objective. Note that with this approach, solutions strong in several objectives will be possibly placed in the wheel in all corresponding sections. Thus, their chances of being selected for reproduction will significantly increase.

Another modification in M-VEGA method is that each chromosome has an additional gene representing its so-called primary objective. This primary objective flag serves a regulatory purpose in the fitness function evaluation of a given potential solution. For instance, it can control some predefined weights assigned to the objectives (*i.e.*, the primary objective can be emphasized in the fitness evaluation process). It can also control the fitness evaluation procedure itself by "switching off" evaluation of all the other objectives but the primary one, which is then equivalent to standard VEGA. Most importantly, however, the primary objective flag is adaptive throughout the process of evolution. In the simplest case, it can change randomly (*e.g.*, for example, in each generation, a potentially new primary objective is assigned to a given chromosome), or it can be implemented as a heuristic that switches evaluation to another objective, once some arbitrary threshold of the fitness evaluation in the initial primary objective has been reached. In both cases, it deals with the serious problem of speciation in the standard VEGA algorithm.

The final modification in the algorithm introduces the concept of "hard" and "soft" modes of fitness computation. The term "hard" computation refers to the situation in which some modifications of the genome, introduced by the fitness estimation procedure itself, are permitted. This can be very advantageous in some cases, as will be shown later. The term "soft" computation refers to the standard fitness function computations, in which no direct changes in the genome are allowed.

Advantages of M-VEGA

The main advantage of M-VEGA is that it is a fairly simple extension of VEGA, and is still easy to implement. However, by applying the aforementioned modifications, the algorithm becomes much more effective in dealing with the problem of speciation.

Disadvantages of M-VEGA

Although M-VEGA deals with the main limitation of VEGA outlined previously, it suffers from the lack of consideration of dominance and elitism (Deb, 2001). Elitism (*i.e.*, the notion that "elite" individuals cannot be expelled from the active gene pool by worse individuals), has recently been indicated as a very important factor in MOEAs that can significantly improve their performance (Laumanns, Zitzler, & Thiele, 2000).

Elitist Nondominated Vector Evaluated Genetic Algorithm (end-VEGA)

This algorithm, originally proposed in Smolinski et al. (2006b) and expanded in Smolinski et al. (2006c), utilizes the improvements introduced in M-VEGA (*i.e.*, evaluation in the light of all objectives, with a designation of the primary objective as well, as the "hard" and "soft" modes of fitness computation), but at the same time deals with its limitations related to the lack of consideration of dominance and elitism. To address dominance, it employs a simple approach based on multiplying the fitness of a given individual by the number of solutions that this individual is dominated by

(+ 1 to ensure that the fitness function of a nondominated solution is not multiplied by 0). Since the fitness function in this project is being minimized, the dominated solutions will be adequately penalized. To include elitism, it utilizes the idea of an external sequential archive (Laumanns et al., 2000) to keep track of the best-so-far (*i.e.*, nondominated) solutions, and to make sure that their genetic material is in the active gene pool.

Advantages of end-VEGA

A very important advantage of end-VEGA is the fact that while it is still a relatively simple extension of VEGA, it includes additional considerations, such as prevention of speciation as well as propagation of dominance and elitism.

Disadvantages of end-VEGA

Although end-VEGA seems to address most of the issues that are considered crucial for a successful multiobjective optimization, there is always room for improvement. One of the main disadvantages of the algorithm is its relatively high computational complexity. This is mostly due to the fact that each chromosome is being evaluated in the light of each objective, as well as to the necessity of maintaining an archive of nondominated solutions that has to be analyzed for dominance and possibly trimmed in every generation.

RS AND MOEA-BASED CLASSIFICATORY SIGNAL DECOMPOSITION

The main concept of classificatory decomposition was motivated by the hybridization of EAs with sparse coding with overcomplete bases (SCOB), introduced in Milanova et al. (2002). Using this approach, the basis functions, as well as the coefficients, are being evolved by optimization of a fitness function that minimizes the reconstruction error and at the same time, maximizes the sparseness of the basis function coding. This methodology produces a set of basis functions and a set of sparse coefficients. This may significantly reduce dimensionality of a given problem but, as any other traditional decomposition technique, does not assure the classificatory usefulness of the resultant model.

In classificatory decomposition, the sparseness term is replaced by a rough-sets-derived data-reduction-driven classification accuracy measure. This should assure that the result will be both "valid" (*i.e.*, via the reconstruction constraint) and useful for the classification task. Furthermore, the classification-related constituent also searches for a reduct, which assures that the classification is done with as few as possible basis functions. Finally, the single-objective EA utilized in the aforementioned technique is replaced by a multi-objective approach in which the EA deals with the reconstruction error and classification accuracy, both at the same time.

It is important to point out that the basis functions in CD are coded as vectors of real numbers, and there are no explicit constraints placed on them as to their shape or any other characteristics, other than that their length has to match that of the input signals. That means that there are no assumptions about the class that the basis functions must come from (*e.g.*, periodic, monotonic, *etc.*). Even though, in practice, it is sensible to impose some constraints on the ranges of the values that the basis functions can assume (as it is plausible to assume that the basis functions' values will be within the same order of magnitude as the values of the observed signals), it still leaves a lot of flexibility. It is often impossible to decompose a signal into a small number of components of a certain characteristic, but it may be possible to do so using fewer, more complex shapes.

Figure 6. Chromosome coding in classificatory decomposition

Basis functions part				Coefficients part		
$m_{11}...m_{1n}$	$m_{21}...m_{2n}$	...	$m_{M1}...m_{Mn}$	$a_{11},a_{12}...a_{1M}$	$a_{21},a_{22}...a_{2M}$	... $a_{N1},a_{N2}...a_{NM}$

Chromosome Coding

Each chromosome forms a complete solution to a given classificatory decomposition problem. It provides descriptions of both the set of basis functions and the coefficients (coded as real numbers) for all the signals in the training data set. For example, if the training set contains N signals with n samples each, and the task is to find M basis functions, the chromosome will be coded as shown in Figure 6.

Obviously, each of the M basis function has the length of the original input series (*i.e.*, n), and there are N vectors of coefficients (*i.e.*, each vector corresponds to one series in the training set) of dimensionality equal to the number of basis functions (*i.e.*, each coefficient corresponds to one basis function).

In the M-VEGA and end-VEGA algorithms, each chromosome is also described by several auxiliary parameters that contain the information about the multiobjective solution. Figure 7 shows an example with O objectives, among which the ith objective is selected as the primary one. Additionally, end-VEGA also stores the information about the number of solutions dominating a given chromosome.

Fitness Evaluation

Reconstruction Error

The problem of minimization of the error resulting from the reconstruction of the signal by a linear combination of the basis functions and coefficients is a relatively simple task. Once a particular distance measure has been decided upon, virtually any optimization algorithm can be used to minimize the distance between the original signal and the reconstructed one. The measure employed in this project, the well-known and widely used 2-norm (Kreyszig, 1978), referred to in signal processing as the signal energy-based measure, is presented in (11).

$$D = \sum_{t=1}^{n}(\mathbf{x}_t - (\mathbf{Ma})_t)^2 \qquad (11)$$

where **x** represents the original time series, **M** is the matrix of basis functions, **a** is a set of coefficients, and $t = 1..n$, where n is the number of samples in the series.

This norm is very simple to calculate and is also differentiable, which proves to be a very useful property.

Figure 7. Chromosome coding in classificatory decomposition – auxiliary genes

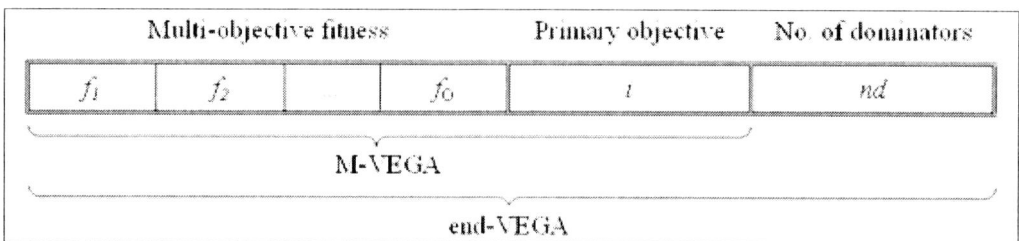

In order to deal with the situation in which the values in the time series are large (thus causing the energy-based distance measure to be large as well), a simple "normalization" by the energy of the original signal is proposed:

$$D_{NORM} = \frac{\sum_{t=1}^{n}(\mathbf{x}_t - (\mathbf{Ma})_t)^2}{\sum_{t=1}^{n}(\mathbf{x}_t)^2} \quad (12)$$

Subsequently, the reconstruction error fitness function f_{REC} for a chromosome p takes the following form:

$$f_{REC}(p) = \frac{\sum_{i=1}^{N} D_{NORM}^i}{N} \quad (13)$$

where D_{NORM}^i is the normalized reconstruction error for the ith signal and N is the total number of input signals.

Classification Accuracy and Reduction in the Number of Coefficients and Basis Functions

The problem of maximizing the classificatory competency of the decomposition scheme, and at the same time reducing the number of computed basis functions, can be dealt with by the application of rough sets.

In this project, two measures were used for the purpose of estimating the classificatory aptitude: (1) the rough-sets-based *accuracy of approximation*, and (2) the *quality of classification*, both presented in the earlier section on the theory of rough sets.

Since these two measures deal directly with interclass discrimination, they work very well in classification problems. It is important to point out, however, that they can be directly applied only to two-class tasks (*i.e.*, objects that belong to some concept X vs. objects that do not belong to X). Therefore, in classificatory decomposition for problems with more than two classes, we will use the average value of the given coefficient as the overall measure of the classification aptitude.

These measures of accuracy of approximation and/or quality of classification are estimated directly on the candidate reduct, which can be computed by any of the existing algorithms/heuristics. Note that therefore, the main objective that deals with the classificatory capability of decomposition can actually be considered a bi-objective optimization problem itself. On one hand, we are looking for the best possible classification accuracy, but on the other, we want to use as few basis functions as possible for that purpose. However, based on previous work done on the application of genetic algorithms in the search for reducts, as described in Wroblewski (1995), it is easy to see that this problem can be dealt with by minimizing a single-objective fitness function that is simply a summation of the classification error and the length of the reduct, as shown in (14):

$$f_{CLASS}(p) = (1 - \gamma_R^{avg.}) + \frac{L(R)}{M} \quad (14)$$

where p is a given representative (*i.e.*, chromosome), $L(R)$ is the length of the potential reduct R (*i.e.*, the number of attributes used in the representative), normalized by the total number of conditional attributes M, and $\gamma_R^{avg.}$ is the *quality of classification* coefficient (averaged over all the decision classes) for the reduct R, as introduced in (5).

The same considerations apply if the coefficient of *accuracy of approximation* is used (15):

$$f_{APPROX}(p) = (1 - \alpha_R^{avg.}) + \frac{L(R)}{M} \quad (15)$$

where p and $L(R)$ are defined as stated, and $\alpha_R^{avg.}$ is the *accuracy of approximation* coefficient (averaged over all the decision classes) for the reduct R, as introduced in (4).

An interesting consideration here is the question of what to do with the coefficients (and the corresponding basis functions) that have not been selected as a part of the reduct. Since we are looking for the best possible classification/approximation accuracy, while we want to use as few basis functions as possible, some mechanism capable of emphasizing the "important" coefficients/basis functions would be advisable. A solution to this problem is possible due to the application of the "hard" fitness computation idea employed by M-VEGA and end-VEGA, which allows the fitness function itself to introduce changes directly to the genetic material of the evaluated chromosome. In this work, two approaches are proposed:

- Coefficients/basis functions annihilation, which simply zeroes-out the "not important" genetic material, and the
- Coefficients/basis functions scaling, which scales the values of the coefficients and the basis functions according to the corresponding attribute significance coefficient (8), introduced earlier.

The idea here is that if we diminish the importance of the basis functions that are not vital in the classification process (or even remove them completely with the annihilation approach), the evolutionary algorithm will improve the remaining basis functions in order to compensate for an increase in the reconstruction error.

Data Preprocessing

It is important to point out that the specification of the chromosome coding and fitness function evaluation does not deal with serious problems of the shift in time between signals and the basis function compression or expansion (*i.e.*, longitudinal scaling) (Keogh, & Pazzani, 1998). Therefore, in order to apply the methodology to that kind of data, we propose to perform a preprocessing procedure to align the signals. Even though this is a quite restricting assumption, many "real-life" datasets actually fulfill it (*e.g.*, the cortical-evoked potentials that the methodology has been applied to), or can be preprocessed in this fashion. On the other hand, due to the application of evolutionary algorithms, this problem could be also dealt with relatively easily by encoding in the chromosome additional numerical values that will represent the shift in time and/or longitudinal scaling. This consideration will be address in further stages of this study.

Decomposition/Classification of "Unseen" Cases

Once the basis functions have been computed from the training dataset, even the simplest optimization algorithm can estimate the coefficients for the "unseen" signals in the testing dataset. In this project, a simple gradient descent algorithm, making use of differentiability of the chosen norm, was used:

$$\frac{\partial D}{\partial \mathbf{a}} = 2(\mathbf{x} - \mathbf{Ma})\mathbf{M}^T \qquad (16)$$

Thus, in order to verify the robustness of the decomposition-based classification, first the coefficients for the generated basis functions and the unseen signals are being computed. Then, those new coefficients are used to build a classifier.

Generation of the Initial Population

As indicated earlier, CD does not impose any strict limitations on the characteristics of the basis functions, which are simply coded as real numbers. This allows for tremendous flexibility in terms of the features of the discovered phenomena underlying a given classification problem, but significantly increases the search space. Under these circumstances, the problem of finding a set of basis functions that allows for a successful reconstruction can be very complex. Therefore, it may be beneficial to initialize the evolutionary algorithm with a population that contains some "educated guess" as to where, roughly, in the search space, the basis functions should be sought. Indeed, our previous research indicates that initializing the algorithm with basis functions, obtained from ICA, significantly improves the effectiveness of CD (Smolinski et al., 2006b; Smolinski et al., 2006c). The result of such a hybridization is a set of basis functions that are both reduced and "improved," as compared to the ICA alone (*i.e.*, less basis functions are needed to successfully reconstruct and classify the data).

In this chapter, we propose another way to initialize the evolutionary algorithm. Rather than utilizing some other decomposition technique to obtain the starting point, we suggest using a sample of randomly selected input signals as the "guesses" for the basis functions. This is a very plausible assumption, as in virtually all "real-life" problems, there will be a very strong relation between the shape of the basis functions and at least some of the raw signals in the database. Once such a sample has been selected; the corresponding coefficients can be found with the gradient descent algorithm described, thus providing the best possible reconstruction at that point.

In this chapter, we will present a comparison of the results obtained with this approach, as well as with a random initial population.

Figure 8. Examples from the Cylinder-Bell-Funnel dataset

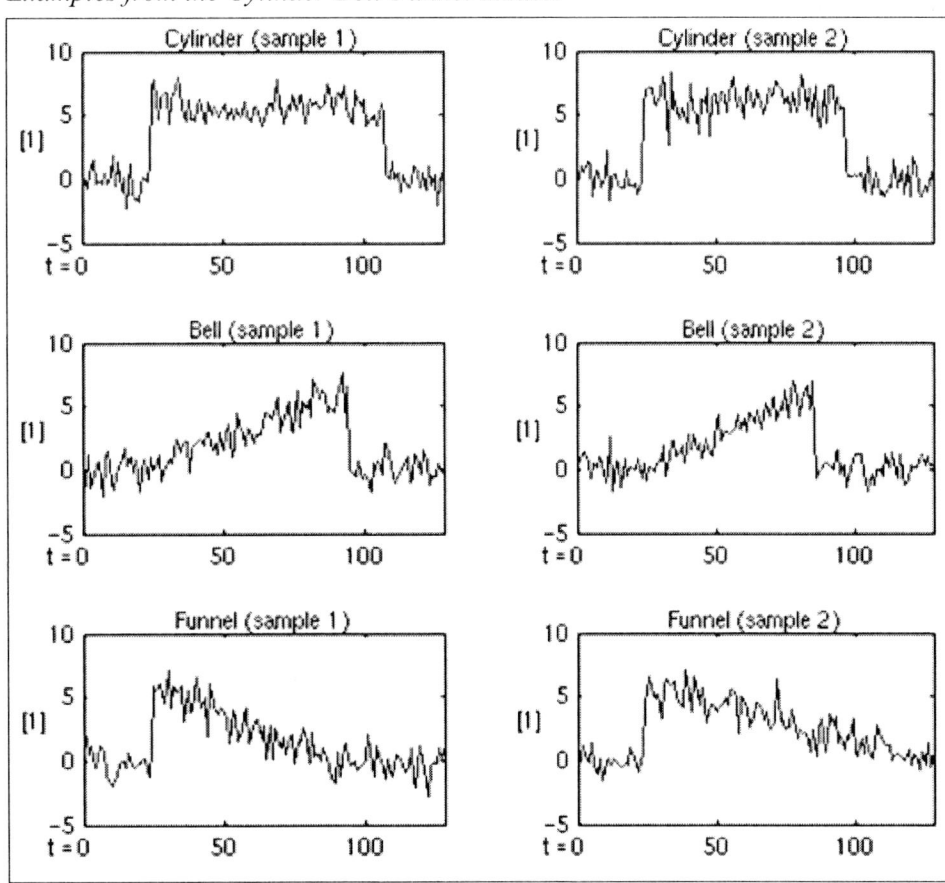

EXPERIMENTAL RESULTS

In the experiments described in this research, two benchmarking datasets were used: the Cylinder-Bell-Funnel task, which was already presented in the introduction, and the Synthetic Control Chart database that will be described later. Both these benchmarks had been previously utilized for a comprehensive study on time-series data mining, performed by Keogh and Kasetty (2002a), and allow for a direct comparison of the results obtained within the present study with the previous findings.

Cylinder-Bell-Funnel

The Cylinder-Bell-Funnel (CBF) is a benchmarking dataset that is one of the most commonly used for time-series classification and clustering experiments. The task is to solve an artificially formed three-class categorization problem with the three classes generated by equations described in the introductory section.

A short Matlab program was implemented (source: Keogh & Folias, 2002b) in order to generate a set of any number of cases (128 data points each) from every class. Figure 8 shows a few examples of the records in the database.

Figure 9. Examples from the Synthetic Control Chart

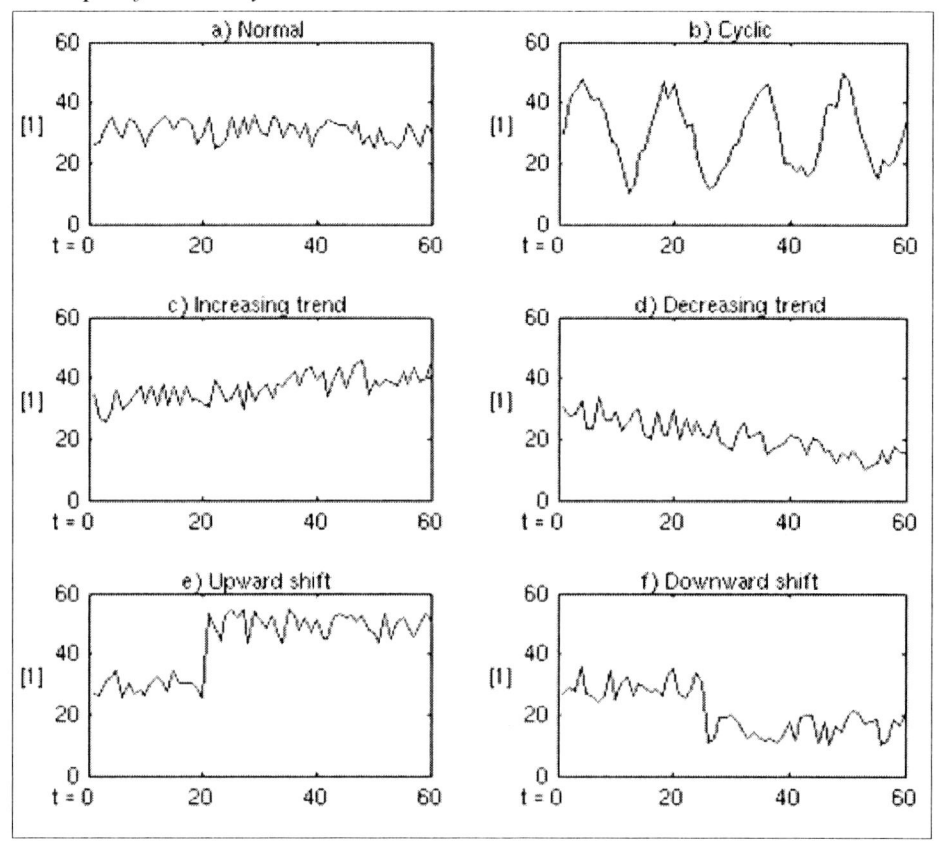

Synthetic Control Chart

The Synthetic Control Chart (SCC) is another benchmarking dataset that is commonly used for time-series classification and clustering. This dataset also contains 600 examples (60 data points each) of control charts synthetically generated by a process designed by Alcock and Manolopoulos (1999), and implemented in Matlab by the authors of this chapter for the purpose of this study. There are six different classes of those control charts: (a) Normal, (b) Cyclic, (c) Increasing trend, (d)

Figure 10. Preprocessed Cylinder-Bell-Funnel dataset

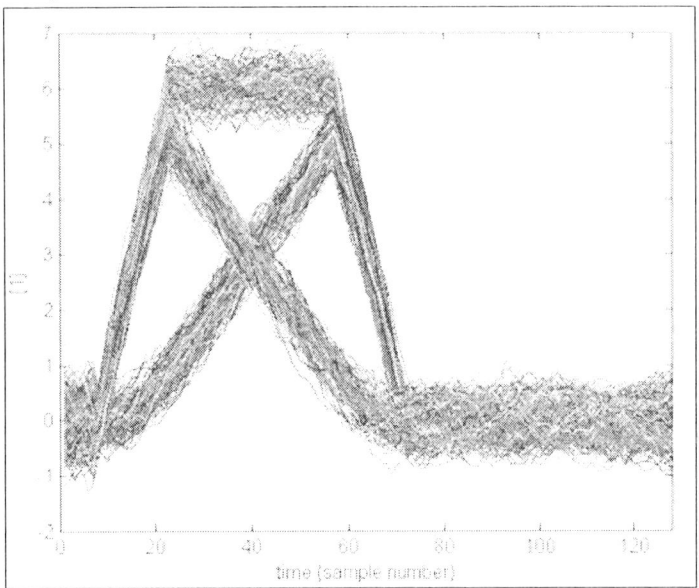

Figure 11. Preprocessed Synthetic Control Chart dataset

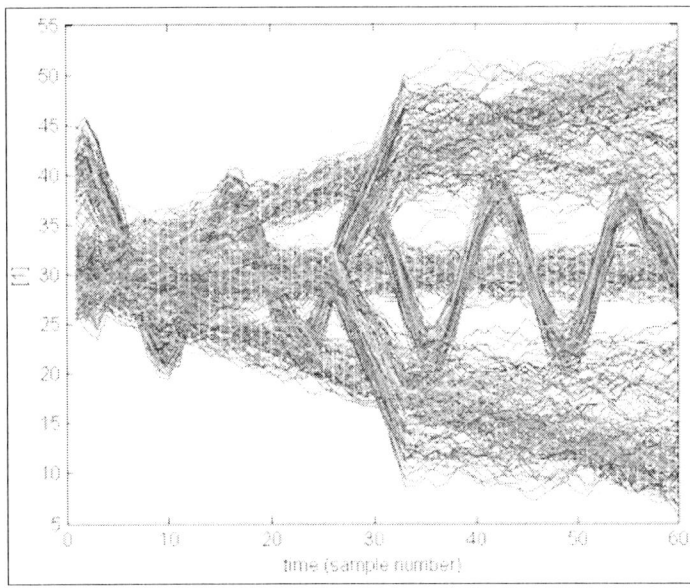

Decreasing trend, (e) Upward shift, and (f) Downward shift. An example from each class in this database is shown in Figure 9.

Setup of Experiments

As indicated earlier, the current specification of the chromosome coding and fitness function evaluation does not allow for shifts in time between signals in the database. Therefore, it is important to first align the sequences with respect to one another, so that a set of consistent basis functions could be computed. In the case of the two benchmarking datasets selected for this study, it is very easy to do so, as the original equations for CBF and SCC have to be only slightly modified to generate an aligned version of the data. By fixing the random variables responsible for the location of the onset of specific events in the signals (*e.g.*, value drop in the bell signal, frequency of the cyclic pattern, *etc.*), a collection of aligned CBF and SCC signals was generated. Then, in order to further simplify the experiments, a mean filter (window of the length of 3% of the signal dimensionality) was applied to smooth the signals in both databases. The complete datasets of aligned and smoothed 600 CBF and SCC signals are shown in Figures 10 and 11, respectively.

In order to assure robustness of our results, a 10-fold cross-validation (CV) scheme was assumed. The algorithm was run 10 times for each of the possible combinations of parameters on each fold of the CV. Thus, each of the investigated combinations was given 100 trials for one of the two datasets. Since there were 28 possible combinations (4 for VEGA: quality of classification

Table 1. Average number of computed basis functions, reconstruction error, and classification accuracy for all three utilized MOEAs with a randomly initialized population: a. CBF, b. SCC

a.

	No change			Annihilation			Scaling		
	#basis funct.	Rec. error	Class error	#basis funct.	Rec. error	Class error	#basis funct.	Rec. error	Class error
Quality of classification									
VEGA	10	23,265.30	0.20	N/A	N/A	N/A	N/A	N/A	N/A
M-VEGA	10	21,449.40	0.19	10	21,443.40	0.19	10	21,449.40	0.19
end-VEGA	10	17,089.60	0.15	10	17,852.50	0.15	10	17,089.60	0.15
Accuracy of approximation									
VEGA	10	23,049.90	0.46	N/A	N/A	N/A	N/A	N/A	N/A
M-VEGA	10	22,781.60	0.45	10	23,030.30	0.45	10	22,598.60	0.45
end-VEGA	10	17,312.60	0.35	10	17,605.00	0.35	10	17,828.80	0.35

b.

	No change			Annihilation			Scaling		
	#basis funct.	Rec. error	Class error	#basis funct.	Rec. error	Class error	#basis funct.	Rec. error	Class error
Quality of classification									
VEGA	6	9,246.65	0	N/A	N/A	N/A	N/A	N/A	N/A
M-VEGA	6	8,343.54	0	6	1,774.31	0	6	1,879.07	0
end-VEGA	5	2,846.90	0	6	596.00	0	6	601.26	0
Accuracy of approximation									
VEGA	6	9,217.64	0	N/A	N/A	N/A	N/A	N/A	N/A
M-VEGA	6	9,076.19	0	7	1,797.02	0.01	6	1,903.12	0.01
end-VEGA	5	2,829.86	0	6	616.00	0.01	6	582.09	0.01

and accuracy of approximation with and without the initial population; 12 for M-VEGA and 12 for end-VEGA: quality of classification and accuracy of approximation with and without the initial population; and three different methods of dealing with the basis functions that are not a part of the reduct), a total number of 9,600 simulations was performed.

For both datasets, the maximum number of allowable basis functions was set to 10. This appeared to be the best trade-off between the minimization of the set of significant basis functions and the maximization of the reconstruction accuracy. The size of the population was set to 30, and the number of generations was set to 100. Mutation probability was initialized with a small random value, and was being adapted along the evolution process (*i.e.*, increased by a small factor if no progress in the fitness functions was observed, and reset when there was an improvement). Crossover probability was set to 0.8. Single-point crossover was utilized.

Results

Tables 1 and 2 present our results in the form of the number of computed basis functions (*i.e.*, reduct size), reconstruction error, and classification accuracy, all averaged over the cross-validation scheme described earlier. The results were obtained for all three utilized MOEAs with different reduct-driven "hard computation" schemes (with exception of VEGA, which does not employ that mechanism), two classification accuracy estimation methods, and two population initialization approaches. For example, end-VEGA, initialized with a random

Table 2. Average number of computed basis functions, reconstruction error, and classification accuracy for all three utilized MOEAs, with the initial population based on a data sample: a. CBF, b. SCC (highlighted combination described in the text)

a.

	No change			Annihilation			Scaling		
	#basis funct.	Rec. error	Class. error	#basis funct.	Rec. error	Class. error	#basis funct.	Rec. error	Class. error
Quality of classification									
VEGA	10	4,918.03	0.21	N/A	N/A	N/A	N/A	N/A	N/A
M-VEGA	8	0.10	0.01	8	0.11	0.01	8	0.10	0.01
end-VEGA	8	0.12	0.01	8	0.12	0.01	8	0.12	0.01
Accuracy of approximation									
VEGA	10	5,027.84	0.46	N/A	N/A	N/A	N/A	N/A	N/A
M-VEGA	8	0.10	0.02	8	0.10	0.02	8	0.11	0.02
end-VEGA	7	0.11	0.02	8	0.11	0.02	8	0.11	0.02

b.

	No change			Annihilation			Scaling		
	#basis funct.	Rec. error	Class. error	#basis funct.	Rec. error	Class. error	#basis funct.	Rec. error	Class. error
Quality of classification									
VEGA	6	0.11	0	N/A	N/A	N/A	N/A	N/A	N/A
M-VEGA	6	0.13	0	6	0.10	0	6	0.10	0
end-VEGA	6	0.14	0	**6**	**0.11**	**0**	6	0.10	0
Accuracy of approximation									
VEGA	6	0.12	0	N/A	N/A	N/A	N/A	N/A	N/A
M-VEGA	6	0.13	0	6	0.10	0	6	0.10	0
end-VEGA	6	0.13	0	6	0.10	0.02	6	0.10	0.02

sample of SCC signals, using the quality of classification coefficient and the annihilation approach (the highlighted combination in Table 2), on average generated 6 basis functions (minimum 5, maximum 8), with the average reconstruction error of 0.11 (minimum 0.02, maximum 0.20), and average classification error of 0.

Based on previously published classification error rates on the CBF and SCC datasets (Kadous, 2002), a conclusion can be drawn that the methodology of classificatory decomposition for signals, in appropriately selected configuration, provides similar capabilities, as several other algorithms used to deal with those problems (*i.e.*, classification accuracy of the order of 90%–100%).

However, the methodology proposed in this work has two major advantages over the other existing techniques. First of all, since the classification process takes place in the space of the coefficients, the computational complexity of the classifier is significantly reduced. In the case of the Cylinder-Bell-Funnel task, for instance, instead of classifying by analyzing time series 128 time stamps each, the categorization process is done by using only 3 (*i.e.*, minimal set yielded by the reduction algorithm) to 10 (*i.e.*, maximal allowable number of basis functions) parameters, *i.e.*, coefficients for the corresponding basis functions. Secondly, since the process of decomposition is being "regulated" by the classification/approximation accuracy measures, the resulting basis functions provide a potentially very useful insight into the mechanisms that govern the underlying classification.

Our results indicate that even though selection of the evolutionary algorithm to be utilized is important, as in most cases, end-VEGA outperformed the two other algorithms, initialization of the evolution is crucial. Although the reconstruction error obtained with end-VEGA initialized with a random population was significantly smaller than in the case of similarly seeded VEGA and M-VEGA (see Table 1), it was still unacceptable. In general, the apparent disproportion in difficulty between the two objectives in CD (*i.e.*, finding a reliable classifier for a problem with a relatively low number of classes seems much easier than reconstructing all the signals in the database using a small number of basis functions), poses a difficult predicament. The algorithm is capable of quickly finding a representation of the data that will assure perfect classification (as in the case of SCC; see Table 1), but will not fulfill the reconstruction constrain. Thankfully, as revealed by our study, the problem can be dealt with in a very straightforward manner. A simple initialization of the starting population in the utilized MOEA, with a randomly selected sample of the data to be decomposed, places the algorithm in such a location in the search space that finding a reliable classificatory decomposition is fast and efficient. Interestingly, however, even with a starting population initialized with "educated guesses," VEGA was still unable to deal with the CBF problem (see Table 2). This supports the claim that selection of the employed EA is important.

The coefficients of quality of classification and accuracy of approximation seem to have comparable capabilities in terms of their usefulness for building a reliable classifier. Even though the quality of classification-based classifier outperformed the accuracy of approximation-driven model of the CBF with random initialization of the population, all the other classification accuracy rates were equivalent.

The influence of the application of the annihilation and scaling schemes was clearly noticeable with the application of a randomly initialized end-VEGA to the SCC problem (see Table 1). Incidentally, this is the dataset for which, even with a random initialization, the algorithm was still able to quickly determine a small number of important basis functions (*i.e.*, an average length of the reduct of 6), with a relatively small reconstruction error. This indicates that for some data, where a small number of basis functions is sufficient, removal of the unnecessary "noise," in the form of nonimportant basis functions, can be beneficial.

SUMMARY

This chapter presented a study of several options for implementing the methodology of classificatory decomposition for signals. The study determined some important aspects of a successful implementation of the method, such as the benefits of initializing the evolutionary algorithm's population with a randomly selected sample of the data to be decomposed. We demonstrated a slight superiority of the coefficient of quality of classification over the coefficient of accuracy of approximation, as used to estimate the classificatory aptitude of the decomposition. We also pointed out that an application of the scheme, based on annihilation or scaling of the "unimportant" basis functions, can significantly improve the reconstruction.

Potential new research directions related to this work may include incorporation of additional parameters in the evolutionary algorithm process (*i.e.* shift in time, basis function contraction or expansion, *etc.*) as well as implementation of time series-sensitive genetic operators that utilize temporal properties of the data; for example, a time-series prediction-driven mutation (*i.e.*, mutation based on the application of a time-series prediction model on several previous samples in the series).

REFERENCES

Alcock, R. J., & Manolopoulos, Y. (1999). Time-series similarity queries employing a feature-based approach. In D.I. Fotiadis, & S.D. Nikolopoulos (Ed.), *7th Hellenic Conference on Informatics* (pp. III.1-9). Ioannina, Greece: University of Ioannina Press.

Chu, S., Keogh, E., Hart, D., & Pazzani, M. (2002). Iterative deepening dynamic time warping for time series. In R. L. Grossman, J. Han, V. Kumar, H. Mannila, & R. Motwani (Ed.), *2nd SIAM International Conference on Data Mining*. Arlington, VA: SIAM. Retrieved December 7, 2006, from http://www.siam.org/meetings/sdm02/proceedings/sdm02-12.pdf

Coello, C.A., Van Veldhuizen, D.A., & Lamont, G.B. (2002). *Evolutionary algorithms for solving multi-objective problems*. Norwell, MA: Kluwer.

Cohon, J.L. (1978). *Multiobjective programming and planning*. New York: Academic Press.

Deb, K. (2001). *Multi-objective optimization using evolutionary algorithms*. New York: John Wiley & Sons.

Fayyad, U.M., Piatetsky-Shapiro, G., Smyth, P., & Uthurusamy, R. (1996). *Advances in knowledge discovery and data mining*. Menlo Park, CA: AAAI Press/The MIT Press.

Flury, B. (1988). *Common principal components and related multivariate models*. New York: John Wiley & Sons.

Gen, M., & Cheng, R. (2000). *Genetic algorithms and engineering optimization*. New York: John Wiley & Sons.

Hyvarinen, A., & Oja, E. (2000). Independent component analysis: Algorithms and applications. *Neural Networks, 13*(4), 411-430.

Kadous, M.W. (2002). *Temporal classification: Extending the classification paradigm to multivariate time series*. Unpublished doctoral dissertation, University of New South Wales, Australia.

Keogh, E., & Folias, T. (2002). *The UCR time series data mining archive*. Riverside, CA, University of California—Computer Science & Engineering Department. Retrieved December 7, 2006, from http://www.cs.ucr.edu/~eamonn/TSDMA/index.html

Keogh, E., & Kasetty, S. (2002). On the need for time series data mining benchmarks: A survey and empirical demonstration. In *Proceedings of the 8th ACM SIGKDD International Conference on Knowledge Discovery and Data Mining* (pp. 102-111). Edmonton, Alberta, Canada: ACM.

Keogh, E., & Pazzani, M. (1998). An enhanced representation of time series which allows fast and accurate classification, clustering and relevance feedback. In R. Agrawal & P. Stolorz (Ed.), *4th International Conference on Knowledge Discovery and Data Mining* (pp. 239-241). New York: AAAI Press.

Komorowski, J., Pawlak, Z., Polkowski, L., & Skowron, A. (1999). Rough sets: A tutorial. In S.K. Pal & A. Skowron (Ed.), *Rough fuzzy hybridization: A new trend in decision-making* (pp. 3-98). Singapore: Springer-Verlag.

Kreyszig, E. (1978). *Introductory functional analysis with applications.* New York: John Wiley & Sons.

Laumanns, M., Zitzler, E., & Thiele, L. (2000). A unified model for multi-objective evolutionary algorithms with elitism. In A. Zalzala et al. (Ed.), *2000 Congress on Evolutionary Computation* (pp. 46-53). San Diego, CA: IEEE Press.

Marek, W., & Pawlak, Z. (1984). Rough sets and information systems. *Fundamenta Informaticae, 17*(1), 105-115.

Milanova, M., Smolinski, T.G., Boratyn, G.M., Zurada, J.M., & Wrobel, A. (2002). Correlation kernel analysis and evolutionary algorithm-based modeling of the sensory activity within the rat's barrel cortex. *Lecture Notes in Computer Science, 2388*, 198-212.

Øhrn, A. (2001). *ROSETTA technical reference manual.* Retrieved December 7, 2006, from http://rosetta.lcb.uu.se/general/resources/manual.pdf

Pawlak, Z. (1982). Rough sets. *International Journal of Computer and Information Sciences 11*, 341-356.

Pawlak, Z. (1991). *Rough sets: Theoretical aspects of reasoning about data.* Norwell, MA: Kluwer.

Rajan, J.J. (1994). *Time series classification.* Unpublished doctoral dissertation, University of Cambridge, UK.

Saito, N. (1994). *Local feature extraction and its application using a library of bases.* Unpublished doctoral dissertation.

Schaffer, J. D. (1984). *Some experiments in machine learning using vector evaluated genetic algorithms.* Unpublished doctoral dissertation, Vanderbilt University, TN.

Smolinski, T.G. (2004). *Classificatory decomposition for time series classification and clustering.* Unpublished doctoral dissertation, University of Louisville, KY.

Smolinski, T. G., Boratyn, G. M., Milanova, M., Buchanan, R, & Prinz, A. A. (2006). Hybridization of independent component analysis, rough sets, and multi-objective evolutionary algorithms for classificatory decomposition of cortical evoked potentials. *Lecture Notes in Artificial Intelligence, 4146* (pp. 174-183).

Smolinski, T.G., Boratyn, G.M., Milanova, M., Zurada, J.M., & Wrobel, A. (2002). Evolutionary algorithms and rough sets-based hybrid approach to classificatory decomposition of cortical evoked potentials. *Lecture Notes in Artificial Intelligence, 2475* (pp. 621-628).

Smolinski, T.G., Buchanan, R., Boratyn, G.M., Milanova, M., & Prinz, A.A. (2006). Independent component analysis-motivated approach to classificatory decomposition of cortical evoked potentials. *BMC Bioinformatics, 7*(Suppl 2), S8.

Smolinski, T.G., Milanova, M., Boratyn, G.M., Buchanan, R., & Prinz, A.A. (2006). Multi-objective evolutionary algorithms and rough sets for decomposition and analysis of cortical evoked potentials. In Y-Q. Zhang & T.Y. Lin (Ed.), *IEEE International Conference on Granular Computing*. Atlanta, GA: IEEE Press.

Swiniarski, R. (1999). Rough sets and principal component analysis and their applications in data model building and classification. In S.K. Pal, & A. Skowron (Ed.), *Rough fuzzy hybridization: A new trend in decision-making* (pp. 275-300). Singapore: Springer-Verlag.

Wroblewski, J. (1995). Finding minimal reducts using genetic algorithms. In P.P. Wang (Ed.), *2nd Annual Joint Conference on Information Sciences* (pp. 186-189). Wrightsville Beach, NC.

Chapter XI
Two Rough Set Approaches to Mining Hop Extraction Data

Jerzy W. Grzymala-Busse
University of Kansas, USA and Institute of Computer Science, PAS, Poland

Zdzislaw S. Hippe
University of Information Technology and Management, Poland

Teresa Mroczek
University of Information Technology and Management, Poland

Edward Roj
Fertilizer Research Institute, Poland

Boleslaw Skowronski
Fertilizer Research Institute, Poland

ABSTRACT

Results of our research on using two approaches, both based on rough sets, to mining three data sets describing bed caking during the hop extraction process, are presented. For data mining we used two methods: direct rule induction by the and generation of belief networks associated with conversion belief networks into rule sets by the BeliefSEEKER system. Statistics for rule sets are presented, including an error rate. Finally, six rule sets were ranked by an expert. Our results show that both our approaches to data mining are of approximately the same quality.

INTRODUCTION

In the past, addition of hop to beer was quite simple: hop was added to the boiling wort, while spent hop formed a filter bed for the wort. Bitterness was controlled by varying the used hop. A few decades ago the process was changed, and the technology of beer production increased in complexity and sophistication.

The following factors are essential for beer production:

- Reduction of hop extraction cost
- Unification of production process

- Environmental protection regulations (pesticide residues, nitrate residues, heavy metals, etc.)
- Requirements of particular consumer groups for special beers

Currently, fresh hop cones are dried in a dryer and then compressed. Hop cones are usually traded in bales that have been compressed to reduce shipping volumes. Some of the main types of the product on the market are either hop pellets, made from unmilled hop cones, or hop powder pellets, made from milled hop cones and presented at different levels of enrichment.

Hop pellets are prepared by freeing hops from foreign matter (stems, rhizomes, metal rods, wires, etc.) and then pelleting them without milling. Hop pellets are not significant in modern beer production, but still can be seen in use in some countries.

Hop powder pellets in various forms are dominant in the brewing industry around the world. Basic preparation consists of removing foreign matter, milling in a hammer mill, blending batches of several hop bales together for product consistency, pelleting through a standardized pellet die, cooling and packing in packs.

Principles of hop extraction have been known for 150 years, when extraction in water and ethanol was, for the first time, successfully applied. An ethanol extraction is still in use, but the predominant extracts are either organic or the CO_2 based extracts (in liquid or supercritical phase). Recently, due to the properties of CO_2, the supercritical extraction is getting more popular among commercial producers.

The soft resins and oils comprise usually about 15% of the hop mass (new varieties sometimes exceed that value) and strongly depend on hop variety (Stevens, 1987). An exemplary composition of hop extract is presented in Table 1 (the example is based on a mid-alpha acid content variety).

There are other commercial methods to extract oil or resins from hop or other commercial plants, such as herbs, flax, hemp, and so forth. Organic solvent-based methods are commonly applied worldwide. But during the last two decades, a

Table 1. Composition of hop extract

Components		Contents
Total resins		15 %
	Soft resins - alpha acids	8 %
	Soft resins - beta acids	4 %
	Hard and uncharacterized soft resins	3 %
Essential oils		1 %
Tannins		4 %
Proteins		15 %
Water		9.5 %
Monosacharides		2 %
Lipids and waxes		3 %
Amino acids		0.1 %
Pectin		2 %
Ash		8 %
Cellulose, lignin, and so forth.		40.4 %

supercritical CO_2 extraction has been used intensively in various applications replacing the classic organic solvent methods (Chassagnes-Mendez, Machado, Araujo, Maia, & Meireles, 2000; Skowronski, 2000; Skowronski & Mordecka, 2001). The extraction process under supercritical conditions is usually carried out in a high-pressure reactor vessel filled with a preprocessed material in the form of tablets, pellets, granules, and so forth., thus, forming a granular bed through which flows a compressed gas. The gas under a supercritical condition is able to dissolve oil products contained in the bed. A mixture of oil products and solvent (CO_2) flows together through a low-pressure separator where liquid and gas fractions can be efficiently separated. The oil mixture is ecologically clean and does not require any additional treatment. The spent gas expands to the atmosphere or is recirculated after compression. Produced liquid resin and oil fractions are directed to the storage tanks.

As we have already mentioned, the extraction process uses pressed organic material (cones or leafs) mainly in the form of tablets or extrusions in order to load more material for each batch. After the bed is loaded, the pressure and temperature are increased to reach a supercritical condition. Once a supercritical condition is reached, the gas flows through the bed at a constant pressure and temperature. After a few hours of running the process, the tank is decompressed to ambient pressure and the batch process is finished. Sometimes during the extraction process, high-pressure drops occur and, as a result, the bed is caked. This event has a negative influence on the extraction process. A pressure drop reduces gas circulation through the circulation loop (pressure tank and the rest of the equipment) and therefore extends the extraction time. This also strongly affects process efficiency by increasing production costs.

The objective of our work was to investigate system variables that would contribute to the bed-caking tendency during the hop extraction process. Commercial experience indicated two parameters determine the bed's state, namely, content of extract and moisture of the granular bed. The content of the extract concentration varies significantly, between 14% and 32%. For that reason, three easy–to-identify input parameters have been identified, namely:

- Variety
- Content of extract (α_{kg})
- Moisture of the bed (m_{wg})

In one of the data sets, two additional variables were used:

- Content of alpha acid in spent hop, that is, remaining concentration (α_{kw}), and
- Moisture of the postprocessed bed (m_{ww}).

Certainly, there are other unknown factors that have not been considered here due to essential measurement problems. Output variables were:

- The caking rate of the bed and
- Extraction time

In fact, there is a strong relationship between the caking rate and time of extraction. If a bed is caked, then extraction time increases. The caking rate description is based on process operator reports. The following four grades of bed caking rates have been used:

1. loose material,
2. loose material with dust,
3. caked and slightly raised bed, and
4. strongly caked and raised bed.

The extraction period has been also quantified according to the following principles:

1. A standard time of extraction over a loose bed.

2. Extension of time by 25% with reference to the standard time.
3. Extension of extraction time by 50% with reference to the standard time.
4. Extension of extraction time by 75% with reference to the standard time.

The experimental data have been processed using two data-mining approaches: rule induction, represented by the MLEM2 algorithm, an option of the LERS (learning from examples based on rough sets) system and generation of Bayesian networks by the BeliefSEEKER system.

A preliminary version of this paper was presented at the ISDA'2005 Conference (Grzymala-Busse, Hippe, Mroczek, Roj, & Skowronski, 2005). In this chapter, we present different results, since we changed our methodology, running MLEM2 directly from raw data instead of running LEM2 on discretized data as in Grzymala-Busse et al. (2005). Here we consider both certain and possible rule sets induced by LEM2. Additionally, we present new results on ranking six rule sets by an expert.

DISCRETIZATION

In our data sets, all attributes, except *Variety*, were numerical. The numerical attributes should be discretized before rule induction. The data-mining system LERS uses for discretization a number of discretization algorithms (Chmielewski & Grzymala-Busse, 1996). To compare both approaches, LERS and BeliefSEEKER, used for data mining, all data sets were discretized using the same LERS discretization method, namely, a polythetic agglomerative method of cluster analysis (Everitt, 1980). *Polythetic* methods use all attributes, while *agglomerative* methods begin with all cases being singleton clusters. Our method was also *hierarchical*, that is, the final structure of all formed clusters was a tree.

More specifically, we selected to use the median cluster analysis method (Everitt, 1980) as a basic clustering method. Different numerical attributes were standarized by dividing the attribute values by the corresponding attribute's standard deviation (Everitt, 1980). Cluster formation was started by computing a distance matrix between every cluster. New clusters were formed by merging two existing clusters that were the closest to each other. When such a pair was founded (clusters b and c), they were fused to form a new cluster d. The formation of the cluster d introduces a new cluster to the space, and hence its distance to all the other remaining clusters must be recomputed. For this purpose, we used the Lance and Williams Flexible Method (Everitt, 1980). Given a cluster a and a new cluster d to be formed from clusters b and c, the distance from d to a was computed as:

$$d_{ad} = d_{da} = 0.5 * d_{ab} + 0.5 * d_{ac} - 0.25 * d_{bc}.$$

At any point during the clustering process, the formed clusters induce a partition on the set of all cases. Cases that belong to the same cluster are indiscernible by the set of numerical attributes. Therefore, we should continue cluster formation until the level of consistency of the partition formed by clusters is equal to or greater than the original data's level of consistency.

Once clusters are formed, the postprocessing starts. First, all clusters are projected on all attributes. Then the resulting intervals are merged to reduce the number of intervals and, at the same time, preserving consistency. Merging of intervals begins from safe merging where, for each attribute, neighboring intervals labeled by the same decision value are replaced by their union. The next step of merging intervals is based on checking every pair of neighboring intervals, whether their merging will result in preserving consistency. If so, intervals are merged permanently. If not, they are marked as unmergeable. Obviously, the order in which pairs of intervals are selected affects the final outcome. In our experiments, we started from attributes with the largest conditional entropy of the decision-given attribute.

LERS

The data system LERS (Grzymala-Busse, 1992, 1997, 2002) induces rules from inconsistent data, that is, data with conflicting cases. Two cases are conflicting when they are characterized by the same values of all attributes, but they belong to different concepts. LERS handles inconsistencies using rough-set theory, introduced by Z. Pawlak in 1982 (1982, 1991). In rough-set theory, lower and upper approximations are computed for concepts involved in conflicts with other concepts.

Rules induced from the lower approximation of the concept *certainly* describe the concept; hence, such rules are called *certain* (Grzymala-Busse, 1988). On the other hand, rules induced from the upper approximation of the concept describe the concept *possibly*, so these rules are called *possible* (Grzymala-Busse, 1988). In general, LERS uses two different approaches to rule induction: one is used in machine learning, the other in knowledge acquisition. In machine learning, the usual task is to learn the smallest set of minimal rules, describing the concept. To accomplish this goal, LERS uses three algorithms: LEM1, LEM2, and MLEM2 (LEM1, LEM2, and MLEM2 stand for learning from examples module, version 1, 2, and modified, respectively). In our experiments, rules were induced using the algorithm MLEM2.

MLEM2

The LEM2 option of LERS is most frequently used for rule induction since, in most cases, it gives better results. LEM2 explores the search space of attribute-value pairs. Its input data set is a lower or upper approximation of a concept, so its input data set is always consistent. In general, LEM2 computes a local covering and then converts it into a rule set. We will quote a few definitions to describe the LEM2 algorithm (Chan & Grzymala-Busse, 1991; Grzymala-Busse, 1997).

The LEM2 algorithm is based on an idea of an attribute-value pair block. Let U be the set of all cases (examples) of the data set. For an attribute-value pair $(a, v) = t$, a *block* of t, denoted by $[t]$, is a set of all cases from U, such that for attribute a have value v. Let B be a nonempty lower or upper approximation of a concept represented by a decision-value pair (d, w). Set B depends on a set T of attribute-value pairs $t = (a, v)$ if and only if:

$$\emptyset \neq [T] = \bigcap_{t \in T}[t] \subseteq B.$$

Set T is a *minimal complex* of B if and only if B depends on T and no proper subset T' of T exists such that B depends on T'. Let **T** be a nonempty collection of nonempty sets of attribute-value pairs. Then **T** is a *local covering* of B if and only if the following conditions are satisfied:

1. Each member T of **T** is a minimal complex of B.

2. $\bigcup_{T \in \mathbf{T}}[T] = B$

3. **T** is minimal, that is, **T** has the smallest possible number of members.

The procedure LEM2, based on rule induction from local coverings, is presented below.

Procedure LEM2
(**input:** a set B;
output: a single local covering **T** of set B);
begin
  $G := B$;
  T$:= \emptyset$;
  while $G \neq \emptyset$ **do**
    begin
    $T := \emptyset$
    $T(G) := \{t \mid [t] \cap G \neq \emptyset\}$;
    while $T = \emptyset$ **or not** $([T] \subseteq B)$ **do**
      begin

select a pair $t \in T(G)$ with the highest attribute priority, if a tie occurs, select a pair $t \in T(G)$ such that $|[t] \cap G|$ is maximum; if another tie occurs, select a pair $t \in T(G)$ with the smallest cardinality of $[t]$; if a further tie occurs, select first pair:

$$T := T \cup \{t\};$$
$$G := [t] \cap G;$$
$$T(G) := \{t|[t] \cap G \neq \varnothing\};$$
$$T(G) := T(G) - T;$$

end; {while}
    for each t in T do
        if $[T-\{t\}] \subseteq B$ then $T := T - \{t\}$;
    T := **T** $\cup$ $\{T\}$;
    $G := B - \cup_{T \in \mathbf{T}} [T]$;
end {while};
    for each $T \in \mathbf{T}$ do
        if $\cup_{S \in \mathbf{T}-\{T\}} [S] = B$ then **T** := **T** $-$ $\{T\}$;
end {procedure}.

For a set X, $|X|$ denotes the cardinality of X.

MLEM2, a modified version of LEM2, processes numerical attributes differently than symbolic attributes. For numerical attributes, MLEM2 sorts all values of a numerical attribute. Then it computes cutpoints as averages for any two consecutive values of the sorted list. For each cutpoint q, MLEM2 creates two blocks: the first block contains all cases for which values of the numerical attribute are smaller than q, the second block contains remaining cases, that is, all cases for which values of the numerical attribute are larger than q. The search space of MLEM2 is the set of all blocks computed this way, together with blocks defined by symbolic attributes. Starting from that point, rule induction in MLEM2 is conducted the same way as in LEM2. At the very end, MLEM2 simplifies rules by merging appropriate intervals for numerical attributes.

Classification System

Rule sets, induced from data sets, are used mostly to classify new, unseen cases. Such rule sets may be used in rule-based expert systems.

There are a few existing classification systems, for example, associated with rule induction systems LERS or AQ. A classification system used in LERS is a modification of the well-known bucket brigade algorithm (Grzymala-Busse, 1997; Stefanowski, 2001). In the rule induction system AQ, the classification system is based on a rule estimate of probability. Some classification systems use a decision list in which rules are ordered; the first rule that matches the case classifies it. In this section, we will concentrate on a classification system used in LERS.

The decision to which concept a case belongs to is made on the basis of three factors: *strength*, *specificity*, and *support*. These factors are defined as follows: *strength* is the total number of cases correctly classified by the rule during training. *Specificity* is the total number of attribute-value pairs on the left-hand side of the rule. The matching rules with a larger number of attribute-value pairs are considered more specific. The third factor, *support*, is defined as the sum of products of strength and specificity for all matching rules indicating the same concept. The concept C for which the support, that is, the following expression:

$$\sum_{R \in Rul} Strength(R) * Specifity(R),$$

is the largest is the winner, and the case is classified as being a member of C, where *Rul* denotes the set of all matching rules R describing the concept C.

In the classification system of LERS, if complete matching is impossible, all partially matching rules are identified. These are rules with at least one attribute-value pair matching the corresponding attribute-value pair of a case. For any partially matching rule R, the additional factor, called *matching_factor* (R), is computed. Matching_factor (R) is defined as the ratio of the number of matched attribute-value pairs with a case to the total number of attribute-value pairs. In partial matching, the concept C for which the following expression is the largest:

$$\sum_{R \in Rul'} Matching\_factor(R) * Strength(R) * Specifity(R),$$

is the winner, and the case is classified as being a member of C, where *Rul'* denotes the set of all matching rules R describing the concept C.

VALIDATION

The most important performance criterion of rule induction methods is the error rate. If the number of cases is less than 100, the *leaving-one-out* method is used to estimate the error rate of the rule set. In leaving-one-out, the number of learn-and-test experiments is equal to the number of cases in the data set. During the *i*-th experiment, the *i*-th case is removed from the data set, a rule set is induced by the rule induction system from the remaining cases, and the classification of the omitted case by rules produced is recorded. The error rate is computed as the ratio of the total number of misclassifications to the number of cases.

On the other hand, if the number of cases in the data set is greater than or equal to 100, the ten-fold cross-validation should be used. This technique is similar to leaving-one-out in that it follows the learn-and-test paradigm. In this case, however, all cases are randomly reordered, and then a set of all cases is divided into 10 mutually disjoint subsets of approximately equal size. For each subset, all remaining cases are used for training, that is, for rule induction, while the subset is used for testing. This method is used primarily to save time at the negligible expense of accuracy.

Ten-fold cross validation is commonly accepted as a standard way of validating rule sets. However, using this method twice, with different preliminary random reordering of all cases yields, in general, two different estimates for the error rate (Grzymala-Busse, 1997).

For large data sets (at least 1,000 cases), a single application of the train-and-test paradigm may be used. This technique is also known as *holdout*. Two thirds of cases should be used for training, one third for testing.

In yet another way of validation, *resubstitution*, it is assumed that the training data set is identical with the testing data set. In general, an estimate for the error rate is here too optimistic. However, this technique is used in many applications. For some applications, it is debatable whether the ten-fold cross-validation is better or not, see, for example, Braga-Neto, Hashimoto, Dougherty, Nguyen, & Carroll, 2004).

BeliefSEEKER

BeliefSEEKER is a computer program generating belief networks for any type of decision tables prepared in the format described in Pawlak (1995). Various algorithms may be applied for learning such networks (Mroczek, Grzymala-Busse, & Hippe, 2004). The development of belief networks is controlled by a specific parameter, representing the maximum dependence between variables, known as marginal likelihood and defined as follows:

$$ML = \prod_{i=1}^{v} \prod_{j=1}^{q_i} \frac{\Gamma(\alpha_{ij})}{\Gamma(\alpha_{ij} + n_{ij})} \prod_{k=1}^{c_i} \frac{\Gamma(\alpha_{ijk} + n_{ijk})}{\Gamma(\alpha_{ijk})},$$

where v is the number of nodes in the network, $i = 1,...,v$, q_i is the number of possible combinations of parents of the node X_i (if a given attribute does not contain nodes of the parent type, then $q_i = 1$), c_i is the number of classes within the attribute X_i, $k = 1,..., c_i$, n_{ijk} is the number of rows in the database, for which parents of the attribute X_i have value j, and this attribute has the value of k, and α_{ijk} and n_{ij} are parameters of the initial Dirichlet's distribution (Heckerman, 1995).

It is worth to mention that the developed learning models (belief networks) can be converted into some sets of belief rules, characterized by a specific parameter called a *certainty factor, CF*, that reveals indirectly the influence of the most significant descriptive attributes on the dependent variable. Also, to facilitate the preliminary evaluation of generated rules, an additional mechanism supports the calculation of their specificity, strength, generality, and accuracy (Hippe, 1996). Characteristic features of the system are:

- Capability to generate various exhaustive learning models (Bayesian networks) for different values of Dirichlet's parameter α and the certainty factor *CF* (Heckerman, 1995).
- Capability to convert generated belief networks into possible rule sets.
- Built-in classification mechanism for unseen cases.

RESULTS OF EXPERIMENTS

Tables 2 and 3 show basic statistics about our three data sets representing granular bed caking during hop extraction. In Tables 4–6, results of our experiments are presented. Note that for every data set, the MLEM2 rule induction algorithm was run once, while BeliefSEEKER generated many rule sets for selected values of α and *CF*. Table 4 shows error rates, computed by ten-fold cross validation for MLEM2 and resubstitution for BeliefSEEKER. Resubstitution as a validation technique for BeliefSEEKER has been selected because of time concerns. Furthermore, Table 4 shows the best results of BeliefSEEKER in terms of error rates. These rule sets were generated with $\alpha = 50$ and $CF = 0.4$ for the first data set, with $\alpha = 10$ and $CF = 0.5$ for the second data set, and with

Table 2. Data sets

Data set	Number of cases	Number of attributes	Number of concepts
Data-set-1	288	5	4
Data-set-2	288	3	4
Data-set-3	179	3	4

Table 3. Number of conflicting cases

Data set	Before discretization	After discretization
Data-set-1	2	6
Data-set-2	56	137
Data-set-3	23	53

Table 4. Error rates

Data set	MLEM2		Belief-SEEKER
	Certain rule set	Possible rule set	
Data-set-1	9.38%	9.72%	6.60%
Data-set-2	11.11%	9.72%	10.76%
Data-set-3	9.50%	8.94%	8.94%

α = 1 and CF = 0.5 for the third data set.

Table 5 presents the cardinalities of all nine rule sets from Table 4. Finally, Table 6 shows ranking of all possible rule sets induced by MLEM2 and BeliefSEEKER. We restricted our attention only to possible rule sets, since the procedure was conducted manually by an expert who ranked rule after rule, judging rules on scale from 1 (unacceptable) to 5 (excellent). Table 6 presents averages of all rule rankings for all six rule sets.

CONCLUSION

This section is divided into three subsections, describing results of MLEM2, BeliefSEEKER, and then general conclusions.

1. Discussion of results obtained by running the MLEM2 system.

The basic results obtained by running the MLEM2 system against our three data sets are:

- *Marynka* variety and its modification contains a lesser amount of extract than *Magnum* variety and its modification.
- The bed state is desirable, 1 or 2, if:
  - Content of extract in a bed is ranged below 22%, and moisture of the bed is ranged below 7%, or a content of extract is below 14% and moisture of the bed is ranged below 10.95%.
  - Content of extract in *Marynka* variety does not exceed 14% and moisture 8.35%,
  - content of extract in *Magnum* variety does nor exceed 20% and moisture 8.15%,
- Both varieties, *Marynka* and *Magnum*, can be caked sometimes, but the absolute content of extract is not a main contributor of caking.
- Moisture of granulated hops plays a stabilizing role at the determined ranges; below the determined value, the extracted bed is loose, and above this value it tends to cake.

2. Discussion of results obtained by using the Bayesian approach.

The basic results obtained by using the Bayesian networks against our three data sets are:

Table 5. Number of rules

Data set	MLEM2		Belief-SEEKER
	Certain rule set	Possible rule set	
Data-set-1	34	34	37
Data-set-2	28	22	9
Data-set-3	12	22	3

Table 6. Ranking of rule sets

Data set	MLEM2	BeliefSEEKER
Data-set-1	4.48	4.62
Data-set-2	4.39	4.22
Data-set-3	3.42	1.00

- *Marynka* variety and its modification contains less extract compared to the *Magnum* variety and its modification.
- Caking ability is a common feature for both varieties and their modifications. In general, the less amount of extract in the granulated hops of the particular variety, the less tendency to caking.
- Higher moisture content in the bed is a favorable caking factor.
- Cake ability usually extends extraction time.

GENERAL CONCLUSIONS

Experts found that the results of both data-mining systems: MLEM2 and BeliefSEEKER are consistent with the current state of the art in the area. In general, possible rule sets were highly ranked by the expert for the first two data sets. The possible rule sets induced from the third data set were ranked lower.

Moreover, stating knowledge in the form of rule sets and Bayesian networks makes it possible to utilize knowledge in a form of an expert system. Thus, results of our research may be considered as a successful first step towards building a system of monitoring the process of hop extraction.

Results of MLEM2 and BeliefSEEKER, in terms of an error rate, indicate that both systems are approximately of the same quality.

REFERENCES

Braga-Neto, U., Hashimoto, R., Dougherty, R. E, Nguyen, D.V., & Carroll, R.J. (2004). Is cross-validation better than resubstitution for ranking genes? *Bioinformatics, 20,* 253-258.

Chan, C.-C., & Grzymala-Busse, J.W. (1991). *On the attribute redundancy and the learning programs ID3, PRISM, and LEM2* (TR-91-14). Lawrence, KS: Department of Computer Science, University of Kansas.

Chassagnez-Mendez, A.L., Machado, N.T., Araujo, M.E., Maia, J.G., & Meireles, M.A. (2000). Supercritical CO_2 extraction of eurcumis and essential eil from the rhizomes of turmeric (Curcuma longa L.), *Ind. Eng. Chem. Res., 39,* 4729-4733.

Chmielewski, M. R., & Grzymala-Busse, J. W. (1996). Global discretization of continuous attributes as preprocessing for machine learning. *International Journal of Approximate Reasoning, 15,* 319-331.

Everitt, B. (1980). *Cluster analysis* (2nd ed.). London: Heinemann Educational Books.

Grzymala-Busse, J.W. (1988). Knowledge acquisition under uncertainty—A rough set approach. *Journal of Intelligent & Robotic Systems, 1,* 3-16.

Grzymala-Busse, J.W. (1992). LERS—A system for learning from examples based on rough sets. In R. Slowinski (Ed.), *Intelligent decision support. Handbook of applications and advances of the rough set theory* (pp. 3-18). Dordrecht, Boston, London: Kluwer Academic Publishers.

Grzymala-Busse, J.W. (1997). A new version of the rule induction system LERS. *Fundamenta Informaticae, 31,* 27-39.

Grzymala-Busse, J.W. (2002). MLEM2: A new algorithm for rule induction from imperfect data. In *Proceedings of the 9th International Conference on Information Processing and Management of Uncertainty in Knowledge-Based Systems* (pp. 243-250). Annecy, France.

Grzymala-Busse, J.W., Hippe, Z.S., Mroczek, T., Roj, E., & Skowronski, B. (2005). Data mining analysis of granular bed caking during hop extraction. In *Proceedings of the ISDA'2005, Fifth International Conference on Intelligent System Design and Applications* (pp. 426-431). IEEE Computer Society, Wroclaw, Poland.

Heckerman, D. (1995). *A tutorial on learning Bayesian networks* (MSR-TR-95-06). Retrieved from heckerman@microsoft.com

Hippe, Z.S. (1996). Design and application of new knowledge engineering tool for solving real world problems, *Knowledge-Based Systems, 9*, 509-515.

Mroczek, T., Grzymala-Busse, J.W., & Hippe, Z.S. (2004). Rules from belief networks: A rough set approach. In S. Tsumoto, R. Slowinski, J. Komorowski, and J.W. Grzymala-Busse (Eds.), *Rough sets and current trends in computing* (pp. 483-487). Berlin, Heidelberg, New York: Springer-Verlag.

Pawlak, Z. (1982). Rough sets. *International Journal of Computer and Information Sciences, 11*, 341-356.

Pawlak, Z. (1991). *Rough sets. Theoretical aspects of reasoning about data*. Dordrecht, Boston, London: Kluwer Academic Publishers.

Pawlak, Z. (1995). Knowledge and rough sets (in Polish). In W. Traczyk (Ed.), *Problems of artificial intelligence* (pp. 9-21). Warsaw, Poland: Wiedza i Zycie.

Skowronski, B. (2000). Interview by L. Dubiel. Hop extract—Polish at last (in Polish), *Przem. Ferment. i Owocowo -Warzywny, 9*, 30-31.

Skowronski B., & Mordecka, Z. (2001). Polish plant for supercritical extraction of hop (in Polish), *Przem. Chem., 80*, 521-523.

Stefanowski, J. (2001). *Algorithms of decision rule induction in data mining*. Poznan, Poland: Poznan, University of Technology Press.

Stevens, R. (1987). *Hops, An introduction to brewing science and technology*. Series II, Vol. I, p. 23. London: Institute of Brewing.

Chapter XII
Rough Sets for Discovering Concurrent System Models from Data Tables

Krzysztof Pancerz
University of Information Technology and Management in Rzeszów and College of Management and Public Administration in Zamosc, Poland

Zbigniew Suraj
University of Rzeszów and State School of Higher Education in Jaroslaw, Poland

ABSTRACT

This chapter constitutes the continuation of a new research trend binding rough-set theory with concurrency theory. In general, this trend concerns the following problems: discovering concurrent system models from experimental data represented by information systems, dynamic information systems, or specialized matrices, a use of rough-set methods for extracting knowledge from data, a use of rules for describing system behaviors, and modeling and analyzing of concurrent systems by means of Petri nets on the basis of extracted rules. Some automated methods of discovering concurrent system models from data tables are presented. Data tables are created on the basis of observations or specifications of process behaviors in the modeled systems. Proposed methods are based on roug- set theory and colored Petri net theory.

INTRODUCTION

Data mining and knowledge discovery are crucial and current research problems in the modern computer sciences (Cios, Pedrycz, & Swiniarski, 1998). Discovering hidden relationships in data is a main goal of machine learning. In a lot of cases, data are generated by concurrent processes. Therefore, discovering concurrent system models is essential from the point of view of understanding the nature of modeled systems, as well as explaining their behaviours. A notion of concurrent systems can be understood widely. In a general case, a concurrent system is a system consisting of some processes whose local states can coexist together and they are partly independent. For example, we can treat systems consisting of economic processes, financial processes, biological processes, genetic processes, meteorological processes, and so forth, as concurrent systems. Subject matter of this chapter concerns methods of concurrent system modelling on the basis of observations or specifications of their behaviours given in the form of different kinds of data tables. Data tables can include results of observations or measurements of specific states of concurrent processes. In this case, created models of concurrent systems are useful for analyzing properties of modeled systems, discovering the new knowledge about behaviours of processes, and so forth. Data tables can also include specifications of behaviours of concurrent processes. Then, created models can be a tool for verification of those specifications; for example, during designing concurrent systems. Methods presented in this chapter can be used, for example, in system designing or analyzing, data analysis, forecasting.

In this chapter, some automatized methods of discovering concurrent system models from data tables are presented. Data tables are created on the basis of observations or specifications of process behaviours in the modeled systems. Proposed methods are based on rough-set theory (Pawlak, 1991) and colored Petri net theory (Jensen, 1997). Rough-set theory, introduced by Z. Pawlak, provides advanced and efficient methods of data analysis and knowledge extraction. Petri nets are the graphical and mathematical tool for modelling of different kinds of phenomena, especially those where actions executed concurrently play a significant role. Model construction is also aided by methods of Boolean reasoning (Brown, 1990). Boolean reasoning makes a base for solving a lot of decision and optimization problems.

Especially, it plays a special role during generation of decision rules. Data describing examined phenomena and processes are represented by means of information systems (in the Pawlak's sense) (Pawlak, 1991), dynamic information systems (Suraj, 1998), or specialized matrices of forbidden states and matrices of forbidden transitions (Suraj & Pancerz, 2005). An information system can include the knowledge about global states of a given concurrent system, understood as vectors of local states of processes making up the concurrent system, whereas a dynamic information system can include, additionally, the knowledge about transitions between global states of the concurrent system. Specialized matrices are designed for specifying undesirable states of a given concurrent system (i.e., those states that cannot hold together), and undesirable transitions between its states. Decomposition of data tables into smaller subtables connected by suitable rules is also possible. Those subtables make up modules of a system. Local states of processes, represented in a given subtable, are linked by means of a functional dependency.

Approaches considered in this chapter are based on the assumption that data collected in data tables include only the partial knowledge about modeled systems. Nevertheless, such partial knowledge is sufficient to construct suitable mathematical models. The remaining knowledge (or, in the sequel, a part of it) can be discovered on the basis of created models.

Two structures of concurrent system models are considered: namely synchronous and asynchronous. In a case of modeling based on information systems, a created synchronous model enables us to generate the so-called maximal consistent extension of a given information system. Such an extension includes all possible global states consistent with all rules extracted from the original data table. An asynchronous model enables us to find all possible transitions between global states of a given concurrent system, for which only one process changes its local state. A model created on the basis of a dynamic information system enables us to generate a maximal consistent extension of that system. In this case, such an extension includes all possible global states consistent with all rules extracted from the original data table, and all possible transitions between global states consistent with all transition rules generated from the original transition system.

Concurrent system models discovered from data tables are represented in the form of colored Petri nets. A use of this kind of Petri nets allows us to obtain compact and readable models, ready for further computer analysis and verification of their properties.

In this chapter, the problems of reconstruction of models (Suraj & Pancerz, 2006) and prediction of their changes in time (Suraj, Pancerz, & Swiniarski, 2005) are also taken up. Those problems occur as a result of appearing the new knowledge about modeled systems and their behaviors. The new knowledge can be expressed by appearing new global states, new transitions between states, new local states of individual processes, or new processes in modeled systems. A prediction method proposed in this chapter points at the character of model changes in time. For representing prediction rules, both prediction matrices and Pawlak's flow graphs are used.

In this chapter, an approach to consistent extensions of information systems and dynamic information systems is presented. Especially, we are interested in partially consistent extensions of such systems (Suraj, 2001; Suraj & Pancerz, 2005; Suraj, Pancerz, & Owsiany, 2005). Methods for computing such extensions are given. In the proposed approach, global states of a modeled system, and also transitions between states (in a case of a dynamic information system) can be consistent only partially with the knowledge included in the original information system or dynamic information system describing a modeled system. The way of computing suitable consistency factors of new global states or new transitions between states with the original knowledge about systems is presented.

In today's computer science development, the usefulness of proposed methods and algorithms for real-life data is conditioned by existing suitable computer tools automating computing processes. Therefore, in this chapter, the ROSECON system is presented (Suraj & Pancerz, 2005). ROSECON is a computer tool supporting users in automatized discovering net models from data tables as well as predicting their changes in time. The majority of methods and algorithms discussed in this chapter are implemented in ROSECON.

At the end of this chapter, there are presented experiments with using proposed methods and algorithms on real-life data coming from finance.

BASIC NOTIONS

In this section, we recall, briefly, basic concepts of rough-set theory (Pawlak, 1991) used in the chapter. Moreover, we present a short informal introduction to colored Petri nets.

Rough-Set Concepts

In rough-set theory, introduced by Z. Pawlak (1982), information systems are used to represent some knowledge about elements of a universe of discourse (Pawlak, 1991). An *information system* is a pair $S=(U, A)$, where U is a nonempty, finite set of objects, called the universe; A is a nonempty, finite set of attributes, that is, $a:U \to V_a$ for $a \in A$, where V_a is called the value set of a. A *decision system* is a pair $S=(U, A)$, where $A=C \cup D$, $C \cap D = \emptyset$, and C is a set of condition attributes, D is a set of decision attributes. Any information (decision) system can be represented as a data table whose columns are labeled with attributes, rows are labeled with objects, and entries of the table are attribute values. Each object in the system $S=(U, A)$ can be represented by a *signature* of the form

$inf_S(u) = \{(a, a(u)) : a \in A\}$. By *temporal information systems,* we understand information systems whose rows (objects) are ordered in time.

For every information (decision) system $S=(U, A)$, a formal language $L(S)$ is associated with S. The alphabet of $L(S)$ consists of A (the set of attribute constants), $V = \bigcup_{a \in A} V_a$ (the set of attribute value constants), and the set $\{\neg, \vee, \wedge, \Rightarrow, \Leftrightarrow\}$ of propositional connectives, called negation, disjunction, conjunction, implication, and equivalence, respectively. An expression of the form (a, v), where $a \in A$ and $v \in V_a$, is called an atomic (elementary) formula of $L(S)$. The set of formulas of $L(S)$ is the least set satisfying the following conditions: (1) (a,v) is a formula of $L(S)$, (2) if φ and ψ are formulas of $L(S)$, then so are $\neg\varphi$, $\varphi \vee \psi$, $\varphi \wedge \psi$, $\varphi \Rightarrow \psi$, and $\varphi \Leftrightarrow \psi$.

The object $u \in U$ satisfies formula φ of $L(S)$, denoted as $u \models \varphi$, if and only if the following conditions are satisfied: $u \models (a,v)$ iff $a(u)=v$, $u \models \neg\varphi$ iff not $u \models \varphi$, $u \models \varphi \vee \psi$ iff $u \models \varphi$ or $u \models \psi$, $u \models \varphi \wedge \psi$ iff $u \models \varphi$ and $u \models \psi$, $u \models \varphi \Rightarrow \psi$ iff $u \models \neg\varphi \vee \psi$, $u \models \varphi \Leftrightarrow \psi$ iff $u \models \varphi \Rightarrow \psi$ and $u \models \psi \Rightarrow \varphi$ If φ is a formula of $L(S)$, then the set $|\varphi|_S = \{u \in U : a(u) = v\}$ is called the meaning of the formula φ in S. For any formula φ of $L(S)$, the set $|\varphi|_S$ can be defined inductively as follows:

$$|(a,v)|_S = \{u \in U : a(u) = v\},$$
$$|\neg\varphi|_S = U - |\varphi|_S, |\varphi \vee \psi|_S = |\varphi|_S \cup |\psi|_S,$$
$$|\varphi \wedge \psi|_S = |\varphi|_S \cap |\psi|_S, |\varphi \Rightarrow \psi|_S = |\neg\varphi|_S \cup |\psi|_S,$$
$$|\varphi \Leftrightarrow \psi|_S = (|\varphi|_S \cap |\psi|_S) \cup (|\neg\varphi|_S \cap |\neg\psi|_S)$$

Any information system contains some knowledge. In our approach, the knowledge included in a given information system S is expressed by rules extracted from S. A rule in the information system S is a formula of the form $\varphi \Rightarrow \psi$, where φ and ψ are referred to as the predecessor and the successor of a rule, respectively. The rule $\varphi \Rightarrow \psi$ is true in S if $|\varphi|_S \subseteq |\psi|_S$. An inhibitor rule in the information system S is a formula of the form $\varphi \Rightarrow \neg\psi$. In this chapter, we consider only true rules for which φ is a conjunction of atomic formulas of $L(S)$ and ψ is an atomic formula of $L(S)$, that is, each rule has the form $(a_1, v_1) \wedge ... \wedge (a_q, v_q) \Rightarrow (a_d, v_d)$. In a case of a decision system $S = (U, C \cup D)$, we can consider decision rules and then $a_d \in D$. A rule is called minimal in S if and only if the removal of any atomic formula from the predecessor of a rule causes that a rule is not true in S. The set of all minimal rules true in S will be denoted as $Rul(S)$. A method for generating the minimal rules in an information system using the rough-set methodology is given in the literature (Skowron & Suraj, 1993; Suraj, 2000). Each rule of the form $(a_1, v_1) \wedge ... \wedge (a_q, v_q) \Rightarrow (a_d, v_d)$, can be transformed into a set of inhibitor rules:

$$\{(a_1, v_1) \wedge ... \wedge (a_q, v_q) \Rightarrow \neg(a_d, v_{d1}), ..., (a_1, v_1) \wedge ... \wedge (a_q, v_q) \Rightarrow \neg(a_d, v_{dk})\}$$

where $v_{d1}, ..., v_{dk} \in V_{a_d} - \{v_d\}$. The set of inhibitor rules true in S corresponding to the set $Rul(S)$ will be denoted as $\overline{Rul}(S)$. Several factors can be associated with each rule $\varphi \Rightarrow \psi$ in the information system $S=(U, A)$. Two of them are interesting to us, namely, a *support factor* (or in short, *support*) $supp(\varphi \Rightarrow \psi) = card(|\varphi \wedge \psi|_S)$ (the number of objects satisfying simultaneously the predecessor and the successor of a rule) and a *strength factor* (or in short, *strength*) $str(\varphi \Rightarrow \psi) = \dfrac{supp(\varphi \Rightarrow \psi)}{card(U)}$ (a relative measure of support). These factors can tell us how a given rule is important (relevant) and, at the same time, more general in the considered information system. Analogous factors can be associated with any set $Rul'(S)$ of rules true in a given information system $S = (U, A)$. Then we have a support factor $supp(Rul'(S)) = card(\{u \in U : u \in |\varphi \wedge \psi|_S$ for any $(\varphi \Rightarrow \psi) \in Rul'(S)\})$ and a strength factor $str(Rul'(S)) = \dfrac{supp(Rul'(S))}{card(U)}$.

Discovering dependencies among attributes is an important issue in data analysis. Let $S=(U, A)$ be an information system. Intuitively, a set $D \subseteq A$ of attributes depends totally on a set $C \subseteq A$ of attributes, if the values of attributes from C uniquely determine all values of attributes from D. In other words, D depends totally on C if and only if there exists a functional dependency between the values of C and D. The fact that D depends totally (functionally) on C will be denoted as $C \Rightarrow D$. The formal approach to dependencies among attributes in information systems using the rough-set theory is given in the literature (Pawlak, 1991; Pawlak, 2004). Sometimes, we can remove some data from a data table, preserving its basic properties. Let $S=(U, A)$ be an information system. Each subset $B \subseteq A$ of attributes determines an equivalence relation on U, called an *indiscernibility relation* $Ind(B)$, defined as $Ind(B) = \{(u,v) \in U \times U : \forall_{a \in B} a(u) = a(v)\}$. Any minimal (with respect to $\subseteq$) subset $B \subseteq A$ of attributes, such that $Ind(B) = Ind(A)$, is called a *reduct* of S. The set of all reducts in a given information system S will be denoted as $Red(S)$.

Let us have the total dependency $C \Rightarrow D$ between two disjoint sets $C, D \subseteq A$ of attributes in a given information system S. It may occur that D depends not on the whole set C but on its subset C'. In order to find this subset, we need the notion of a *relative reduct*. We can consider a special case of a relative reduct called a *functional relative reduct*. Let $C, D \subseteq A$ be two disjoint sets of attributes in S and $C \Rightarrow D$. Then $C' \subseteq C$ is a functional D-reduct of C if, and only if, C' is a minimal (with respect to $\subseteq$) subset of C such that $C' \Rightarrow D$.

Rudiments of Colored Petri Nets

In general, Petri nets are a graphical and mathematical tool for modeling processes acting concurrently. Colored Petri nets (or in short, CP-nets) were introduced by K. Jensen (1981). They constitute generalization and extension of place-transition nets (Reisig, 1985). For this reason, they are called high-level nets. Colored Petri nets enable us to obtain more general description of a modeled system, where some details are hidden and a structure is more transparent (readable). States of a CP-net are represented by places (drawn as circles), whereas actions are represented by transitions (drawn as rectangles). Places are connected with transitions by arcs (and also, inversely, transitions with places). For each place p, a nonempty set of tokens, called a *color set,* is associated with p. In a given place p, tokens belonging to the color set associated with p can only appear. A marking of the net determines distribution of tokens on all places. A marking of a single place p is a multiset over the color set associated with p. Expressions on arcs determine which tokens are removed from input places and which tokens are placed in output places of transitions with which arcs are connected. For each transition t, a Boolean expression, called a guard expression, is associated with t. If this expression is fulfilled, then a transition can be fired. Firing of a given transition t causes removing suitable tokens from input places of t and placing suitable tokens in output places of t. So, transitions are responsible for state (marking) changes. We say that a given marking M' is reachable from a marking M if and only if there exists a sequence of transitions that can be fired, causing changes of markings from M to M' through some intermediate markings. Arc expressions and guard expressions may include, among others, variables that are replaced with suitable values during firing transitions. A theory of colored Petri nets is very rich. For a formal description of such nets, we refer the reader to Jensen (1997).

DATA TABLES FOR DESCRIPTION OF CONCURRENT SYSTEMS

In order to describe or specify concurrent systems, we can use different forms of data tables. In our approach, data tables can represent information systems, dynamic information systems, decomposed information systems. or specialized matrices of forbidden states and forbidden transitions. In this subsection, we present issues concerning descriptions or specifications of concurrent systems by means of such data tables. The idea of a concurrent system representation by means of information systems is due to Z. Pawlak (1992). We assume that data tables include only a part of the knowledge about the behaviors of described or specified systems (the so-called *Open World Assumption*). The remaining knowledge (or still a part of it) can be discovered by means of determining consistent or partially consistent extensions of information systems or dynamic information systems represented by those data tables. Consistent and partially consistent extensions will be considered in the next section. Some specialized data tables like, for example, forbidden state matrices and forbidden transition matrices (Suraj & Pancerz, 2005) not considered here can also be used to specify concurrent systems.

Description of Concurrent Systems by Means of Information Systems

Let $S=(U, A)$ be an information system. Elements of the set U can be interpreted as global states of a given concurrent system CS, whereas attributes (elements of the set A) as processes in CS. For every process a from A, the set V_a of local states is associated with a. If the information system is represented by a data table, then the columns of a data table are labeled with names of processes. Each row of a data table (labeled with an element from the set U) includes a record of local states of processes of CS. Each record can be treated as a global state of CS. Therefore, in the sequel, for an information system S describing a concurrent system, we can interchangeably use the following terminology: attributes of S are called *processes* in S, objects of S are called *global states* in S, and values of attributes of S are called *local states* of processes in S.

If a data table is treated as a specification of a given system then one of the problems is to determine whether this specification defines a concurrent system or a sequential system and what are the rules describing the system behavior. Rough sets can be helpful to solve this problem (Pawlak, 1992).

Definition 1 (Partially concurrent system). If $S=(U, A)$ is an information system and $B \subset A$ is a reduct in S, then S is called a partially concurrent system.

For reducts of a given information system $S=(U, A)$ we have the following property. If $R \in Red(S)$, then there exists a functional dependency between the sets of attributes R and $A-R$, that is, $R \Rightarrow A-R$. Hence, we have that local states of processes from the set $(A-R)$ depend functionally on local states of processes from the reduct R. In this case, behavior of processes from R forces specific local states of processes outside of R. Taking a given reduct R into consideration, we can say about *primary processes* and *secondary processes* in S with respect to R. Then, primary processes are pair-wise independent of each other, whereas secondary processes change their local states depending on local states of primary processes. It is easy to see that notions of primary and secondary processes are relative in a given information system S because they depend on a chosen reduct in S. A formal definition is given next.

Definition 2 (Primary processes and secondary processes). Let $S=(U, A)$ be an information system describing a concurrent system and R be a reduct in S. Processes from R are called primary processes in S w.r.t. R, whereas processes from $A-R$ are called secondary processes in S w.r.t. R.

Example 1. Let us assume that an information system describes an economic system consisting of five processes (macroeconomics indexes): an inflation rate (marked with *infl*), a turnover of a foreign trade-export (marked with *exp*), a turnover of a foreign trade-import (marked with *imp*), a dollar exchange rate (marked with *usd*), a euro exchange rate (marked with *euro*). The global states of an economic system observed by us are collected in Table 1, representing an information system $S=(U, A)$, for which:

- A set of objects (global states) $U = \{u_1, u_2, ..., u_{10}\}$,
- A set of attributes (processes) $A = \{infl, exp, imp, usd, euro\}$,
- Sets of attribute values (local states of processes):
$V_{infl} = V_{exp} = V_{imp} = V_{usd} = V_{euro} = \{-1, 0, 1\}$

The meaning of local states is the following: -1 denotes a decrease, 0 - no change, and 1 - an increase.

The information system S has two reducts: $Red(S) = \{R_1, R_2\}$, where $R_1 = \{exp, usd, euro\}$, and $R_2 = \{imp, usd, euro\}$. Hence, the following functional dependencies are valid in S: $\{exp, usd, euro\} \Rightarrow \{infl, imp\}$,

Table 1. An information system S describing economic processes

U/A	infl	exp	imp	usd	euro
u_1	1	1	1	0	0
u_2	0	0	1	-1	-1
u_3	0	-1	-1	0	0
u_4	0	0	0	0	1
u_5	-1	1	1	1	0
u_6	0	0	0	0	0
u_7	-1	-1	-1	1	1
u_8	-1	1	1	0	1
u_9	-1	1	1	-1	0
u_{10}	-1	-1	-1	0	-1

$\{imp, usd, euro\} \Rightarrow \{infl, exp\}$. Taking the reduct R_1 into consideration, we obtain that *exp, usd,* and *euro* are primary processes, whereas *infl* and *imp* are secondary processes in S w.r.t. R_1.

The set $Rul(S)$ of all minimal rules true in the system S includes over 90 rules. Table 2 contains only such rules for which a strength factor is not smaller than 0.2. For each rule ρ, two factors (support $supp(\rho)$ and strength $str(\rho)$) are shown. In this case, rules represent dependencies among local states of processes in the economic system.

Description of Concurrent Systems by Means of Dynamic Information Systems

A description of concurrent systems by means of information systems does not take their dy-

Table 2. The set of minimal rules true in S for which the strength factor is not smaller than 0.2

Rule ρ	$supp(\rho)$	$str(\rho)$
$(exp, 0) \Rightarrow (infl, 0)$	3	0.3
$(imp, 0) \Rightarrow (infl, 0)$	2	0.2
$(usd, 1) \Rightarrow (infl, -1)$	2	0.2
$(usd, 0) \wedge (imp, 1) \Rightarrow (exp, 1)$	2	0.2
$(euro, 0) \wedge (imp, 1) \Rightarrow (exp, 1)$	3	0.3
$(imp, -1) \Rightarrow (exp, -1)$	3	0.3
$(imp, 0) \Rightarrow (exp, 0)$	2	0.2
$(imp, 1) \wedge (infl, -1) \Rightarrow (exp, 1)$	3	0.3
$(euro, 0) \wedge (infl, -1) \Rightarrow (exp, 1)$	2	0.2
$(exp, 1) \Rightarrow (imp, 1)$	4	0.4
$(usd, -1) \Rightarrow (imp, 1)$	2	0.2
$(exp, -1) \Rightarrow (imp, -1)$	3	0.3
$(usd, 0) \wedge (exp, 0) \Rightarrow (imp, 0)$	2	0.2
$(euro, 0) \wedge (infl, -1) \Rightarrow (imp, 1)$	2	0.2
$(euro, 0) \wedge (infl, 0) \Rightarrow (usd, 0)$	2	0.2
$(imp, 0) \Rightarrow (usd, 0)$	2	0.2

namic behaviors into consideration. In Suraj (1998), dynamic information systems have been proposed for a description of concurrent systems. Let $DS = (U, A, E, T)$, where $S=(U, A)$ and $TS = (U, E, T)$, be a dynamic information system. In this case, a set $A = \{a_1, a_2, ..., a_m\}$ can be treated as a set of processes, like previously. For every process $a \in A$, a set V_a of local states is associated with a. Behavior of a given concurrent system CS is presented by means of two linked subtables marked with S and TS, respectively. The first subtable S represents global states of CS observed by us, and it is called an underlying system of DS. The second subtable TS represents a transition relation T between global states of CS, and it is called a transition system of DS. Each row of the first subtable includes a record of local states of processes from A. Each record is labeled with an element from the set U of global states. The second subtable represents a transition system. Columns of the second subtable are labeled with events from the set E, whereas rows, with global states from U, like previously. Elements of this subtable for a given global state u make up successor states of u. The first global state of an underlying system can be an initial state of a transition system. Next, we give formal definitions.

Definition 3 (Transition system). A transition system is a triple $TS = (S, E, T)$, where S is a nonempty set of states, E is a nonempty set of events, and $T \subseteq S \times E \times S$ is a transition relation.

Each ordered triple $t = (s, e, s') \in T$ is called a transition in TS, occurring due to an event e. The state s is called a *previous state* of t, whereas s' - a *next state* of t. In a transition system, we can distinguish the so-called terminal states. By a terminal state, we understand a state for which there does not exist any event causing a transition from this state to another state.

Definition 4 (Dynamic information system). A dynamic information system is a quadruple $DS = (U, A, E, T)$, where $S=(U, A)$ is an information system called an underlying system, and $TS = (U, E, T)$ is a transition system.

A transition relation in a given dynamic information system DS can be presented in the form of a directed labeled graph $TG(DS) = (N, B)$ over E, where N is a set of nodes such that $N = U$, and B is a set of directed labeled arcs such that:

$$B = \{(u, e, v) \in U \times E \times U : \exists_{e \in E} (u, e, v) \in T\}.$$

Nodes of $TG(DS)$ represent global states from the set U of DS, whereas arcs represent transitions between global states from U determined by the transition relation T.

Example 2. Let us assume that a dynamic information system represents an economic system consisting of three processes (macroeconomics indexes): a dollar exchange rate (marked with *usd*), a euro exchange rate (marked with *euro*), and a stock exchange index (marked with *wig*). The observation of the behavior of processes is collected in Table 3 representing a dynamic information system $DS = (U, A, E, T)$, for which:

- A set of objects (global states) $U = \{u_1, u_2, ..., u_{10}\}$,
- A set of attributes (processes) $A = \{usd, euro, wig\}$,
- Sets of attribute values (local states of processes): $V_{usd} = V_{euro} = \{-1, 0, 1, 2\}$, $V_{wig} = \{-1, 0, 1\}$,
- A set of events $E = \{e_1, e_2, ..., e_9\}$,
- A transition relation:
$T = \{(u_1, e_1, u_2), (u_2, e_2, u_3), (u_3, e_3, u_4),$
$(u_4, e_4, u_5), (u_5, e_5, u_6), (u_6, e_6, u_7),$
$(u_7, e_7, u_8), (u_8, e_8, u_9), (u_9, e_9, u_{10})\}$

The meaning of local states is as follows: -1 denotes a decrease of 1%, 0 - no change, 1 - an increase of 1%, 2 - an increase of 2%. A transition graph $TG(DS)$ for DS is shown in Figure

1. Objects u_4 and u_5 are indiscernible (i.e., their signatures are the same) and they are represented by one node.

In our consideration, dynamic information systems are presented by means of data tables representing information systems in the Pawlak's sense. In this case, each dynamic information system DS is depicted by means of two data tables. The first data table represents an underlying system S of DS that is, in fact, an information system. The second one represents a decision system that is further referred to as a decision transition system. This table represents transitions determined by a transition relation. It is possible to distinguish two approaches to constructing a decision transition system. Let $DS = (U, A, E, T)$ be a dynamic information system, where $S = (U, A)$ is its underlying system. The first approach is called a weak form of a decision transition system. In this approach, we construct a decision table $S_T^w = (U_T^w, A \cup A')$, where $A = \{a_1,...,a_m\}$ is a set of condition attributes, whereas $A' = \{a'_1,...,a'_m\}$ is a set of decision attributes. Each attribute $a' \in A'$ corresponds exactly to one attribute $a \in A$. Such a table includes a part of all possible pairs of global states from the underlying system S. Each row of S_T^w corresponds to one transition between global states $u, u' \in U$ determined by a transition relation T, that is, when there exists an event $e \in E$ such that $(u, e, u') \in T$. Condition attributes $a_1,...,a_m$ determine the previous global states of transitions, whereas decision attributes $a'_1,...,a'_m$ - the next global states of transitions.

Example 3. Suppose we are given a dynamic information system DS describing an economic process from Example 2 An underlying system S of DS is shown in Table 4.

A decision transition system $S_T^w = (U_T^w, A \cup A')$ in the weak form is presented in Table 5. For S_T^w

*Figure 1. A transition graph TG(DS) for **DS***

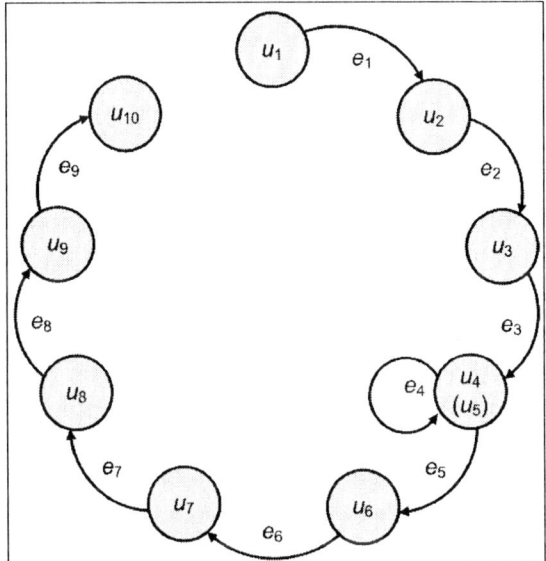

*Table 3. A dynamic information system **DS** describing economic processes*

U/A	usd	euro	wig	U/E	e_1	e_2	e_3	e_4	e_5	e_6	e_7	e_8	e_9
u_1	0	0	1		u_2								
u_2	1	0	-1			u_3							
u_3	0	0	-1				u_4						
u_4	2	1	0					u_5					
u_5	2	1	0						u_6				
u_6	-1	-1	0							u_7			
u_7	0	0	0								u_8		
u_8	-1	-1	1									u_9	
u_9	-1	0	2										u_{10}
u_{10}	1	0	0										

*Table 4. An underlying system **S** of a dynamic information system **DS** describing economic processes*

U/A	usd	euro	wig
u_1	0	0	1
u_2	1	0	-1
u_3	0	0	-1
u_4	2	1	0
u_5	2	1	0
u_6	-1	-1	0
u_7	0	0	0
u_8	-1	-1	1
u_9	-1	0	2
u_{10}	1	0	0

*Table 5. A decision transition system S_T^w in the weak form for a dynamic information system **DS** describing economic processes*

$U_T^w / A \cup A'$	usd	euro	wig	usd'	euro'	wig'
t_1	0	0	1	1	0	-1
t_2	1	0	-1	0	0	-1
t_3	0	0	-1	2	1	0
t_4	2	1	0	2	1	0
t_5	2	1	0	-1	-1	0
t_6	-1	-1	0	0	0	0
t_7	0	0	0	-1	-1	1
t_8	-1	-1	1	-1	0	2
t_9	-1	0	2	1	0	0

we have: a set of objects $U_T^w = \{t_1, t_2, ..., t_9\}$, a set of condition attributes $A = \{usd, euro, wig\}$, a set of decision attributes $A' = \{usd', euro', wig'\}$. The decision attribute *usd'* corresponds to the condition attribute *usd* and, analogously, *euro'* corresponds to *euro*, *wig'* corresponds to *wig*.

The second approach is called *a strong form* of a decision transition system. This approach has been proposed in Suraj (1998). In this approach, we construct a decision table $S_T^s = (U_T^s, A^* \cup \{d\})$, where $A^* = A \cup A'$ is a set of condition attributes, $A = \{a_1, ..., a_m\}$, and $A' = \{a'_1, ..., a'_m\}$, whereas d is a decision attribute. Each attribute $a' \in A'$ corresponds exactly to one attribute $a \in A$. Such a table contains all the possible pairs (u, u') of global states from the underlying system S. Attributes $a_1, ..., a_m$ determine the previous states, whereas attributes $a'_1, ..., a'_m$ determine the next states of transitions. A value of the decision d is 1 if and only if there exists an event $e \in E$ such that $(u, e, u') \in T$, or 0 otherwise.

Example 4. Suppose we are given a dynamic information system *DS* describing an economic system from Example 2. A fragment of a decision transition system $S_T^s = (U_T^s, A^* \cup \{d\})$, in the strong form, is shown in Table 6. For the system S_T^s we have: a set of objects $U_T^s = \{t'_1, t'_2, ..., t'_{100}\}$, a set of condition attributes $A \cup A' = \{usd, euro, wig, usd', euro', wig'\}$, a decision attribute d. The attribute *usd'* corresponds to the attribute *usd* and, analogously, *euro'* corresponds to *euro*, *wig'* corresponds to *wig*.

It is easy to see that the first approach is, in a certain sense, a particular case of the second one. A decision table S_T^w contains only such pairs of global states from a table S_T^s, for which a decision attribute d is equal to 1. A choice of the form of a decision transition system influences the new knowledge discovered for the described system. A strong form of a decision transition system includes more information about the behavior of the system. Therefore, it limits the possibility of adding new behavior of the system consistent with the original knowledge. In the sequel, objects in decision transition systems can also be called, for simplicity, *transitions*.

In a dynamic information system, we can consider two kinds of rules:

- Rules in an underlying system called, in short, underlying rules.
- Rules in a decision transition system called, in short, transition rules.

Table 6. A fragment of a decision transition system S_T^s in the strong form for a dynamic information system DS describing economic processes

$U_T^s / A \cup A' \cup \{d\}$	usd	euro	wig	usd'	euro'	wig'	d
t'_1	0	0	1	0	0	0	0
t'_2	0	0	1	1	0	-1	1
t'_3	0	0	1	0	0	-1	0
...	...	...	...	...	...	...	...
t'_{11}	1	0	-1	0	0	0	0
t'_{12}	1	0	-1	1	0	-1	0
t'_{13}	1	0	-1	0	0	-1	1
...	...	...	...	...	...	...	...
t'_{98}	1	0	0	-1	-1	1	0
t'_{99}	1	0	0	-1	0	2	0
t'_{100}	1	0	0	1	0	0	0

Underlying rules represent dependencies among local states of processes in global states, whereas transition rules represent dependencies among local states of processes at transitions between global states. Transition rules are computed as decision rules in a decision transition system (each decision attribute is taken separately during extracting rules using rough-set methodology). Both kinds of rules can be considered as positive rules as well as inhibitor rules. Further, we will consider only minimal underlying rules and minimal transition rules true in a given dynamic information system.

Example 5. Let us consider a dynamic information system DS describing an economic system from Example 2 A set $\overline{UndRul(S)}$ of inhibitor underlying rules true in DS contains over 30 rules. Table 7 presents only such underlying rules for which a strength factor is not smaller than 0.2. For each rule ρ, two factors (support $supp(\rho)$ and strength $str(\rho)$) are shown.

A set $\overline{TranRul(DS)}$ of inhibitor transition rules (extracted from the decision transition system S_T^w) true in DS contains over 90 rules. Table 8 presents only such rules for which a strength factor is not smaller than 0.2. For each rule ρ, two factors (support $supp(\rho)$ and strength $str(\rho)$) are shown.

Description of Concurrent Systems by Means of Decomposed Information Systems

In general, a decomposition is a division of a given information system into smaller, relatively independent subsystems. Components obtained

Table 7. A set of inhibitor underlying rules true in the dynamic information system DS, for which a strength factor is not smaller than 0.2.

Rule ρ	$supp(\rho)$	$str(\rho)$
$(euro,-1) \Rightarrow \neg(usd,2)$	2	0.2222
$(euro,-1) \Rightarrow \neg(usd,1)$	2	0.2222
$(euro,-1) \Rightarrow \neg(usd,0)$	2	0.2222
$(usd,0) \Rightarrow \neg(euro,1)$	3	0.3333
$(usd,0) \Rightarrow \neg(euro,-1)$	3	0.3333
$(usd,1) \Rightarrow \neg(euro,1)$	2	0.2222
$(usd,1) \Rightarrow \neg(euro,-1)$	2	0.2222
$(wig,-1) \Rightarrow \neg(euro,1)$	2	0.2222
$(wig,-1) \Rightarrow \neg(euro,-1)$	2	0.2222

*Table 8. A set of inhibitor transition rules true in a dynamic information system **DS**, for which a strength factor is not smaller than 0.2.*

Rule ρ	$supp(\rho)$	$str(\rho)$
$(wig, 1) \Rightarrow \neg(euro', 1)$	2	0.2222
$(wig, 1) \Rightarrow \neg(euro', -1)$	2	0.2222
$(usd, -1) \Rightarrow \neg(euro', 1)$	3	0.3333
$(usd, -1) \Rightarrow \neg(euro', -1)$	3	0.3333
$(euro, -1) \Rightarrow \neg(euro', 1)$	2	0.2222
$(euro, -1) \Rightarrow \neg(euro', -1)$	2	0.2222
$(usd, 2) \Rightarrow \neg(wig', 2)$	2	0.2222
$(usd, 2) \Rightarrow \neg(wig', 1)$	2	0.2222
$(usd, 2) \Rightarrow \neg(wig', -1)$	2	0.2222
$(euro, 1) \Rightarrow \neg(wig', 2)$	2	0.2222
$(euro, 1) \Rightarrow \neg(wig', 2)$	2	0.2222
$(euro, 1) \Rightarrow \neg(wig', 1)$	2	0.2222

in this way represent, in a certain sense, modules of an information system linked inside by means of functional dependencies. We can consider two kinds of problems. The first one can be expressed as follows. An information system S is given. The aim is to decompose this system into components with the set of the so-called links (i.e., suitable rules) binding these components. The second one we can formulate as follows. For a given information system S, it is needed to find the family of the so-called coverings of S. Informally, we can say that any covering of a given information system S is the minimal set of components (with respect to the number of them) that, put together, make up an original system S. Decomposition of information systems has been considered, among others, in (Suraj, 1996; Suraj, 2000). There have been presented methods for computing components of information systems with respect to their reducts. Those methods are based on rough-set theory. Here, we present the so-called global decomposition of information systems.

Let $S = (U, A)$ be an information system. There may exist a minimal functional dependency $B \Rightarrow C$ between sets of attributes B and C defining a component of S, where $B, C \subseteq A$ and $B \cap C = \emptyset$. Each component of $S = (U, A)$ is a subsystem $S' = (U', A')$ of S satisfying the following requirements: $U' \subseteq U$, $A' = \{a': a \in B \subseteq A\}$, $a'(u) = a(u)$ for $u \in U'$ and $V_{a'} = V_a$ for $a \in A$. We can distinguish two kinds of components, namely *normal components* and *degenerated components*. Normal components contain attributes linked by means of functional dependencies among their values. Degenerated components contain attributes such that there do not exist any functional dependencies among values of these attributes and values of other attributes. In general, each information system can be decomposed into components. In a trivial case, a given information system S is decomposed into only one component that is the same as the system S. Components of information systems are defined by functional relative reducts in these systems. We are interested in functional $\{a\}$-reducts of the set of attributes $A-\{a\}$ (or in short, functional $\{a\}$-reducts) in a given information system $S=(U, A)$. The set of all $\{a\}$-reducts of $A-\{a\}$ in an information system $S=(U, A)$ will be denoted as $FunRed^a(S)$. In order to compute these reducts, we can use the rough-set methods given in Suraj (1996, 2000).

A global decomposition of a given information system $S=(U, A)$ is based on computing functional relative reducts for each attribute $a \in A$. Then, we can define components of S as follows.

Definition 5 (Normal component). Let $S=(U, A)$ be an information system. An information system $S^* = (U^*, A^*)$ is called a normal component of S if and only if S^* is a subsystem of S and $A^* = X \cup Y$, where $Y = \{a \in A : X \in FunRed^a(S)\}$.

Definition 6 (Degenerated component). Let $S = (U, A)$ be an information system. An information system $S^* = (U^*, A^*)$ is called a degenerated component of S if and only if S^* is a subsystem of S and $A^* = \{a\}$ and there does not exist any normal component $S' = (U', X' \cup Y')$ of S such that $a \in X' \cup Y'$.

By components we understand both normal components and degenerated components.

Values of attributes belonging to a given component S^* of S are linked by means of rules called *internal links* within S^*. There also exist rules linking values of attributes from the component S^* with values of attributes outside of S^*. Such rules are called *external links* outside of S^*. Both kinds of links can be considered as positive rules as well as inhibitor rules. In our approach, we consider only minimal rules true in a given decomposed information system.

Let $S=(U,A)$ be an information system. We say that S is *coverable* by components if and only if there exists a set of components $S_1 = (U_1, X_1 \cup Y_1)$, $S_2 = (U_2, X_2 \cup Y_2), \ldots, S_k = (U_k, X_k \cup Y_k)$ of S such that $X_1 \cup X_2 \cup \ldots \cup X_k \cup Y_1 \cup Y_2 \cup \ldots \cup Y_k = A$, yielding a *covering* of S. A pair $cov(S) = (\{S_1, S_2, \ldots, S_k\}, C)$ is called a covering of an information system S. $S_1, S_2, \ldots, S_k$ are its components (normal or degenerated), whereas C is a set of links (internal and external). A problem of computing coverings of a given information system is equivalent to a problem of computing coverings of a given set by their subsets. So, it is NP hard.

An information system S describing a concurrent system CS can be decomposed into components. Then, each normal component of S contains a subset of all processes of CS, which local states are linked by means of a functional dependency. Therefore, we can say that normal components of an information system represent modules of CS having tightly linked internal structures. Moreover, there exist external links binding modules with one another. Let $S^* = (U^*, X \cup Y)$ be a normal component of an information system S describing a concurrent system CS. For each normal component, there exists a functional dependency between sets of attributes X and Y, that is, $X \Rightarrow Y$. In that case, we can say about primary processes and secondary processes in the component S^*. Local states of processes from the set X of S^* determine uniquely local states of processes from the set Y of S^*.

Definition 7 (Primary processes and secondary processes in a component). Let $S = (U, A)$ be an information system describing a concurrent system and let $S^* = (U^*, X \cup Y)$ be a normal component of S. Processes from the set X are called primary processes and processes from the set Y are called secondary processes in S^*.

Primary processes are partially independent of each other, so they can act concurrently.

Example 6. Let us consider an information system S describing economic processes from Example 1.

*Table 9. Components of an information system **S** describing economic processes*

Component	Type	Set X (primary processes)	Set Y (secondary processes)
S_1	normal	$\{exp, usd, euro\}$	$\{infl\}$
S_2	normal	$\{imp, usd, euro\}$	$\{infl\}$
S_3	normal	$\{infl, imp\}$	$\{exp\}$
S_4	normal	$\{imp, euro\}$	$\{exp\}$
S_5	normal	$\{exp, usd\}$	$\{imp\}$
S_6	normal	$\{exp, euro\}$	$\{imp\}$

A set of components of S obtained after a global decomposition of S is shown in Table 9. All components are normal. Individual components are presented in Tables 10 (a component S_1), 11 (a component S_2), 12 (a component S_3), 13 (a component S_4), 14 (a component S_5), and 15 (a component S_6). For example, the component S_1 defines a module of a concurrent system containing processes *exp*,

Table 10. A component S_1 of **S**

U_1 / A_1	exp	usd	euro	infl
u_1	1	0	0	1
u_2	0	-1	-1	0
u_3	-1	0	0	0
u_4	0	0	1	0
u_5	1	1	0	-1
u_6	0	0	0	0
u_7	-1	1	1	-1
u_8	1	0	1	-1
u_9	1	-1	0	-1
u_{10}	-1	0	-1	-1

Table 11. A component S_2 of **S**

U_2 / A_2	imp	usd	euro	infl
u_1	1	0	0	1
u_2	1	-1	-1	0
u_3	-1	0	0	0
u_4	0	0	1	0
u_5	1	1	0	-1
u_6	0	0	0	0
u_7	-1	1	1	-1
u_8	1	0	1	-1
u_9	1	-1	0	-1
u_{10}	-1	0	-1	-1

Table 12. A component S_3 of **S**

U_3 / A_3	infl	imp	exp
u_1	1	1	1
u_2	0	1	0
u_3	0	-1	-1
u_4	0	0	0
u_5	-1	1	1
u_7	-1	-1	-1

Table 13. A component S_4 of **S**

U_4 / A_4	imp	euro	exp
u_1	1	0	1
u_2	1	-1	0
u_3	-1	0	-1
u_4	0	1	0
u_6	0	0	0
u_7	-1	1	-1
u_8	1	1	1
u_{10}	-1	-1	-1

Table 14. A component S_5 of **S**

U_5 / A_5	exp	usd	imp
u_1	1	0	1
u_2	0	-1	1
u_3	-1	0	-1
u_4	0	0	0
u_5	1	1	1
u_7	-1	1	-1
u_9	1	-1	1

Table 15. A component S_6 of **S**

U_6 / A_6	exp	euro	imp
u_1	1	0	1
u_2	0	-1	1
u_3	-1	0	-1
u_4	0	1	0
u_6	0	0	0
u_7	-1	1	-1
u_8	1	1	1
u_{10}	-1	-1	-1

usd, euro, and *infl*. The processes *exp, usd, euro* are primary processes whereas the process *infl* is a secondary process, that is, local states of the process *infl* depend functionally on current local states of processes *exp, usd, euro*.

Sets of links are shown for the component S_2. Table 16 contains internal links within S_2, whereas Table 17 contains external links outside of S_2.

Table 16. A set of internal links within the component S_2

$(exp,0) \Rightarrow (infl,0)$	$(usd,1) \Rightarrow (infl,-1)$
$(exp,1) \wedge (usd,-1) \Rightarrow (infl,-1)$	$(exp,0) \wedge (usd,-1) \Rightarrow (euro,-1)$
$(exp,1) \wedge (usd,1) \Rightarrow (euro,0)$	$(exp,-1) \wedge (usd,1) \Rightarrow (euro,1)$
$(exp,1) \wedge (usd,-1) \Rightarrow (euro,0)$	

Table 17. A set of external links outside of the component S_2

$(exp,1) \Rightarrow (imp,1)$	$(usd,-1) \Rightarrow (imp,1)$
$(exp,-1) \Rightarrow (imp,-1)$	$(exp,0) \wedge (usd,0) \Rightarrow (imp,0)$

Table 18. Coverings of an information system S describing economic processes

covering	components
$cov_1(S)$	S_3, S_5, S_6
$cov_2(S)$	S_3, S_4, S_5
$cov_3(S)$	S_2, S_6
$cov_4(S)$	S_2, S_5
$cov_5(S)$	S_2, S_4
$cov_6(S)$	S_2, S_3
$cov_7(S)$	S_1, S_6
$cov_8(S)$	S_1, S_5
$cov_9(S)$	S_1, S_4
$cov_{10}(S)$	S_1, S_3
$cov_{11}(S)$	S_1, S_2

From the family of components of the system S we can choose coverings of S. All coverings of S are collected in Table 18.

EXTENSIONS OF INFORMATION SYSTEMS

Extensions of information systems have been considered, among others, in Rzasa and Suraj (2002) and Suraj (2001). Any extension S^* of a given information system S is created by adding to the system S new objects whose signatures contain only values of attributes that appeared in S. Especially, an important role among extensions of a given information system S is played by the so-called *consistent extension* of S. Such an extension is obtained if all the new objects added to the system S satisfy each minimal rule true in S. Among all possible consistent extensions of a given information system S, there exists the so-called *maximal consistent extension* (with respect to the number of objects). If an information system S describes a concurrent system CS, then its maximal consistent extension contains all the global states consistent with all the minimal rules representing dependencies among local states of processes of CS. These rules are extracted from the information system S. In some cases, a maximal consistent extension can contain new global states of CS that have not been earlier observed in CS, but these new states are consistent with rules extracted from S. Then, we have a nontrivial maximal consistent extension of S. New global states can be treated as the new knowledge of the concurrent system CS.

An approach to consistent extensions of information systems can be generalized using the so-called partially consistent extensions. In a case of a partially consistent extension S^* of a given information system S, we admit a situation that new objects added to S satisfy only a part of all minimal rules true in S. Then, an essential thing is to determine a consistency factor of a new object added to S, with the knowledge $K(S)$ included in the original information system S. Here, we propose some method of computing consistency factors. This method is based on rough-set theory. Computing a consistency factor for a given object is based on determining importance (relevance) of rules, extracted from the system S, that are not satisfied by the new object. We assume that if the importance of these rules is greater, the consistency factor of a new object with the knowledge $K(S)$ is smaller. The importance of a set of rules not satisfied by the new object is determined by means of a strength factor of this set of rules in S. It is worth noting that different approaches to determining a consistency factor are also possible. Next, we present some formal definitions concerning extensions of information systems.

Definition 8 (Extension). Let $S=(U, A)$ be an information system. An information system $S^* = (U^*, A^*)$ is called an extension of S if and only if $U \subseteq U^*$ and $card(A) = card(A^*)$ and $\forall_{a \in A} \exists_{a^* \in A^*} V_{a^*} = V_a$ and $a^*(u) = a(u)$ for each $u \in U$

It is easy to see that any extension of a given information system S includes only such objects whose attribute values appeared in the original data table representing S. Moreover, all objects that appear in S also appear in S^*. In the sequel, a set A^* of attributes in the extension S^* of S will be denoted as A (i.e., like in the original system S).

Example 7. Let us assume that an information system $S=(U, A)$ describes some genetic system consisting of three genes marked with g_1, g_2, and g_3. Global states observed in our system are collected in Table 19 representing the information system $S=(U, A)$, for which:

- A set of objects (global states) $U = \{u_1, u_2, ..., u_{11}\}$
- A set of attributes (processes) $A = \{g_1, g_2, g_3\}$

- Sets of attribute values (local states of processes): $V_{g1} = V_{g2} = V_{g3} = \{A, C, G\}$.

Here, global states can be interpreted as chromosomes, whereas attribute values represent allele: A (Adenine), C (Cytosine), G (Guanine). Note

Table 19. An information system S describing a genetic system

U/A	g_1	g_2	g_3
u_1	A	G	C
u_2	C	A	G
u_3	C	G	A
u_4	C	C	C
u_5	A	G	A
u_6	C	G	C
u_7	A	C	A
u_8	A	A	C
u_9	A	C	G
u_{10}	G	C	A
u_{11}	G	C	C

Table 20. An exemplary extension S^* of the information system S describing a genetic system

U/A	g_1	g_2	g_3
u_1	A	G	C
u_2	C	A	G
u_3	C	G	A
u_4	C	C	C
u_5	A	G	A
u_6	C	G	C
u_7	A	C	A
u_8	A	A	C
u_9	A	C	G
u_{10}	G	C	A
u_{11}	G	C	C
u_{12}	A	C	C
u_{13}	A	A	A
u_{14}	G	A	G

that, unfortunately, A denotes the set of attributes and one of the values of attributes. However, it should not confuse the reader.

In Table 20, an exemplary extension $S^* = (U^*, A)$ of the information system S is shown. It is easy to see that the system S^* satisfies all requirements given in Definition 8.

A maximal extension of a given information system S is also called a *Cartesian extension* of S and it is denoted as S^{max}.

Definition 9 (Cartesian extension). Let $S = (U, A)$ be an information system where $A = \{a_1, a_2, ..., a_m\}$, and let $\{V_{ai}\}_{i=1,2,...,m}$ be a family of sets of attribute values in S. An information system $S^* = (U^*, A)$ is called a Cartesian extension of S if and only if $U^* = \{u : a_1(u) \in V_{a1}, a_2(u) \in V_{a2}, ..., a_m(u) \in V_{am}\}$.

Let $S = (U, A)$ be an information system, $S^* = (U^*, A)$ - its extension, and $u \in U^*$. For the object u, we can compute its consistency factor $\xi_S(u)$ with the knowledge $K(S)$ included in the original information system S. Let us assume that this knowledge is expressed by the set $Rul(S)$ of all minimal rules true in S. Then, the consistency factor $\xi_S(u)$ can be defined as follows.

Definition 10 (Consistency factor). Let $S=(U, A)$ be an information system, $S^* = (U^*, A)$ - its extension, and $u \in U^*$. A consistency factor $\xi_S(u)$ of the object u with the knowledge included in S is a number defined as:

$$\xi_S(u) = 1 - \sigma$$

where σ is a strength factor of a set of rules from $Rul(S)$ not satisfied by the object u.

A consistency factor satisfies inequalities $0 \leq \xi_S(u) \leq 1$ for each $u \in U^*$. Obviously, if $u \in U$, then $\xi_S(u) = 1$, because the set of rules not satisfied by the object u from the original system is empty.

Briefly, in order to compute a consistency factor $\xi_S(u)$ for a given object u, we should make two main steps:

1. Determine a set $Rul''(S)$ of rules from the set $Rul(S)$ of all minimal rules, true in the information system S, that are not satisfied by the object u.
2. Compute a strength factor of the set $Rul''(S)$ of rules in the information system S.

An efficient algorithm for determining which rules are not satisfied by the new object in a given information system S is presented in Suraj and Pancerz (2006). This algorithm does not involve computing any rules in the system S.

Example 8. Let us consider an information system S describing a genetic system and its extension S^* from Example 7. On the basis of the knowledge $K(S)$ included in the original system S, we will compute consistency factors of objects u_{12} and u_{14} from the extension S^* with the knowledge $K(S)$. The knowledge $K(S)$ is expressed by the set $Rul(S)$ of all the minimal rules true in S (see Table 21).

A set of rules from $Rul(S)$ not satisfied by the object u_{12} is empty, whereas a set of rules from $Rul(S)$ not satisfied by the object u_{14} is shown in Table 22. Moreover, Table 22 contains objects from the system S supporting individual rules.

A strength factor for a set of rules not satisfied

*Table 21. A set Rul(S) of all minimal rules true in the information system **S** describing a genetic system*

$(g_2, A) \wedge (g_3, G) \Rightarrow (g_1, C)$	$(g_2, A) \wedge (g_3, C) \Rightarrow (g_1, A)$
$(g_2, C) \wedge (g_3, G) \Rightarrow (g_1, A)$	$(g_1, C) \wedge (g_3, G) \Rightarrow (g_2, A)$
$(g_1, C) \wedge (g_3, A) \Rightarrow (g_2, G)$	$(g_1, A) \wedge (g_3, G) \Rightarrow (g_2, C)$
$(g_1, G) \Rightarrow (g_2, C)$	$(g_1, C) \wedge (g_2, A) \Rightarrow (g_3, G)$
$(g_1, C) \wedge (g_2, C) \Rightarrow (g_3, C)$	$(g_1, A) \wedge (g_2, A) \Rightarrow (g_3, C)$

Table 22. A set of rules from Rul(S) not satisfied by the object u_{14}

Rule	Objects from S supporting a rule
$(g_2, A) \wedge (g_3, G) \Rightarrow (g_1, C)$	$\{u_2\}$
$(g_1, G) \Rightarrow (g_2, C)$	$\{u_{10}, u_{11}\}$

by u_{12} is equal to 0, whereas a strength factor of a set of rules not satisfied by u_{14} is equal to 0.2727. Hence, $\xi(u_{12}) = 1$ and $\xi(u_{14}) = 0.7273$. We can say that the object u_{12} is consistent with the knowledge $K(S)$ to the degree 1 (totally consistent), whereas the object u_{14} - to the degree 0.7273 (partially consistent).

After determining a consistency factor for each object of any extension of a given information system S, we can say about consistent or partially consistent extensions of S.

Definition 11 (Consistent extension and partially consistent extension). Let $S = (U, A)$ be an information system and $S^* = (U^*, A)$ its extension. S^* is called a consistent extension of S if and only if $\xi_S(u) = 1$ for each $u \in U^*$. Otherwise, S^* is called a partially consistent extension.

Analogously, we can consider extensions of dynamic information systems. Any extension DS^* of a given dynamic information system DS is created by adding to an underlying system S of DS new global states whose signatures contain only such attribute values that appeared in S, and by adding to a transition system TS new transitions between global states from the extension of the underlying system S. In this case, an important role is also played by consistent extensions of a given dynamic information system DS. Any consistent extension of DS is obtained if each new global state added to the underlying system S of DS satisfies each minimal underlying rule true in S, and each new transition added to the transition system TS of DS satisfies each minimal transition rule true in a decision transition system S_t representing transition relation T of DS. Among all possible consistent extensions of a given dynamic information system DS, there exists a maximal consistent extension of DS with respect to the number of global states in the underlying system S of DS and the number of transitions in the decision transition system S_t of DS. If a dynamic information system DS describes a concurrent system CS, then its maximal consistent extension represents maximal set of global states of CS consistent with all minimal rules representing dependencies among local states of processes in CS, and a set of all possible transitions between global states consistent with all minimal transition rules. If the maximal consistent extension of DS includes new global states or new transitions, then we obtain, in a certain sense, the new knowledge of the behavior of CS that has not been observed yet, but which is consistent with the original knowledge included in DS. For a dynamic information system, we can also consider partially consistent extensions. A way to compute a consistent extension or a partially consistent extension of a given dynamic information system DS is analogous to that for information systems. Now, we additionally take into consideration a decision transition system and minimal transition rules extracted from it.

NET MODELS

In this section, we give only a general outline of the problem of constructing net models of concurrent systems described by means of information systems or dynamic information systems. A process of creating a net model of a concurrent system CS, on the basis of its description or specification in the form of a data table DT, is carried out according to the following steps:

1. Construct a net representing the set of processes of CS described in DT.

2. Add to the net, obtained in the previous step, connections representing constraints among local states of processes that must be satisfied when these processes act simultaneously in CS.
3. Describe the elements (places, transitions, and arcs) of the net defined in previous steps according to the definition of a colored Petri net.

A structure of a model is determined on the basis of a description (specification) of CS (e.g., in the form of an information system, a dynamic information system, or a decomposed information system). Guard expressions for transitions are computed on the basis of suitable sets of rules (positive or inhibitor) extracted from the description (specification) of CS using rough-set methods. One of the relevant elements in this process is a transformation from the rules to guard expressions for transitions of a created model. Rules can be treated as Boolean algebra expressions; therefore, for a transformation, we can use the following Boolean algebra laws: $[x \Rightarrow \neg y] \Leftrightarrow [\neg(x \wedge y)]$, $[(\neg x_1) \wedge ... \wedge (\neg x_i)] \Leftrightarrow \neg[x_1 \vee ... \vee x_i]$ and $x \vee x \Leftrightarrow x$ (or dual).

We consider two kinds of net models, called here synchronous models and asynchronous models. In the asynchronous model, each process has one's own transition changing its local state. A change of a local state of a given process appears asynchronously in relation to changes of local states of remaining processes. In the synchronous model, we have only one transition changing simultaneously (synchronously) local states of all the processes of a concurrent system. Examples given here should explain the main idea of discovering these two kinds of net models from data tables considered by us. We show only net models created on the basis of information systems. However, analogously, net models can be created on the basis of dynamic information systems or decomposed information systems. In a case of a dynamic information system, we additionally take transition rules into consideration. In a case of a decomposed information system, the structure of the net is determined on the basis of components, whereas guard expressions are constructed on the basis of internal and external links.

Example 9. Let us consider an information system S, given in Example 7, describing a genetic system. For this system, we obtain a synchronous model in the form of the colored Petri net $CPN^S(S)$ shown in Figure 2.

We have for the net $CPN^S(S)$:

- The set of places $P = \{pg1, pg2, pg3\}$.
- The set of transitions $T = \{t\}$.

The place $pg1$ represents the process g_1, the place $pg2$ represents the process g_2, whereas the place $pg3$ represents the process g_3. Color sets associated with places stand for local states of processes represented by these places. A Boolean expression, determined on the basis of the set $\overline{Rul}(S)$ of inhibitor rules true in the system S, has the form:

$B = g\{[(g_2, A) \wedge (g_1, G)] \vee [(g_2, G) \wedge (g_1, G)] \vee [(g_1, C) \wedge (g_3, C) \wedge (g_2, A)] \vee$
$\vee [(g_1, G) \wedge (g_3, G) \wedge (g_2, C)] \vee [(g_1, C) \wedge (g_3, G) \wedge (g_2, C)] \vee [(g_2, G) \wedge (g_3, G) \wedge (g_1, C)] \vee \vee [(g_2, C) \wedge (g_3, A) \wedge (g_1, C)] \vee [(g_2, A) \wedge (g_3, A) \wedge (g_1, C)] \vee [(g_2, A) \wedge (g_3, G) \wedge (g_1, A)] \vee \vee [(g_2, G) \wedge (g_3, G) \wedge (g_1, A)] \vee [(g_3, A) \wedge (g_2, A) \wedge (g_1, A)]\}$

After assigning a variable associated with an appropriate output arc of the transition t to each atomic formula in the expression B, we obtain a guard expression as shown in the figure. This expression is written in the CPN ML language syntax (Jensen, 1997). The logical operators used there have the following meaning: *not* – negation, *andalso* – conjunction, and *orelse* – disjunction.

Figure 2. A synchronous net model CPN^S(S) for the information system S describing a genetic system

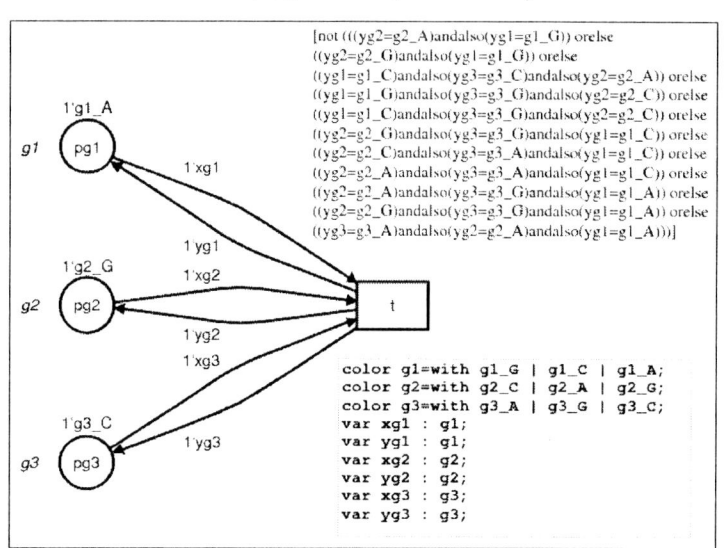

An initial marking M_0 of $CPN^S(S)$ represents a global state u_1 of S having the following signature $inf_A(u_1) = \{(g_1, A), (g_2, G), (g_3, C)\}$. A set of markings reachable from the initial marking M_0 in the net $CPN^S(S)$ is shown in Table 23.

This set includes 12 markings (from M_0 to M_{11}). One can be found out that the markings from M_0 to M_2 and from M_4 to M_{11} correspond to global states that appeared in the system S, whereas the marking M_3 represents a new global state that did not appear in the system S, but such a global state satisfies all the minimal rules true in the system S. There exists only one new global state with this property. Thus, a set of all the reachable

Table 23. A set of markings reachable from the initial marking M_0 for the net $CPN^S(S)$

$[M_0\rangle/P$	pg1	pg2	pg3
M_0	1'g1_A	1'g2_G	1'g3_C
M_1	1'g1_A	1'g2_A	1'g3_C
M_2	1'g1_A	1'g2_C	1'g3_A
M_3	1'g1_A	1'g2_C	1'g3_C
M_4	1'g1_A	1'g2_C	1'g3_G
M_5	1'g1_A	1'g2_G	1'g3_A
M_6	1'g1_C	1'g2_A	1'g3_G
M_7	1'g1_C	1'g2_C	1'g3_C
M_8	1'g1_C	1'g2_G	1'g3_A
M_9	1'g1_C	1'g2_G	1'g3_C
M_{10}	1'g1_G	1'g2_C	1'g3_A
M_{11}	1'g1_G	1'g2_C	1'g3_C

Table 24. A maximal consistent extension of the information system S describing a genetic system

U/A	g_1	g_2	g_3
u_1	A	G	C
u_2	C	A	G
u_3	C	G	A
u_4	C	C	C
u_5	A	G	A
u_6	C	G	C
u_7	A	C	A
u_8	A	A	C
u_9	A	C	G
u_{10}	G	C	A
u_{11}	G	C	C
u_{12}	A	C	C

markings in the net $CPN^s(S)$ defines a maximal consistent extension $\underline{S}^{max}$ of the system S shown in Table 24.

In Example 8, we computed for the object u_{12} a consistency factor $\xi(u_{12})$ with the knowledge included in the system S. A value of this factor was equal to 1, which confirms the fact that the object u_{12} belongs to the consistent extension of the system S.

Example 10. Let us consider the information system S from Example 7 describing a genetic system. We obtain an asynchronous model in the form of the colored Petri net $CPN^a(S)$ shown in Figure 3 for this system.

We have for the net $CPN^a(S)$:

- The set of places $P = \{pg1, pg2, pg3\}$,
- The set of transitions $T = \{tg1, tg2, tg3\}$.

The place $pg1$ represents the process g_1, the place $pg2$ represents the process g_2, whereas the place $pg3$ represents the process g_3. Transitions $tg1$, $tg2$, and $tg3$ represent changes of local states of processes g_1, g_2, and g_3, respectively. A guard expression for the transition t_{gi}, where $i = 1, 2, 3$, is determined on the basis of such rules from the set $Rul(S)$ that contain an attribute g_i in atomic formulas of their predecessors or successors. The initial marking M_0 of $CPN^a(S)$ represents a global state u_1 of S having the following signature $inf_A(u_1) = \{(g_1, A), (g_2, G), (g_3, C)\}$. The set of markings reachable from the initial marking M_0 is shown in Table 25.

Figure 3. The asynchronous net model $CPN^a(S)$ for the information system S describing a genetic system

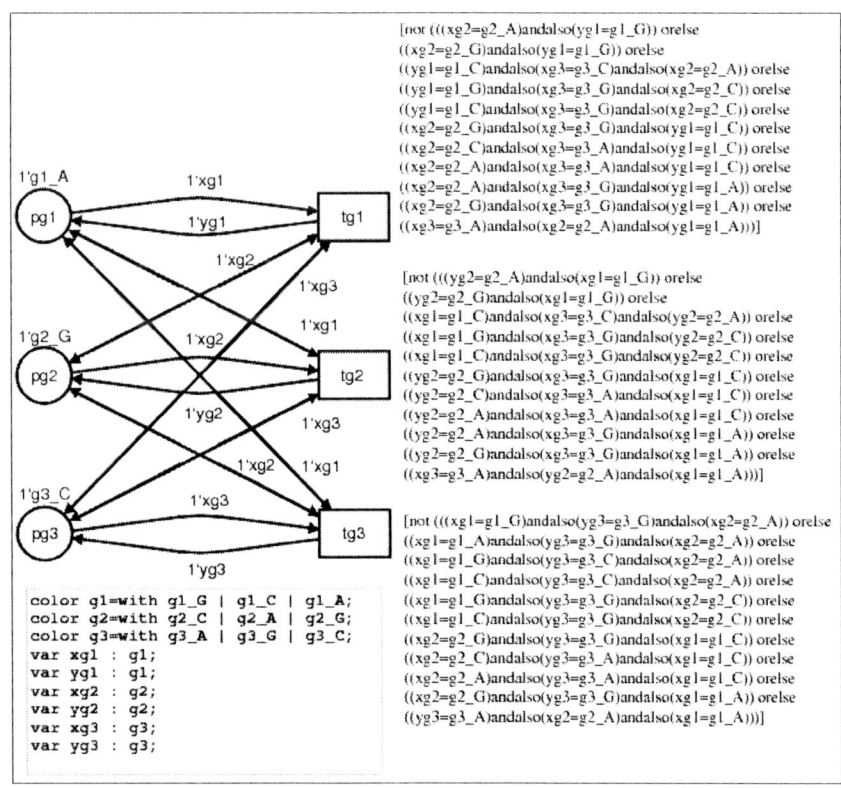

Table 25. The set of markings reachable from the initial marking M_0 for the net $CPN^a(S)$

$[M_0\rangle/P$	pg1	pg2	pg3
M_0	1'g1_A	1'g2_G	1'g3_C
M_1	1'g1_C	1'g2_G	1'g3_C
M_2	1'g1_A	1'g2_A	1'g3_C
M_3	1'g1_A	1'g2_C	1'g3_C
M_4	1'g1_A	1'g2_G	1'g3_A
M_5	1'g1_C	1'g2_C	1'g3_C
M_6	1'g1_C	1'g2_G	1'g3_A
M_7	1'g1_G	1'g2_C	1'g3_C
M_8	1'g1_A	1'g2_C	1'g3_A
M_9	1'g1_A	1'g2_C	1'g3_G
M_{10}	1'g1_G	1'g2_C	1'g3_A

Table 26. Successor markings for individual markings reachable in the net $CPN^a(S)$

Marking	Successor markings
M_0	M_0, M_1, M_2, M_3, M_4
M_1	M_0, M_1, M_5, M_6
M_2	M_0, M_2, M_3
M_3	$M_0, M_2, M_3, M_5, M_7, M_8, M_9$
M_4	M_0, M_4, M_6, M_8
M_5	M_1, M_3, M_5, M_7
M_6	M_1, M_4, M_6
M_7	M_3, M_5, M_7, M_{10}
M_8	$M_3, M_4, M_8, M_9, M_{10}$
M_9	M_3, M_8, M_9
M_{10}	M_7, M_8, M_{10}

It is easy to see that no marking in Table 25 represents the global state u_2 from the system S. Let us denote a marking that represents the global state u_2 as M_{11}. A set of markings reachable from the marking M_{11} is shown in Table 27. A successor marking for M_{11} is only M_{11}.

Markings reachable from the given marking M can be obtained by computing the so-called occurrence graph $OG(M)$ (Jensen, 1997) of a colored Petri net for the marking M. Nodes of

Table 27. A set of markings reachable from the marking M_{11} for the net $CPN^a(S)$

$[M_{11}\rangle/P$	pg1	pg2	pg3
M_{11}	1'g1_C	1'g2_A	1'g3_G

$OG(M)$ represent markings reachable from M, whereas each arc of $OG(M)$ is labeled with a pair containing a transition and evaluation of variables, for which a net changes its marking. On the basis of occurrence graphs computed for markings M_0 and M_{11}, we can determine an asynchronous transition graph $ATG(\underline{S}^{max})$ for the maximal consistent extension $\underline{S}^{max}$ of S. This graph has the form as in Figure 4.

Each two nodes u, v of $ATG(\underline{S}^{max})$ are connected by arcs directed from u to v and from v to u. Therefore, each pair of arcs is presented in the figure as a bidirectional arc, which reflects a symmetry of the asynchronous transition relation. Each directed arc represents an asynchronous transition between two global states differing only in a value of one gene. So, in the case of our

Figure 4. An asynchronous transition graph for the maximal consistent extension of the information system S

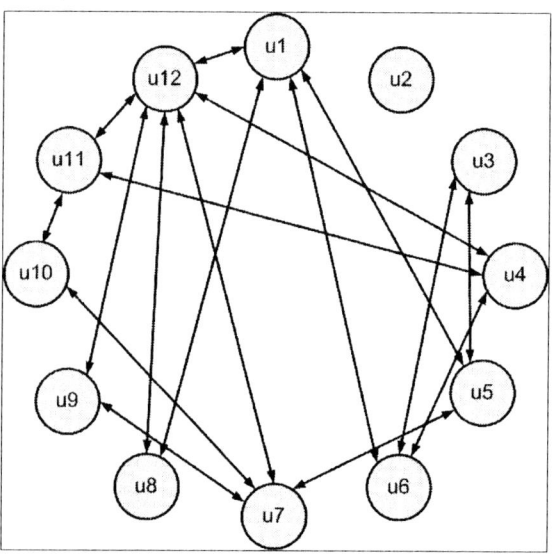

genetic system, each transition represents a point mutation. Analysis of the obtained graph enables us to determine, among others, which chromosomes are reachable from other chromosomes by sequences of point mutations.

RECONSTRUCTION OF NET MODELS

In our approach, the net model can be built on the basis of a decomposed information system S describing a given concurrent system. If the description of a concurrent system changes (i.e., a new information system S^* appears), we have to reconstruct the net model representing the concurrent system. The structure of a constructed net is determined on the basis of components of an information system. So, changing reducts and components in S can lead to a change in the structure of a net model. In that case, we would like to know how the reducts and components change when the new information about the system behavior appears. The idea of the reconstruction of a net model constructed for S can be presented graphically using a block diagram as in Figure 5.

The renewed computation of reducts and components of information systems is time-consuming because algorithms are NP hard. So, it is important to compute new reducts and components in an efficient way, that is, without the necessity of renewed computations. Some method has been proposed in Suraj and Pancerz (2005). In the approach presented there, a particular case has been considered, when the new description (in the form of an information system S^*) of a modeled system includes one new object (global state) with relation to the old description (in the form of an information system S).

PREDICTION

One of the important aspects of data mining is the analysis of data changing in time (temporal data). Many of the systems change their properties as time goes. Then, models constructed for one period of time must be reconstructed for another period of time. In this section, we assume that concurrent systems are described by temporal information systems (data tables include consecutive global states). In our approach, we observe behavior of modeled systems in consecutive time windows that temporal information systems are split into. Observation of changes enables us to determine the so-called prediction rules that can be used to predict future changes of models. Models of concurrent systems can be built on the basis of decomposed data tables. Components of information systems are defined on the basis of functional relative reducts. Therefore, an important thing is to determine how functional relative reducts change in consecutive time windows. On the basis of the obtained knowledge, we can predict future changes of functional relative reducts. At the beginning, we assume that all the time windows (which a given temporal information system S

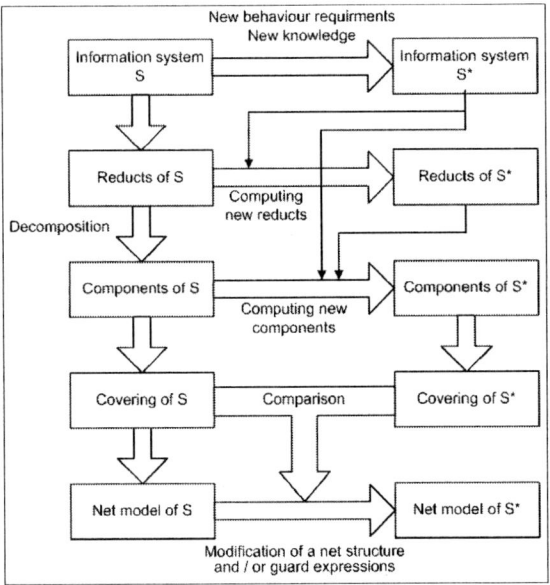

Figure 5. Reconstruction problem

is split into) have the same length (include the same number of objects), and the shift between any two consecutive time windows is constant. In order to determine prediction rules of model changes, we carry out the following steps, briefly described:

1. We split a given temporal information system $S = (U, A)$ into time windows of the same length, preserving a constant shift between two consecutive time windows. We obtain a set **S** of all the time windows.
2. For each time window from the set **S** and each attribute $a \in A$, we compute a set of all functional $\{a\}$-reducts of $A - \{a\}$. We obtain a data table (called a temporal table of functional reducts) whose columns are labeled with attributes from A, whereas rows, with consecutive time windows, from **S**. The cells of such a table contain sets of functional relative reducts.
3. For each attribute $a \in A$, we build a temporal decision system. Attributes of this system are labeled with the consecutive time windows (the last attribute is treated as a decision). The number of consecutive time windows taken into consideration is set by us. Each row represents a sequence of sets of functional relative reducts that appeared in consecutive time windows.
4. For each attribute $a \in A$, we compute prediction rules from the temporal decision system. In order to represent such rules, we can use flow graphs proposed by Z. Pawlak (2005), as it will be shown in the example, as well as prediction matrices (Suraj et al., 2005).

Example 11. Let us consider an information system describing weather processes: temperature (marked with *temp*), dew point (marked with *dew*), humidity (marked with *hum*), pressure (marked with *press*), and wind speed (marked with *wind*). Global states observed in our system are collected in Table 28, representing an information system $S = (U, A)$, for which:

- A set of objects (global states) $U = \{d_1, d_2, ..., d_{21}\}$,
- a set of attributes (processes) $A = \{temp, dew, hum, press, wind\}$,
- sets of attribute values (local states of processes): $V_{temp} = \{38, 50, 58\}$ [F], $V_{dew} = \{31, 40, 47\}$ [F], $V_{hum} = \{40, 61, 70, 77\}$ [%], $V_{press} = \{2906, 2982, 2995, 3004, 3016\}$ [$100 \times in$], $V_{wind} = \{0, 4, 6, 8, 10\}$ [mph].

We assume the following values of parameters: the length of time windows is equal to 7, the shift between two consecutive time windows

*Table 28. An information system **S** describing weather processes*

U/A	temp	dew	hum	press	wind
d_1	50	31	40	2982	8
d_2	58	31	61	2906	10
d_3	50	47	77	2906	0
d_4	58	47	70	2906	0
d_5	58	47	61	2906	10
d_6	58	47	70	2906	10
d_7	58	47	77	2906	10
d_8	50	47	77	2906	6
d_9	50	40	70	2906	6
d_{10}	50	40	77	2906	8
d_{11}	50	40	77	2906	6
d_{12}	50	47	70	2982	6
d_{13}	50	40	77	2982	4
d_{14}	38	31	61	3004	8
d_{15}	38	31	70	3004	0
d_{16}	50	40	70	2995	10
d_{17}	50	31	61	3016	4
d_{18}	50	40	70	3004	0
d_{19}	50	40	61	2995	8
d_{20}	50	40	61	2982	10
d_{21}	50	31	61	2995	0

is equal to 1, and the number of consecutive time windows taken for constructing temporal decision systems is equal to 4. A fragment of a temporal table of functional relative reducts is shown in Table 29. A fragment of a temporal decision system for the attribute *temp* is shown in Table 30. The empty cells in this table mean that in a given time window, there is a lack of functional relative reducts. Each cell with ∅ means that the value of a suitable attribute is constant in a given time window (so naturally, it does not depend on the values of remaining attributes).

Some prediction rules expressed in the form of a Pawlak's flow graph for the attribute *temp* are shown in Figure 6.

The number of layers in the flow graph is equal to the number of attributes in the temporal decision system. The layer labeled with t_i, where $i = 1, 2, 3, 4$, in the flow graph contains nodes representing sets of functional relative reducts that appear as values in the column t_i of the temporal decision system. Each arc connecting two nodes (representing two sets of functional reducts $FunRed_k$ and $FunRed_l$) from the neighboring layers t_i and t_j is labeled with three factors:

- A strength factor $str = \dfrac{card(X \cap Y)}{card(U_T)}$,

- A certainty factor $cer = \dfrac{card(X \cap Y)}{card(X)}$,

Table 29. A temporal table of functional relative reducts (fragment)

	temp	dew	hum	press	wind
S_1	{hum, wind}			{hum} {wind} {temp, dew}	
S_2	{hum, wind}			∅	
S_3	{hum, wind}	{temp, hum} {hum, wind}		∅	
S_4	{wind}	{hum, wind}		∅	
S_5	{wind}			∅	
...	...	...	...	...	...
S_{15}	{dew, hum} {dew, press}	{temp, hum, wind} {temp, press, wind}	{press, wind}	{hum, wind}	{dew, hum, press}

*Table 30. A temporal decision system for the attribute **temp** (fragment)*

	t_1	t_2	t_3	t_4
s_1	{hum, wind}	{hum, wind}	{hum, wind}	{wind}
s_2	{hum, wind}	{hum, wind}	{wind}	{wind}
...	...	...	...	...
s_{12}	{dew, press} {dew, wind}	{dew, press} {dew, wind}	{dew, press} {dew, wind}	{dew, hum} {dew, press}

*Figure 6. A Pawlak's flow graph expressing prediction rules for the attribute **temp** (fragment)*

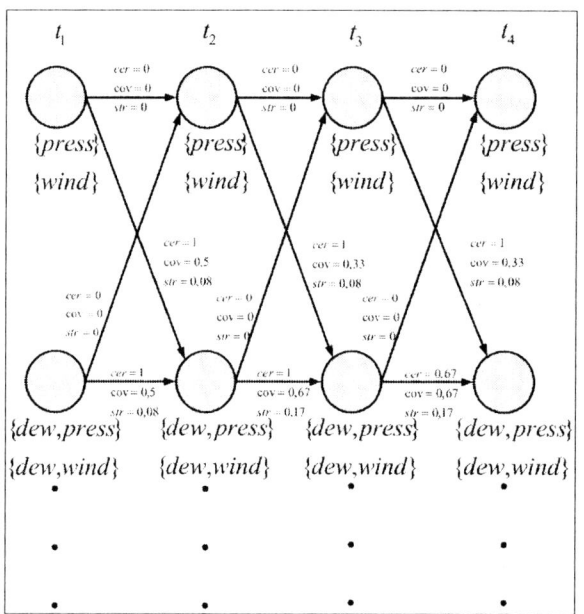

- A coverage factor $str = \dfrac{card(X \cap Y)}{card(Y)}$,

where X is a set of all the objects, for which $FunRed_k$ is a value of the attribute t_l, Y is a set of all the objects, for which $FunRed_l$ is a value of the attribute t_j, and U_T is a set of all the objects in the temporal decision system.

THE ROSECON SYSTEM

The ROSECON system is a computer tool enabling users to automatize discovering-net models of concurrent systems from data tables, as well as predicting their changes in time. All approaches presented in this chapter have been implemented in this system. The acronym ROSECON comes from *ROugh SEts and CONcurrency*. Algorithms implemented in ROSECON use methods of rough-set theory. The models created have the form of colored Petri nets. Such models enable users to determine different properties concerning structures and behaviors of modeled systems. ROSECON is destined for working on PC computers in Windows 95/2000/XP operating systems. It offers a user-friendly GUI environment for triggering computations and displaying results. It is available on the Web site http://www.rsds.wsiz.rzeszow.pl. Currently, ROSECON can be used, among others, for:

- Discovering synchronous and asynchronous net models of concurrent systems described by information systems and dynamic information systems.
- Discovering net models of concurrent systems described by decomposed information systems, with the decomposition of these systems, if necessary.
- Generating occurrence graphs of colored Petri net models.

- computing consistent and partially consistent extensions of information systems and dynamic information systems.
- Computing costs of reconstruction of net models changing in time.
- Predicting model changes in time.

All input and output data have the form of suitable XML structures, which is consistent with modern trends in software tools. ROSECON can cooperate with professional computer tools for working with colored Petri nets, namely Design/CPN (http: www.daimi.au.dk/designCPN) and CPN Tools (http://www.daimi.au.dk/CPNtools). All the net models generated by the ROSECON can be exported to the XML formats accepted by these tools.

ROSECON is also equipped with the auxiliary tools. A specialized editor provides facilities for creating and modifying data tables, recording data tables in an XML format accepted by ROSECON, import and export of data from/to other tools (MS Excel, Matlab). A statistics module gives users some statistical information about data tables, like attribute value statistics: mean, standard deviation, median, sample range, variation coefficient, correlation, attribute value distribution. If a data table represents an information system describing a given concurrent system, then a part of statistical information can be helpful to see the structures of individual processes.

EXPERIMENTS

We can apply computing a partially consistent extension of a given temporal information system describing an economic system consisting of financial processes to predicting future states that can appear in this system. In our experiments, we use the ROSECON system. As experimental data, we take daily exchange rates among the Polish zloty and important currencies like the US dollar (USD), the euro (EUR), the Japanese yen (JPY), the Swiss franc (CHF), the Pound sterling (GBP). We have five data tables (for years 2000, 2001, 2002, 2003, and 2004). Each data table consists of daily exchange rates for each business day in a given year. For each table, we build an information system in the following way. Attributes correspond to currencies, whereas objects correspond to consecutive business days. For each attribute, its value set consists of three values: -1 (decrease), 0 (no change), and 1 (increase). A fragment of one of the information systems is shown in Table 31.

Having information systems prepared in this way, we carry out our experiments for each information system separately, according to the following steps:

- We split an information system into two parts. The first part (consisting of objects from d_1 to d_{100}) is a training system S_{train}. The second part (consisting of remaining objects) is a test system S_{test}.
- For the training system S_{train} we determine its Cartesian extension S_{train}^{max}.
- For each new object from the Cartesian extension S_{train}^{max} we compute its consistency factor with the knowledge included in S_{train} and expressed by the set of rules from $Rul(S_{train})$, which have the minimal strength 0.04.

Table 31. An exemplary information system (fragment)

U/a	Usd	Eur	Jpy	Chf	Gbp
D_1	0	1	-1	1	1
D_2	0	1	-1	1	1
D_3	-1	-1	-1	-1	-1
D_4	-1	-1	-1	-1	-1
D_5	-1	-1	0	-1	-1
...	...	...	...	...	...
D_{251}	0	0	-1	1	1

- A partially consistent extension obtained in this way can be applied to predicting future states that can appear in the system, that is, we can assume that objects with the greater consistency factor should appear more often in the future. We check our prediction on the test system S_{test}. For each object from the Cartesian extension S_{train}^{max} of S_{train} we count its appearances in the test system S_{test}.

Results of prediction tests (for two chosen years) are shown in Figure 7. The graphs present percentage appearances of objects in the test systems having suitable values of consistency factors computed for training systems. It is easy to see that objects from extensions of training systems having greater values of consistency factors appeared significantly more often in the test systems. This regularity appeared in each year taken into consideration.

CONCLUSION AND OPEN PROBLEMS

This chapter concentrates on discovering models of concurrent systems from data tables, as well as their reconstruction and prediction of changes in time. The proposed approach is based mainly on rough-set theory, Boolean reasoning, and colored Petri nets. Rough-set theory with Boolean reasoning has been used as a universal and useful tool for extracting knowledge from data (in our case, in the form of rules). Constructed models have the form of colored Petri nets. This kind of Petri net makes obtaining coherent and readable models possible. Both theories (rough-set theory as well as colored Petri net theory) characterize well-elaborated theoretical foundations and existence of professional computer tools, aiding us in computing large data tables. The proposed approach can be used for solving many problems appearing in the contemporary computer science, for example, design and analysis of systems, data analysis, forecasting, and so on. Among main applications, we can distinguish the following: system engineering, medicine, molecular biology, economics, finance, meteorology. So, the approach proposed can be applied any time we are faced with the problems of analysis and modeling of processes making up a certain system. Moreover, in such a case, the behavior of the system is presented in the form of a data table (or data tables).

One of the limitations seems to be the exponential complexity of algorithms for generating

Figure 7. Results of prediction test for years 2003 and 2004

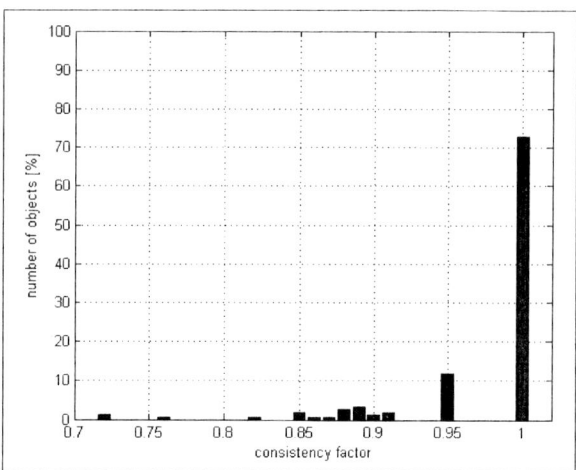
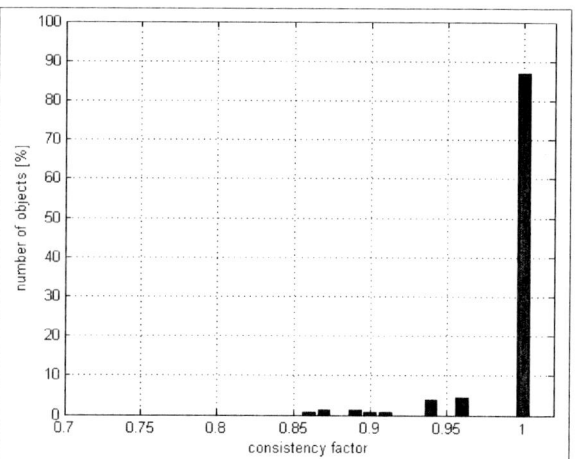

reducts and rules in information systems. Reducts and rules make up a basic way to represent the knowledge of behaviors of analyzed and modeled systems. Another problem is a huge number of rules obtained in the case of large data tables and sometimes their excess.

The main directions of further work are the following:

- Building hierarchical models of concurrent systems described by data tables,
- Discovering models from continuous data.
- Transformation from obtained models to computer programs carrying out the control of modeled systems (e.g., for PLC controllers).

We can also point at many open problems. It is necessary to list the following:

- A problem of computing consistent and partially consistent extensions of information systems and dynamic information systems without necessity of computing rules from data tables.
- A problem of equivalence (in the sense of representation of the same maximal consistent extensions of given information systems) of models of systems built on the basis of different coverings determined by components of information systems.
- A problem of synthesis of concurrent systems specified by dynamic information systems with respect to chosen properties of transition relations between global states of systems.

ACKNOWLEDGMENT

This work has been partially supported by the grant No. 3 T11C 005 28 from Ministry of Scientific Research and Information Technology of the Republic of Poland.

REFERENCES

Brown, E.M. (1990). *Boolean reasoning.* Dordrecht: Kluwer Academic Publishers.

Cios, K., Pedrycz, W., & Swiniarski, R. (1998). *Data mining methods for knowledge discovery.* Dordrecht: Kluwer Academic Publishers.

Jensen, K. (1981). Coloured Petri nets and the invariant-method. *Theoretical Computer Science, 14,* 317-336.

Jensen, K. (1997). *Coloured Petri nets. Vol. 1: Basic concepts.* Berlin, Heidelberg: Springer-Verlag.

Jensen, K. (1997). *Coloured Petri nets. Vol. 2: Analysis methods.* Berlin, Heidelberg: Springer-Verlag.

Pancerz, K., & Suraj, Z. (2003). Synthesis of Petri net models: A rough set approach. *Fundamenta Informaticae, 55*(2), 149-165.

Pawlak, Z. (1982). Rough sets. *International Journal of Information and Computer Sciences, 11,* 341-356.

Pawlak, Z. (1991). *Rough sets—Theoretical aspects of reasoning about data.* Dordrecht: Kluwer Academic Publishers.

Pawlak, Z. (1992). Concurrent versus sequential: The rough sets perspective. *Bulletin of the EATCS, 48,* 178-190.

Pawlak, Z. (2004). Some issues on rough sets. *Lecture Notes in Computer Sciences, 3100,* 1-58.

Pawlak, Z. (2005). Flow graphs and data mining. *Lecture Notes in Computer Sciences, 3400,* 1-36.

Peters, J.F., Skowron, A., & Suraj, Z. (2000). An application of rough set method to control dsesign. *Fundamenta Informaticae, 43*(1-4), 269-290.

Reisig, W. (1985). *Petri nets. An introduction.* Berlin: Springer-Verlag.

Rzasa, W., & Suraj, Z. (2002). A new method for determining of extensions and restrictions of information systems. *Lecture Notes in Artificial Intelligence, 2475,* 197-204.

Skowron, A., & Suraj, Z. (1993). Rough sets and concurrency. *Bulletin of the Polish Academy of Sciences, 41*(3), 237-254.

Suraj, Z. (1996). Discovery of concurrent data models from experimental tables: A rough set approach. *Fundameta Informaticae, 28*(3-4), 353-376.

Suraj, Z. (1998). The synthesis problem of concurrent systems specified by dynamic information systems. In L. Polkowski & A. Skowron (Eds.), *Rough sets in knowledge discovery 2* (pp. 418-448). Berlin: Physica-Verlag.

Suraj, Z. (2000). Rough set methods for the synthesis and analysis of concurrent processes. In L. Polkowski, S. Tsumoto, & T.Y. Lin (Eds.), *Rough set methods and applications* (pp. 379-488). Berlin: Physica-Verlag.

Suraj, Z. (2001). Some remarks on extensions and restrictions of information systems. *Lecture Notes in Artificial Intelligence, 2005,* 204-211.

Suraj, Z., & Pancerz, K. (2005). Restriction-based concurrent system design using the rough set formalism. *Fundamenta Informaticae, 67*(1-3), 233-247.

Suraj, Z., & Pancerz, K. (2005, September). *Some remarks on computing consistent extensions of dynamic information systems.* Paper presented at the 5th International Conference on Intelligent Systems Design and Applications, Wroclaw, Poland.

Suraj, Z., & Pancerz, K. (2005, November). *The ROSECON system - A computer tool for modelling and analysing of processes.* Paper presented at the International Conference on Computational Intelligence for Modelling, Control and Automation, Vienna, Austria.

Suraj, Z., & Pancerz, K. (2006). Reconstruction of concurrent system models described by decomposed data tables. *Fundamenta Informaticae, 71*(1), 101-119.

Suraj, Z., & Pancerz, K. (2006, September). *Efficient computing consistent and partially consistent extensions of information systems.* Paper presented at the Workshop on Concurrency, Specification and Programming, Wandlitz, Germany.

Suraj, Z., Pancerz, K., & Owsiany, G. (2005). On consistent and partially consistent extensions of information systems. *Lecture Notes in Artificial Intelligence, 3641,* 224-233.

Suraj, Z., Pancerz, K., & Swiniarski R.W. (2005, June). *Prediction of model changes of concurrent systems described by temporal information systems.* Paper presented at the 2005 World Congress in Applied Computing, Las Vegas, Nevada, USA.

Compilation of References

Adomavicius, G., & Tuzhilin, A. (1997). Discovery of actionable patterns in databases: The action hierarchy approach. In *Proceedings of KDD97 Conference*. Newport Beach, CA: AAAI Press

Agrawal R., Imielinski, T., & Swami A.N. (1993). Mining association rules between sets of items in large databases. In P. Buneman & S. Jajodia (Eds.), *ACM SIGMOD International Conference on Management of Data* (pp. 207–216). Washington, DC.

Agrawal, R, & Srikant, R. (1994). Fast algorithms for mining association rules. In J.B. Bocca, M. Jarke, & C. Zaniolo (Eds.), *Twentieth International Conference on Very Large Data Bases VLDB* (pp. 487-499). Morgan Kaufmann.

Agrawal, R., Mannila H., Srikant, R., Toivonen, H., & Verkamo, A. I. (1996). Fast discovery of association rules. In *Advances in knowledge discovery and data mining* (pp. 307-328). Menlo Park, CA: AAAI Press/The MIT Press.

Ahlqvist, O., Keukelaar, J., & Oukbir, K. (2000). Rough classification and accuracy assessment. *International Journal of Geographical Information Science, 14*(5), 475-496.

Alcock, R. J., & Manolopoulos, Y. (1999). Time-series similarity queries employing a feature-based approach. In D.I. Fotiadis, & S.D. Nikolopoulos (Ed.), *7th Hellenic Conference on Informatics* (pp. III.1-9). Ioannina, Greece: University of Ioannina Press.

Alexandroff, P. (1937). Diskrete räume. *Matematičeskij Sbornik, 2*, 501-518.

Atanassov, K. (1986). Intuitionistic fuzzy sets. *Fuzzy Sets and Systems, 20*, 87-96.

Atanassov, K. (1999). *Intuitionistic fuzzy sets: Theory and applications*. Heidelberg, NY: Physica-Verlag, A Springer-Verlag Company.

Bakar, A.A., Sulaiman, M.N., Othman, M., & Selamat, M.H. (2002). Propositional satisfiability algorithm to find minimal reducts for data mining. *International Journal of Computer Mathematics, 79*(4), 379-389.

Banerjee M., & Chakraborty, M. (1993). Rough algebra. *Bulletin of the Polish Academy of Sciences, Mathematics, 41*(4), 293-297.

Banerjee M., & Chakraborty, M. (1996). Rough sets through algebraic logic. *Fundamenta Informaticae, 28*(3-4), 211-221.

Banerjee M., & Chakraborty, M. (2003). Algebras from rough sets. In S. K. Pal, L. Polkowski, & A. Skowron, (Eds.), *Rough-neural computing: Techniques for computing with words* (pp.157-184). Heidelberg, Germany: Springer Verlag.

Banzhaf, J. F. (1965). Weighted voting doesn't work: A mathematical analysis. *Rutgers Law Review, 19*, 317-343.

Barto, A. G., Sutton, R. S., & Anderson, C. W. (1983). Neuronlike elements that can solve difficult problems. *IEEE Trans. on Systems, Man, and Cybernetics, 13*, 834-846.

Bazan J., & Szczuka. M. (2001). RSES and RSESlib—A collection of tools for rough set computations. In W. Ziarko & Y. Yao (Eds.), *Proceedings of the 2nd International Conference on Rough Sets and Current Trends in Computing (RSCTC2000),* Banff, Canada, *Lecture notes in artificial intelligence, 2005* (pp. 106-113). Heidelberg, Germany: Springer.

Bazan, J. (1998). A comparison of dynamic and non-dynamic rough set methods for extracting laws from decision tables. In *Rough sets in knowledge discovery* (pp. 321-365). Heidelberg: Physica-Verlag.

Bazan, J., Nguyen, H.S., Skowron, A , &. Szczuka. M., (2003). A view on rough set concept approximation. In G. Wang, Q. Liu, Y. Yao, & A. Skowron (Eds.), *Proceedings of the Ninth International Conference on Rough Sets, Fuzzy Sets, Data Mining and Granular Computing (RSFDGrC2003),* Chongqing, China (pp. 181-188). Heidelberg: Springer Germany.

Bazan, J., Skowron, A., & Synak, P. (1994). Dynamic reducts as a tool for extracting laws from decision tables. In *Proceedings of the 8th Symposium on Methodologies for Intelligent Systems, Lecture Notes in Artificial Intelligence 869* (pp. 346-355).

Bazan. J. (1998). A comparison of dynamic and non-dynamic rough set methods for extracting laws from decision tables. In L. Polkowski & A. Skowron (Eds.), *Rough sets in knowledge discovery 1: Methodology and applications. Studies in fuzziness and soft computing,* 18 (pp. 321-365). Heidelberg, Germany: Springer.

Beaubouef, T., & Petry F. (2002). Fuzzy set uncertainty in a rough object oriented database. In *Proceedings of North American Fuzzy Information Processing Society (NAFIPS-FLINT'02),* New Orleans, LA (pp. 365-370).

Beaubouef, T., & Petry, F. (1994). Rough querying of crisp data in relational databases. In *Proceedings of the Third International Workshop on Rough Sets and Soft Computing (RSSC'94),* San Jose, California.

Beaubouef, T., & Petry, F. (2000). Fuzzy rough set techniques for uncertainty processing in a relational database. *International Journal of Intelligent System,s 15*(5), 389-424.

Beaubouef, T., & Petry, F. (2004a). Rough functional dependencies. In *Proceedings of the 2004 Multiconferences: International Conference On Information and Knowledge Engineering (IKE'04)* (pp. 175-179). Las Vegas, NV.

Beaubouef, T., & Petry, F. (2005a). Normalization in a rough relational database. In *Proceedings of Tenth International Conference on Rough Sets, Fuzzy Sets, Data Mining, and Granular Computing (RSFDGrC'05)* (pp. 257-265). Regina, Canada.

Beaubouef, T., & Petry, F. (2005b). Representation of spatial data in an OODB using rough and fuzzy set modeling. *Soft Computing Journal, 9*(5), 364-373.

Beaubouef, T., Ladner, R., & Petry, F. (2004b). Rough set spatial data modeling for data mining. *International Journal of Intelligent Systems, 19*(7), 567-584.

Beaubouef, T., Petry, F., & Buckles, B. (1995). Extension of the relational database and its algebra with rough set techniques. *Computational Intelligence, 11*(2), 233-245.

Beaubouef, T., Petry, F., & Ladner, R. (In Press). Spatial data methods and vague regions: A rough set approach. *Applied Soft Computing Journal.*

Berenji, H. R. (2003). A convergent actor-critic-based FRL algorithm with application to power management of wireless transmitters. *IEEE Trans. on Fuzzy Systems, 11*(4), 478-485.

Berger, R. P. (1995). *Fur, feathers & transmission lines.* Retrieved from http://www.hydro.mb.ca/environment/publications/fur_feathers.pdf

Bertsekas, D. P., & Tsitsiklis, J. N. (1996). *Neuro-dynamic programming.* Belmont, MA: Athena Scientific.

Beynon, M.J. (2000). An investigation of β-reduct selection within the variable precision rough sets model. In *Proceedings of the Second International Conference on Rough Sets and Current Trends in Computing (RSCTC 2000)* (pp. 114–122).

Beynon, M.J. (2001). Reducts within the variable precision rough sets model: A further investigation. *European Journal of Operational Research, 134*(3), 592-605.

Birkhoff, G. (1940) *Lattice theory* (1st ed.). Providence, RI: Colloquium Publications XXV, American Mathematical Society.

Birkhoff, G. (1967) *Lattice theory* (3rd ed.). Providence, RI: Colloquium Publications XXV, American Mathematical Society.

Birkhoff, G. (1967). *Lattice theory* (3rd ed.). Providence, RI: American Mathematical Society Colloquium Publications.

Bittner, T. (2000). Rough sets in spatio-temporal data mining. In *Proceedings of International Workshop on Temporal, Spatial and Spatio-Temporal Data Mining, Lyon, France. Lecture Notes in Artificial Intelligence* (pp. 89-104). Berlin-Heidelberg: Springer-Verlag.

Bittner, T., & Stell, J. (2003). Stratified rough sets and vagueness, in spatial information theory: Cognitive and computational foundations of geographic information science. In W. Kuhn, M. Worboys, & S. Timpf (Eds.), *International Conference (COSIT'03)* (pp. 286-303).

Bjorvand, A.T., & Komorowski, J. (1997). Practical applications of genetic algorithms for efficient reduct computation. In *Proceedings of the 15th IMACS World Congress on Scientific Computation, Modelling and Applied Mathematics, 4* (pp. 601-606).

Blake, A. (1973). *Canonical expressions in Boolean algebra*. PhD thesis, University of Chicago.

Bonabeau, E., Dorigo, M., & Theraulez, G. (1999). *Swarm intelligence: From natural to artificial systems*. New York: Oxford University Press Inc.

Bonikowski, Z. (1994). Algebraic structures of rough sets. In *Rough sets, fuzzy sets and knowledge discovery (RSKD'93). Workshops in computing* (pp. 242-247). Berlin, London: Springer-Verlag and British Computer Society.

Boole. G. (1854). *The law of thought*. MacMillan (also Dover Publications, New-York).

Braga-Neto, U., Hashimoto, R., Dougherty, R. E, Nguyen, D.V., & Carroll, R.J. (2004). Is cross-validation better than resubstitution for ranking genes? *Bioinformatics, 20*, 253-258.

Breiman, L., Friedman, J., Olshen, R., & Stone, C. (1984). *Classification and regression trees*. Monterey, CA: Wadsworth and Brooks.

Brown, E.M. (1990). *Boolean reasoning*. Dordrecht: Kluwer Academic Publishers.

Brown, F. (1990). *Boolean reasoning*. Dordrecht, Germany: Kluwer Academic Publishers.

Buckles, B., & Petry, F. (1982). A fuzzy representation for relational data bases. *International Journal of Fuzzy Sets and Systems, 7*(3), 213-226.

Burris, S., & Sankappanavar, H. P. (1981). *A course in universal algebra*. Berlin-Heidelberg, Germany: Springer-Verlag.

Buszkowski, W. (1998). Approximation spaces and definability for incomplete information systems. In *Proceedings of 1st International Conference on Rough Sets and Current Trends in Computing (RSCTC'98)* (pp. 115-122).

Butz, C. J., Yan, W., & Yang, B. (2005). The computational complexity of inference using rough set flow graphs. *The Tenth International Conference on Rough Sets, Fuzzy Sets, Data Mining, and Granular Computing, 1* (pp. 335-344).

Butz, C. J., Yan, W., & Yang, B. (2006). An efficient algorithm for inference in rough set flow graphs. *LNCS Transactions on Rough Sets, 5*, (to appear).

Cattaneo, G. (1997). Generalized rough sets. Preclusivity fuzzy-intuitionistic (BZ) lattices. *Studia Logica, 58*, 47-77.

Cattaneo, G. (1998). Abstract approximation spaces for rough theories. In L. Polkowski, & A. Skowron, (Eds.). *Rough sets in knowledge discovery* (pp. 59-98). Heidelberg: Physica-Verlag.

Cattell, R., Barry, D., Bartels, D., Berler, M., Eastman, J., Gamerman, S., Jordan, D., Springer, A., Strickland, H., & Wade, D. (1997). *The object database standard: ODMG2.0*. San Francisco: Morgan Kaufmann.

Chan, C.-C., & Grzymala-Busse, J.W. (1991). *On the attribute redundancy and the learning programs ID3, PRISM, and LEM2* (TR-91-14). Lawrence, KS: Department of Computer Science, University of Kansas.

Chanas, S., & Kuchta, D. (1992). Further remarks on the relation between rough and fuzzy sets. *Fuzzy Sets and Systems, 47*, 391-394.

Chassagnez-Mendez, A.L., Machado, N.T., Araujo, M.E., Maia, J.G., & Meireles, M.A. (2000). Supercritical CO_2 extraction of eurcumis and essential eil from the rhizomes of turmeric (Curcuma longa L.), *Ind. Eng. Chem. Res., 39*, 4729-4733.

Chen, Y.H., & Yao, Y.Y. (2006). Multiview intelligent data analysis based on granular computing. In *Proceedings of IEEE International Conference on Granular Computing (Grc'06)* (pp. 281-286).

Chmielewski, M. R., & Grzymala-Busse, J. W. (1996). Global discretization of continuous attributes as preprocessing for machine learning. *International Journal of Approximate Reasoning, 15*, 319-331.

Chouchoulas, A., & Shen, Q. (2001). Rough set-aided keyword reduction for text categorisation. *Applied Artificial Intelligence, 15*(9), 843-873.

Chu, S., Keogh, E., Hart, D., & Pazzani, M. (2002). Iterative deepening dynamic time warping for time series. In R. L. Grossman, J. Han, V. Kumar, H. Mannila, & R. Motwani (Ed.), *2nd SIAM International Conference on Data Mining*, Arlington, VA: SIAM. Retrieved December 7, 2006, from http://www.siam.org/meetings/sdm02/proceedings/sdm02-12.pdf

Chuchro, M. (1994). On rough sets in topological Boolean algebras. In W. Ziarko (Ed.), *Rough sets, fuzzy sets and knowledge discovery. Proceedings of the International Workshop on Rough Sets and Knowledge Discovery (RSKD'93)* (pp 157-160). Berlin, Heidelberg: Springer-Verlag.

Cios, K., Pedrycz, W., & Swiniarski, R. (1998). *Data mining methods for knowledge discovery*. Dordrecht: Kluwer Academic Publishers.

Coello, C.A., Van Veldhuizen, D.A., & Lamont, G.B. (2002). *Evolutionary algorithms for solving multi-objective problems*. Norwell, MA: Kluwer.

Cohn, P.M. (1965). *Universal algebra*. New York: Harper and Row Publishers.

Cohon, J.L. (1978). *Multiobjective programming and planning*. New York: Academic Press.

Comer, S. (1991). An algebraic approach to the approximation of information. *Fundamenta Informaticae, 14*, 492-502.

Comer, S. (1995). Perfect extensions of regular double Stone algebras. *Algebra Universalis, 34*, 96-109.

Cordón, O., del Jesus, M. J., & Herrera, F. (1999). Evolutionary approaches to the learning of fuzzy rule-based classification systems. In L.C. Jain (Ed.), *Evolution of engineering and information systems and their applications* (pp. 107-160). CRC Press.

Dash, M., & Liu, H. (1997). Feature selection for classification. *Intelligent Data Analysis, 1*(3), 131-156.

Davey, B.A., & Priestley, H.A. (2001). *Introduction to lattices and order*. Cambridge, UK: Cambridge University Press.

De Tré, G., & De Caluwe, R. (1999). A generalized object-oriented database model with generalized constraints. In *Proceedings of North American Fuzzy Information Processing Society* (pp. 381-386). New York.

Deb, K. (2001). *Multi-objective optimization using evolutionary algorithms*. New York: John Wiley & Sons.

Dempster, A.P. (1967). Upper and lower probabilities induced by a multivalued mapping. *Annals of Mathematical Statistics, 38*, 325-339.

Demri, S. (1999). A logic with relative knowledge operators. *Journal of Logic, Language and Information, 8*(2), 167-185.

Compilation of References

Demri, S. (2000). The nondeterministic information logic NIL is PSPACE-complete. *Fundamenta Informaticae, 42*(3-4), 211-234.

Demri, S., & Orłowska, E. (1998). Informational representability of models for information logics. In E. Orłowska (Ed.). *Incomplete information: Rough set analysis* (pp. 383-409). Heidelberg, Germany, New York: Springer-Verlag.

Demri, S., & Orłowska, E. (2002). *Incomplete information: Structures, inference, complexity.* Berlin, Heidelberg, Germany: Springer-Verlag.

Dubois, D., & Prade, H. (1990). Rough fuzzy sets and fuzzy rough sets. *International Journal of General Systems, 17*, 191-209.

Dubois, D., & Prade, H. (1992). Putting rough sets and fuzzy sets together. *Intelligent decision support* (pp. 203-232). Dordrecht: Kluwer Academic Publishers.

Dubois, D., Godo, L., Prade, H., & Esteva, F. (2005). An information-based discussion of vagueness. In H. Cohen & C. Lefebvre (Eds.), *Handbook of categorization in cognitive science* (pp. 892-913). Elsevier.

Düntsch, I. (1997). A logic for rough sets. *Theoretical Computer Science, 179*, 427-436.

Düntsch, I., & Gediga, G. (1997). Algebraic aspects of attribute dependencies in information systems. *Fundamenta Inforamticae, 29*, 119-133.

Düntsch, I., & Gediga, G. (2000). *Rough set data analysis: A road to non-invasive knowledge discovery.* Bangor, UK: Methodos Publishers.

Düntsch, I., & Gediga, G. (2003). Approximation operators in qualitative data analysis. In H. de Swart, E. Orlowska, G. Schmidt, & M. Roubens (Eds.), *Theory and application of relational structures as knowledge instruments* (pp. 216-233). Heidelberg: Springer.

Düntsch, I., & Orłowska, E. (2001). Boolean algebras arising from information systems. Proceedings of the Tarski Centenary Conference, Banach Centre, Warsaw, June 2001. *Annals of Pure and Applied Logic, 217*(1-3), 77-89.

Elan. (2006). *Elan SC520 Microcontroller User's Manual.* Retrieved from http://www.embeddedarm.com/

Everitt, B. (1980). *Cluster analysis* (2nd ed.). London, UK: Heinemann Educational Books.

Fayyad, U.M., & Irani, K.B. (1993). Multi-interval discretization of continuous-valued attributes for classification learning. *IJCAI*, 1022-1029.

Fayyad, U.M., Piatetsky-Shapiro, G., Smyth, P., & Uthurusamy, R. (1996). *Advances in knowledge discovery and data mining.* Menlo Park, CA: AAAI Press/The MIT Press.

Flury, B. (1988). *Common principal components and related multivariate models.* New York: John Wiley & Sons.

Frege, G. (1903). *Grundgesetzen der Arithmetik, 2.* Jena, Germany: Verlag von Herman Pohle.

Ganter, B. & Wille, R. (1999). *Formal concept analysis: Mathematical foundations.* New York: Springer-Verlag.

Garey, M.R., & Johnson, D.S. (1979). *Computers and intractability: A guide to the theory of NP completeness.* New York: W.H. Freeman & Co.

Gaskett, C. (2002). *Q-learning for robot control.* PhD thesis, Supervisor: A. Zelinsky, Department of Systems Engineering, The Australian National University.

Gaskett, C., Ude, A., & Cheng, G. (2005). Hand-eye coordination through endpoint closed-loop and learned endpoint open-loop visual servo control. In *Proceedings of the International Journal of Humanoid Robotics, 2*(2), 203-224.

Gediga, G., & Düntsch, I. (2002). Modal-style operators in qualitative data analysis. In *Proceedings of IEEE International Conference on Data Mining* (pp. 155-162).

Gediga, G., & Düntsch, I. (2003). On model evaluation, indices of importance and interaction values in rough set analysis. In S.K. Pal, L. Polkowski, & A. Skowron (Eds.), *Rough-neuro computing: A way for computing with words* (pp. 251-276). Heidelberg: Physica Verlag.

Geffner, H., & Wainer, J. (1998). Modeling action, knowledge and control. In H. Prade (Ed.), *ECAI 98, 13th European Conference on AI* (pp. 532-536). John Wiley & Sons.

Gehrke, M., & Walker, E. (1992). On the structure of rough sets. *Bulletin of the Polish Academy of Sciences, Mathematics, 40*, 235-245.

Gen, M., & Cheng, R. (2000). *Genetic algorithms and engineering optimization.* New York: John Wiley & Sons.

Gratzer, G. (1971). *Lattice theory: First concepts and distirbutive lattices.* San Francisco: W. H. Freeman and Company.

Gratzer, G. (1978). *General lattice theory.* Basel, Switzerland: Birkhauser.

Greco, A., Matarazzo, B., & Słowiński, R. (2002). Rough approximation by dominance relations. *International Journal of Intelligent Systems, 17*, 153-171.

Greco, S., Matarazzo, B., & Slowinski, R. (1998). Fuzzy measures as a technique for rough set analysis. In *Proceedings of the. 6th European Congress on Intelligent Techniques & Soft Computing, Aachen, 1* (pp. 99-103).

Greco, S., Matarazzo, B., & Slowinski, R. (1999). Rough approximation of a preference relation by dominance relations. *European Journal of Operational Research, 117*, 63-83.

Greco, S., Matarazzo, B., Pappalardo, N., & Slowinski, R. (2005). Measuring expected effects of interventions based on decision rules. *Journal of Experimental and Theoretical Artificial Intelligence, Taylor Francis, 17*(1-2).

Greco, S., Pawlak, Z., & Słowiński, R. (2002). Generalized decision algorithms, rough inference rules and flow graphs. In J.J. Alpigini, J.F. Peters, A. Skowron, N. Zhong (Eds.), *Rough sets and current trends in computing, 3rd International Conference (RSCTC2002). Lectures Notes in Artificial Intelligence, 2475* (pp. 93-104). Berlin, Heildelberg, Germany: Springer-Verlag.

Grzymala-Busse, J. (1988). Knowledge acquisition under uncertainty—a rough set approach, *Journal of Intelligent and Robotics Systems, 1*, 3-16.

Grzymala-Busse, J. (1991). *Managing uncertainty in expert systems.* Boston: Kluwer Academic Publishers.

Grzymala-Busse, J. (1997). A new version of the rule induction system LERS. *Fundamenta Informaticae, 31*(1), 27-39.

Grzymała-Busse, J. W. (1988). Knowledge acquisition under uncertainty - A rough set approach. *Journal of Intelligent and Robotic Systems, 1*(1), 3-16.

Grzymała-Busse, J. W., & Stefanowski, J. (1999). Two approaches to numerical attribute discretization for rule induction. In *Proceedings: the 5th International Conference of the Decision Sciences Institute (ICDCI'99)* (pp. 1377-1379). Athens, Greece.

Grzymala-Busse, J.W. (1992). LERS—A system for learning from examples based on rough sets. In R. Slowinski (Ed.), *Intelligent decision support. Handbook of applications and advances of the rough set theory* (pp. 3-18). Dordrecht, Boston, London: Kluwer Academic Publishers.

Grzymala-Busse, J.W. (2002). MLEM2: A new algorithm for rule induction from imperfect data. In *Proceedings of the 9th International Conference on Information Processing and Management of Uncertainty in Knowledge-Based Systems* (pp. 243-250). Annecy, France.

Grzymała-Busse, J.W., & Hu, M. (2001). A comparison of several approaches to missing attribute values in data mining. In W. Ziarko & Y. Yao (Eds.), *Proceedings of the 2nd International Conference on Rough Sets and Current Trends in Computing (RSCTC2000). Lectures Notes in Artificial Intelligence, 2005* (pp. 340-347). Heildelber, Germany: Springer-Verlag.

Grzymała-Busse, J.W., & Stefanowski, J. (2001). Three discretization methods for rule induction. *International Journal of Intelligent Systems, 16*(1), 29-38.

Grzymala-Busse, J.W., Hippe, Z.S., Mroczek, T., Roj, E., & Skowronski, B. (2005). Data mining analysis of granular bed caking during hop extraction. In *Proceedings of the ISDA'2005, Fifth International Conference on Intelligent System Design and Applications* (pp. 426-431). IEEE Computer Society, Wroclaw, Poland,.

Grzymała-Busse, J.W., Stefanowski, J., & Ziarko, W. (1996). Rough sets: Facts versus misconceptions. *Informatica, 20*, 455-465.

Gyseghem, N., & De Caluwe, R. (1997). Fuzzy and uncertain object-oriented databases: Concepts and models. *Advances in Fuzzy Systems—Applications and Theory, 13*, 123-177.

Han, J., Hu, X., & Lin, T.Y. (2004). Feature subset selection based on relative dependency between attributes. *Rough Sets and Current Trends in Computing: 4th International Conference, RSCTC 2004* (pp. 176-185). Uppsala, Sweden.

Han, J., Pei, J., & Yin, Y. (2000). Mining frequent patterns without candidate generation. In W. Chen, J. Naughton, & P.A. Bernstein (Eds.), *ACM SIGMOD International Conference on Management of Data* (pp. 1-12). ACM Press.

Heckerman, D. (1995). *A tutorial on learning Bayesian networks* (MSR-TR-95-06). Retrieved from heckerman@microsoft.com

Hempel, C.G. (1939). Vagueness and logic. *Philosophy of Science, 6*, 163-180.

Hippe, Z.S. (1996). Design and application of new knowledge engineering tool for solving real world problems, *Knowledge-Based Systems, 9*, 509-515.

Ho, T.B. (1997). Acquiring concept approximations in the framework of rough concept analysis. In *Proceedings of 7th European-Japanese Conference on Information Modelling and Knowledge Bases* (pp.186-195).

Höhle, U. (1988). Quotients with respect to similarity relations. *Fuzzy Sets and Systems, 27*(1), 31-44.

Holland, J. (1975). *Adaptation in natural and artificial systems*. Ann Arbour: The University of Michigan Press.

Hoos, H.H., & Stützle, T. (1999). Towards a characterisation of the behaviour of stochastic local search algorithms for SAT. *Artificial Intelligence, 112*, 213-232.

Hu, K., Sui, Y., Lu, Y., Wang, J., & Shi, C. (2001). Concept approximation in concept lattice. In *Proceedings of 5th Pacific-Asia Conference on Knowledge Discovery and Data Mining (PAKDD'01)* (pp. 167-173).

Huntington, E.V. (1933). Boolean algebra. A correction. *Transactions of AMS, 35*, 557-558.

Hyvarinen, A., & Oja, E. (2000). Independent component analysis: Algorithms and applications. *Neural Networks, 13*(4), 411-430.

Internet Engineering Task Force (IEFT). (2006). *RFC 1332 The PPP Internet Protocol Control Protocol (IPCP)*. Retrieved from http://www.ietf.org/rfc/rfc1332.txt

Inuiguchi, M., Hirano, S., & Tsumoto, S. (Eds.). (2003). *Rough set theory and granular computing*. Berlin, Heidelberg, Germany: Springer Verlag.

Iwiński, T. B. (1987). Algebraic approach to rough sets. *Bulletin of the Polish Academy of Sciences, Mathematics, 35*, 673-683.

Järvinen, J. (1999). *Knowledge representation and rough sets*. Doctoral dissertation. University of Turku: Turku Center for Computer Science.

Järvinen, J. (2006). *Lattice theory of rough sets*. Unpublished manuscript.

Jensen, K. (1981). Coloured Petri nets and the invariant-method. *Theoretical Computer Science, 14*, 317-336.

Jensen, K. (1997). *Coloured Petri nets. Vol. 1: Basic concepts*. Berlin, Heidelberg: Springer-Verlag.

Jensen, R. (2006). Performing feature selection with ACO. To appear in *Swarm intelligence and data mining*. Springer SCI book series.

Jensen, R., & Shen, Q. (2004). Fuzzy-rough attribute reduction with application to Web categorization. *Fuzzy Sets and Systems, 141*(3), 469-485.

Jensen, R., & Shen, Q. (2004). Semantics-preserving dimensionality reduction: Rough and fuzzy-rough based approaches. *IEEE Transactions on Knowledge and Data Engineering, 16*(12), 1457-1471.

Jensen, R., & Shen, Q. (2005). Fuzzy-rough data reduction with ant colony optimization. *Fuzzy Sets and Systems, 149*(1), 5-20.

Jensen, R., Shen, Q., & Tuson, A. (2005). Finding rough set reducts with SAT. In *Proceedings of the 10th International Conference on Rough Sets, Fuzzy Sets, Data Mining and Granular Computing*, LNAI 3641 (pp. 194-203).

Jin, Y. (2000). Fuzzy modeling of high-dimensional systems: Complexity reduction and interpretability improvement. *IEEE Transactions on Fuzzy Systems, 8*(2), 212-221.

Kadous, M.W. (2002). *Temporal classification: Extending the classification paradigm to multivariate time series*. Unpublished doctoral dissertation, University of New South Wales, Australia.

Katriniak, T. (1973). The structure of distributive double p-algebras. Regularity and Congruences. *Algebra Universalis, 3*, 238-246.

Kautz, H.A., & Selman B. (1992). Planning as satisfiability. In *Proceedings of the Tenth European Conference on Artificial Intelligence (ECAI'92)* (pp. 359-363).

Kautz, H.A., & Selman. B. (1996). Pushing the envelope: Planning, propositional logic, and stochastic search. In *Proceedings of the Twelfth National Conference on Artificial Intelligence (AAAI'96)* (pp. 1194-1201).

Keefe, R. (2000). *Theories of vagueness*. Cambridge, UK: Cambridge University Press.

Kennedy, J., & Eberhart, R.C. (1995). Particle swarm optimization. In *Proceedings of IEEE International Conference on Neural Networks* (pp. 1942-1948). Piscataway, NJ.

Kent, R.E. (1996). Rough concept analysis. *Fundamenta Informaticae, 27*, 169-181.

Keogh, E. & Folias, T. (2002). The UCR time series data mining archive. *Riverside, CA, University of California—Computer Science & Engineering Department*. Retrieved December 7, 2006, from http://www.cs.ucr.edu/~eamonn/TSDMA/index.html

Keogh, E., & Kasetty, S. (2002). On the need for time series data mining benchmarks: A survey and empirical demonstration. In *Proceedings of the 8th ACM SIGKDD International Conference on Knowledge Discovery and Data Mining* (pp. 102-111). Edmonton, Alberta, Canada: ACM.

Keogh, E., & Pazzani, M. (1998). An enhanced representation of time series which allows fast and accurate classification, clustering and relevance feedback. In R. Agrawal, & P. Stolorz (Ed.), *4th International Conference on Knowledge Discovery and Data Mining* (pp. 239-241). New York: AAAI Press.

Koczkodaj, W. W., Orlowski, M., & Marek, V. W. (1998). Myths about rough set theory. *Communications of the ACM, 41*(11), 102-103.

Komorowski, J., Pawlak, Z., Polkowski, L., & Skowron, A. (1999). Rough sets: A tutorial. In S.K. Pal, & A. Skowron (Ed.), *Rough fuzzy hybridization: A new trend in decision-making* (pp. 3-98). Singapore: Springer-Verlag.

Konda, V. R., & Tsitsiklis, J. N. (1995). Actor--critic algorithms. *Adv. Neural Inform. Processing Sys.*, 345-352.

Kondrad, E., Orłowska, E., & Pawlak, Z. (1981). *Knowledge representation systems*. Technical Report 433, Institute of Computer Science, Polish Academy of Sciences.

Koppelberg, S. (1989). General theory of Boolean algebras. In J.D. Monk & R. Bonett, (Eds.), *Handbook of Boolean algebras*. Amsterdam: North Holland.

Kosko, B. (1986). Fuzzy entropy and conditioning. *Information Sciences, 40*(2), 165–174.

Kreyszig, E. (1978). *Introductory functional analysis with applications*. New York: John Wiley & Sons.

Kryszkiewicz, M. (1994). Maintenance of reducts in the variable precision rough sets model. *ICS Research Report* 31/94, Warsaw University of Technology.

Kryszkiewicz, M. (1997). Maintenance of reducts in the variable precision rough set model. In T.Y. Lin & N. Cercone (Eds.), *Rough sets and data mining – Analysis of imperfect data* (pp. 355-372). Boston, Kluwer Academic Publishers.

Kudo, M., & Skalansky, J. (2000). Comparison of algorithms that select features for pattern classifiers. *Pattern Recognition, 33*(1), 25-41.

Langley, P. (1994). Selection of relevant features in machine learning. In *Proceedings of the AAAI Fall Symposium on Relevance* (pp. 1-5).

Laumanns, M., Zitzler, E., & Thiele, L. (2000). A unified model for multi-objective evolutionary algorithms with elitism. In A. Zalzala et al. (Ed.), *2000 Congress on Evolutionary Computation* (pp. 46-53). San Diego, CA: IEEE Press.

Lin, T.Y. (1998). Granular computing on binary relations. I: Data mining and neighborhood systems. In L. Polkowski, & A. Skowron, (Eds.). *Rough sets in knowledge discovery* (pp. 107-121). Heidelberg: Physica-Verlag.

Lin, T.Y. (1999). Granular computing: Fuzzy logic and rough sets. In L.A. Zadeh & J. Kacprzyk (Eds.), *Computing with words in information/intelligent systems* (pp. 183-200). Berlin, Heidelberg, Germany: Springer-Verlag.

Lin, T.Y. (2006). A roadmap from rough set theory to granular computing. In G. Wang, J.F. Peters, A. Skowron, & Y.Y. Yao, (Eds.), *Rough sets and knowledge technology. Proceedings of the First International Conference, RSKT2006, Chonging, China. Lectures Notes in Computer Science, 4062* (pp. 33-41). Berlin, Heidelberg, Germany: Springer-Verlag.

Lin, T.Y., & Cercone, N. (Eds.). (1997). *Rough sets and data mining*. Dodrecht: Kluwer Academic Publisher.

Lipski, W. (1976). Informational systems with incomplete information. In *3rd International Symposium on Automata, Languages and Programming* (pp. 120-130). Edinburgh, Scotland.

Liu, B., Hsu, W., & Chen, S. (1997). Using general impressions to analyze discovered classification rules. In *Proceedings of KDD97 Conference*. Newport Beach, CA: AAAI Press.

Liu, H., & Motoda, H. (Eds.), (1999). *Feature selection for knowledge discovery and data mining*. Kluwer Academic Publishers.

Mac Parthaláin, N., Jensen, R., & Shen, Q. (2006). Fuzzy entropy-assisted fuzzy-rough feature selection. To appear in *Proceedings of the 15th International Conference on Fuzzy Systems (FUZZ-IEEE'06)*.

Makinouchi, A. (1977). A consideration on normal form of not-necessarily normalized relation in the relational data model. In *Proceedings of the Third International Conference on Very Large Databases* (pp. 447-453).

Manquinho, V.M., Flores, P.F., Silva, J.P.M.., & Oliveira A.L. (1997). Prime implicant computation using satisfiability algorithms. In the *International Conference on Tools with Artificial Intelligence (ICTAI '97)* (pp. 232-239).

Marek, W., & Pawlak, Z. (1984). Rough sets and information systems. *Fundamenta Informaticae, 17*(1), 105-115.

McKinsey, J.C.C., & Tarski, A. (1946). On closed elements in closure algebras, *The Annals of Mathematics, 47*, 122-162.

Mi, J.S., Wu, W.Z. & Zhang, W.X. (2003). Approaches to approximation reducts in inconsistent decision tables. In *Proceedings of 9th International Conference on Rough Sets, Fuzzy Sets, Data Mining, and Granular Computing (RSFDGrC'03)* (pp. 283-286).

Mi, J.S., Wu, W.Z. & Zhang, W.X. (2004). Approaches to knowledge reduction based on variable precision rough set model, *Information Sciences, 159*, 255-272.

Miao, D., & Wang, J. (1999). An information representation of the concepts and operations in rough set theory, *Journal of Software, 10*, 113-116.

Milanova, M., Smolinski, T.G., Boratyn, G.M., Zurada, J.M., & Wrobel, A. (2002). Correlation kernel analysis and evolutionary algorithm-based modeling of the sensory activity within the rat's barrel cortex. *Lecture Notes in Computer Science, 2388*, 198-212.

Modrzejewski, M. (1993). Feature selection using rough sets theory. In *Proceedings of the 11th International Conference on Machine Learning* (pp. 213-226).

Mroczek, T., Grzymala-Busse, J.W., & Hippe, Z.S. (2004). Rules from belief networks: A rough set approach. In S. Tsumoto, R. Slowinski, J. Komorowski, & J.W. Grzymala-Busse (Eds.), *Rough sets and current trends in computing* (pp. 483-487). Berlin, Heidelberg, New York: Springer-Verlag.

Nakata, M., & Sakai, H. (2005). Rough sets handling missing values probabilistically interpreted. In D. Ślęzak G. Wang, M. S. Szczuka, I. Düntsch, & Y.Yao (Eds.) *Rough sets, fuzzy sets, data mining, and granular computing, 10th International Conference, Regina, Canada (RSFDGrC 2005), Lectures Notes in Artificial Intelligence 3641* (pp. 325-334). Berlin, Heidelberg: Springer-Verlag.

Nanda, S,. & Majumdar, S. (1992). Fuzzy rough sets. *Fuzzy Sets and Systems, 45*, 157-160.

Ngo, C.L., & Nguyen H.S. (2005). A method of web search result clustering based on rough sets. In *Proceedings of 2005 IEEE/WIC/ACM International Conference on Web Intelligence* (pp. 673-679). IEEE Computer Society Press.

Nguyen, H.S. (1997). *Discretization of real value attributes, Boolean reasoning approach.* PhD thesis, Warsaw University, Warsaw, Poland.

Nguyen, H.S. (1998). Discretization problems for rough set methods. In L. Polkowski, & A. Skowron, (Eds.), *Rough sets and current trends in computing, 1st International Conference (RSCTC'98), Warsaw, Poland. Lectures Notes in Artificial Intelligence, 1424* (pp. 545-552). Berlin, Heidelberg, Germany: Springer-Verlag.

Nguyen, H.S. (1999). Efficient SQL-querying method for data mining in large data bases. In *Proceedings of Sixteenth International Joint Conference on Artificial Intelligence, IJCAI-99* (pp. 806-811). Stockholm, Sweden: Morgan Kaufmann.

Nguyen, H.S. (2001). On efficient handling of continuous attributes in large data bases. *Fundamenta Informaticae, 48*(1), 61-81.

Nguyen, H.S. (2006). Approximate Boolean reasoning: Foundations and applications in data mining. Transactions on rough sets V. *Lecture notes on computer science, 4100*, 334-506.

Nguyen, H.S. (2006). Knowledge discovery by relation approximation: A rough set approach. In Proceedings of the International Conference on Rough Sets and Knowledge Technology (RSKT), Chongqing, China. *LNAI, 4062* (pp. 103-106). Heidelberg, Germany: Springer.

Nguyen, H.S., & Nguyen S.H. (1998). Discretization methods for data mining. In L. Polkowski & A. Skowron (Eds.), *Rough sets in knowledge discovery* (pp. 451-482). Heidelberg; New York: Springer.

Nguyen, H.S., &. Nguyen, S.H. (2005). Fast split selection method and its application in decision tree construction from large databases. *International Journal of Hybrid Intelligent Systems, 2*(2), 149-160.

Nguyen, H.S., Łuksza, M., Mąkosa, E., & Komorowski J. (2005). An approach to mining data with continuous decision values. In M.A. Kłopotek, S.T. Wierzchoń, & K. Trojanowski (Eds.), *Proceedings of the International IIS: IIPWM'05 Conference held in Gdansk, Poland. Advances in Soft Computing* (pp. 653-662). Heidelberg, Germany: Springer.

Nguyen, H.S., & Skowron, A. (1997). Boolean reasoning for feature extraction problems. In *Proceedings of the 10th International Symposium on Methodologies for Intelligent Systems* (pp. 117-126).

Nguyen, H.S., & Ślęzak D. (1999). Approximate reducts and association rules—Correspondence and complexity results. In A. Skowron, S. Ohsuga, & N. Zhong (Eds.), New directions in rough sets, data mining and granular-soft computing, Proceedings of RSFDGrC'99, Yamaguchi, Japan, *LNAI, 1711*, 137-145., Heidelberg, Germany: Springer.

Nguyen, S.H. (2000). Regularity analysis and its applications in Data Mining. In L. Polkowski, T.Y. Lin, &

S. Tsumoto (Eds.), Rough set methods and applications: New developments in knowledge discovery in information systems. *Studies in fuzziness and soft computing 56*, 289-378. Heidelberg, Germany: Springer.

Nguyen, S.H., Bazan, J., Skowron, A., & Nguyen, H.S. (2004). Layered learning for concept synthesis. Transactions on rough sets I, *Lecture Notes on Computer Science, 3100*, 187-208. Heidelberg, Germany: Springer.

Nguyen, S.H., & Nguyen H.S. (1996). Some efficient algorithms for rough set methods. In *Sixth International Conference on Information Processing and Management of Uncertainty on Knowledge Based Systems IPMU1996*, volume III (pp. 1451–1456). Granada, Spain.

Nguyen, S.H., & Skowron, A. (1997a). Searching for relational patterns in data. In *Proceedings of the First European Symposium on Principles of Data Mining and Knowledge Discovery* (pp. 265-276), Trondheim, Norway, June 25-27.

Nguyen, S.H., Skowron, A., & Synak P. (1998). Discovery of data patterns with applications to decomposition and classification problems. In L. Polkowski & A. Skowron (Eds.), Rough sets in knowledge discovery 2: Applications, case studies and software systems. *Studies in fuzziness and soft computing, 19*, 55-97. Heidelberg, Germany: Springer.

Obtułowicz, A. (1987). Rough sets and Heyting algebra valued sets. *Bulletin of the Polish Academy of Sciences, Mathematics, 35*(9-10), 667-671.

Øhrn, A. (1999). *Discernibility and rough sets in medicine: Tools and applications*. Department of Computer and Information Science. Trondheim, Norway, Norwegian University of Science and Technology: 239.

Øhrn, A. (2001). *ROSETTA technical reference manual*. Retrieved December 7, 2006, from http://rosetta.lcb.uu.se/general/resources/manual.pdf

Ohrn, A., Komorowski, J., Skowron, A., & Synak P. (1998). The ROSETTA software system. In L. Polkowski & A. Skowron (Eds.), Rough sets in knowledge discovery 2: Applications, case studies and software systems. *Studies in fuzziness and soft computing, 19*, 572-576. Heidelberg, Germany: Springer.

Ola, A., & Ozsoyoglu, G. (1993). Incomplete relational database models based on intervals. *IEEE Transactions on Knowledge and Data Engineering, 5*(2), 293-308.

Orłowska, E. (1988). Representation of vague information. *Information Systems, 13*, 167-174.

Orłowska, E. (1989). Logic for reasoning about knowledge. *Zeitschrift fur Mathematische Logik und Grundlagen der Mathematik, 35*, 559-568.

Orłowska, E. (1993). Reasoning with incomplete information: Rough set based information logics. In V. Algar, S. Bergler, & F.Q. Dong (Eds.). *Incompleteness and Uncertainty in Information Systems Workshop* (pp. 16-33). Berlin, Springer-Verlag.

Orłowska, E. (1993). Rough semantics for non-classical logics. In W. Ziarko (Ed.), *2nd International Workshop on Rough Sets and Knowledge Discovery,* Banff, Canada (pp. 143-148).

Orłowska, E. (1995). Information algebras. In V. S. Alagar, & M. Nivat (Eds.), *Algebraic methodology and software technology, Lectures Notes in Computer Science, 936*, 50-65.

Orłowska, E. (1997). Studying incompleteness of information: A class of information logics. In K. Kijania-Placek, & J. Woleński (Eds.). *The Lvov-Warsaw School and contemporary philosophy* (pp.303-320). Dordrecht: Kluwer Academic Publishers.

Orlowska, E. (1998). Introduction: What you always wanted to know about rough sets. In E. Orlowska (Ed.), *Incomplete information: Rough set analysis.* New York: Physica-Verlag Heidelberg.

Orłowska, E. (Ed.). (1998). *Incomplete information: Rough set analysis.* Heidelberg, Germany, New York: Springer-Verlag.

Orłowska, E., & Pawlak, Z. (1984). Representation of nondeterministic information. *Theoretical Compter Science, 29*, 27-39.

Pagliani, P. (1993). From concept lattices to approximation spaces: Algebraic structures of some spaces of partial objects. *Fundamenta Informaticae, 18,* 1-25.

Pagliani, P. (1996). Rough sets and Nelson algebras. *Fundamenta Inforamticae, 27*(2-3), 205-219.

Pagliani, P. (1998a). Rough set theory and logic-algebraic structures. In E. Orłowska (Ed.). *Incomplete information: Rough set analysis* (pp. 109-192). Heidelberg, Germany, New York: Springer-Verlag.

Pagliani, P. (1998b). Intrinsic co-Heyting boundaries and information incompleteness in rough set analysis. In L. Polkowski & A. Skowron, (Eds.), *Rough sets and current trends in Computing, 1st International Conference (RSCTC'98),* Warsaw, Poland. *Lectures Notes in Artificial Intelligence, 1424* (pp. 123-130). Berlin, Heidelberg, Germany: Springer-Verlag.

Pagliani, P., & Chakraborty, M.K. (2005). Information quanta and approximation spaces. I: Nonclassical approximation operators. In *Proceedings of IEEE International Conference on Granular Computing* (pp. 605-610).

Pal, S.K., Polkowski, L., & Skowron, A. (Eds.). (2004). *Rough-neural computing: Techniques for computing with words.* Heidelberg, Germany, Springer Verlag.

Pancerz, K., & Suraj, Z. (2003). Synthesis of Petri net models: A rough set approach. *Fundamenta Informaticae, 55*(2), 149-165.

Pawlak, Z. (1973). *Mathematical foundations of informational retrieval* (Research Report 101). Institute of Computer Science, Polish Academy of Sciences.

Pawlak, Z. (1981). Information systems - theoretical foundations. *Information Systems, 6,* 205-218.

Pawlak, Z. (1981). Classification of objects by means of attributes. *Institute for Computer Science, Polish Academy of Sciences Report 429.*

Pawlak, Z. (1981). Rough sets. *Institute for Computer Science, Polish Academy of Sciences Report 431.*

Pawlak, Z. (1982). Rough sets. *International Journal of Computer and Information Sciences, 11,* 341-356.

Pawlak, Z. (1984). Rough sets. *International Journal of Man-Machine Studies, 21,* 127-134.

Pawlak, Z. (1985). Rough sets and fuzzy sets. *Fuzzy Sets and Systems, 17,* 99-102.

Pawlak, Z. (1991). Information systems—theoretical foundations. *Information Systems Journal, 6,* 205-218.

Pawlak, Z. (1991). *Rough sets. Theoretical aspects of reasoning about data.* Dodrecht: Kluwer Academic Publisher.

Pawlak, Z. (1992). Concurrent versus sequential: The rough sets perspective. *Bulletin of the EATCS, 48,* 178-190.

Pawlak, Z. (1995). Knowledge and rough sets (in Polish). In W. Traczyk (Ed.), *Problems of artificial intelligence* (pp. 9-21). Warsaw, Poland: Wiedza i Zycie.

Pawlak, Z. (2002). In pursuit of patterns in data reasoning from data - The rough set way. *The Third International Conference on Rough Sets, and Current Trends in Computing* (pp. 1-9).

Pawlak, Z. (2003). Elementary rough set granules: toward a rough set processor. In S.K. Pal, L. Polkowski, & A. Skowron (Eds.), *Rough-neural computing: Techniques for computing with words* (pp. 5-13). Heidelberg, Germany, Springer Verlag.

Pawlak, Z. (2003). Flow graphs and decision algorithms. *The Ninth International Conference on Rough Sets, Fuzzy Sets, Data Mining, and Granular Computing* (pp. 1-10).

Pawlak, Z. (2004). Some issues on rough sets. *Lecture Notes in Computer Sciences, 3100,* 1-58.

Pawlak, Z. (2005). Flow graphs and data mining. *Lecture Notes in Computer Sciences, 3400,* 1-36.

Pawlak, Z., & Skowron, A. (2007). Rudiments of rough sets. *Sciences, 177,* 3-27.

Pawlak, Z., & Skowron, A. (2007). Rough sets: Some extensions. *Sciences, 177,* 28-40.

Pawlak, Z., & Skowron, A. (2007). Rough sets and Boolean reasoning. *Sciences, 177*, 41-73.

Pei, D.W., & Xu, Z.B. (2004). Rough set models on two universes. *International Journal of General Systems, 33*, 569-581.

Peters, J. F. (2005b). Rough ethology: Towards a biologically-inspired study of collective behavior in intelligent systems with approximation spaces. *Transactions on Rough Sets, III, LNCS 3400*, 153-174.

Peters, J. F., & Henry, C. (2005). Reinforcement learning in swarms that learn. In *Proceedings of the IEEE/WIC/ACM International Conference on Intelligent Agent Technology (IAT 2005)*, Compiègne Univ. of Tech., France (pp. 400-406).

Peters, J. F., & Pawlak, Z. (2007). Zdzisław Pawlak life and work (1906-2006). *Information Sciences, 177*, 1-2.

Peters, J. F., Henry, C., & Ramanna, S. (2005). Rough ethograms: Study of intelligent system behavior. In: M. A. Kłopotek, S. Wierzchoń, & K. Trojanowski (Eds.), *New trends in intelligent information processing and Web mining (IIS05)*, Gdańsk, Poland (pp. 117-126).

Peters, J. F., Skowron, A., Synak, P., & Ramanna, S. (2003). Rough sets and information granulation. In T. Bilgic, D. Baets, & O. Kaynak (Eds.), Tenth Int. Fuzzy Systems Assoc. World Congress IFSA, Instanbul, Turkey. *Lecture Notes in Artificial Intelligence 2715*, 370-377. Heidelberg: Physica-Verlag.

Peters, J.F. (2007). Near sets. Special theory about nearness of objects. *Fundamenta Informaticae, 75*, in press.

Peters, J.F., Skowron, A., & Suraj, Z. (2000). An application of rough set method to control dsesign. *Fundamenta Informaticae, 43*(1-4), 269-290.

Pizzuti, C. (1996). Computing prime implicants by integer programming. In *Eighth International Conference on Tools with Artificial Intelligence (ICTAI '96)* (pp. 332-336).

Pogonowski, J. (1981). *Tolerance spaces with applications to linguistics*. Poznań: Adam Mickiewicz University Press.

Polkowski, L. (2002). *Rough sets. Mathematical foundations*. Heidelberg: Springer-Verlag.

Polkowski, L., & Skowron, A. (1996). Rough mereology: A new paradigm for approximate reasoning. *International Journal of Approximate Reasoning, 15*, 333-365.

Polkowski, L., & Skowron, A. (Eds.). (1998). *Rough Sets and Current Trends in Computing, 1st International Conference (RSCTC'98), Warsaw, Poland. Lectures Notes in Artificial Intelligence, 1424*. Berlin, Heidelberg, Germany: Springer-Verlag.

Polkowski, L., & Skowron, A. (Eds.). (1998). *Rough sets in knowledge discovery*. Heidelberg: Physica-Verlag.

Polkowski, L., Lin, T.Y., & Tsumoto, S. (Eds.). (2000). Rough set methods and applications: New developments in knowledge discovery in information systems. *Studies in Fuzziness and Soft Computing, 56*. Heidelberg, Germany: Physica-Verlag.

Polkowski, L., Skowron, A., & Żytkow, J. (1995). Rough foundations for rough sets. In T.Y. Lin, & A.M. Wildberger (Eds.), *Soft computing: Rough sets, fuzzy logic, neural networks, uncertainty management, knowledge discovery* (pp. 55-58). San Diego, CA: Simulation Councils, Inc.

Pomykała, J., & Pomykała, J.A. (1988). The Stone algebra of rough sets. *Bulletin of the Polish Academy of Sciences, Mathematics, 36*, 451-512.

Qi, J.J., Wei, L., & Li, Z.Z. (2005). A partitional view of concept lattice. In *Proceedings of 10th International Conference on Rough Sets, Fuzzy Sets, Data Mining, and Granular Computing (RSFDGrC'05)*, Part I (pp. 74-83).

Qiu, G. F., Zhang, W. X. & Wu, W. Z. (2005). Characterizations of attributes in generalized approximation representation spaces. In *Proceedings of 10th International Conference on Rough Sets, Fuzzy Sets, Data Mining, and Granular Computing (RSFDGrC'05)* (pp. 84-93).

Quine, W.V.O. (1952). The problem of simplifying truth functions. *American Mathematical Monthly, 59*, 521-531.

Quine, W.V.O. (1959). On cores and prime implicants of truth functions. *American Mathematical, 66*, 755-760.

Quine, W.V.O. (1961). *Mathematical logic*. Cambridge, MA: Harvard University Press.

Quinlan, J.R. (1993). C4.5: Programs for machine learning. *The Morgan Kaufmann series in machine learning*. San Mateo, CA: Morgan Kaufmann Publishers.

Quinlan, R. (1986). Induction of decision trees. *Machine Learning, 1*, 81-106.

Rajan, J.J. (1994). *Time series classification*. Unpublished doctoral dissertation, University of Cambridge, UK.

Ras, Z. (1999). Discovering rules in information trees In J. Zytkow & J. Rauch (Eds.), *Principles of Data Mining and Knowledge Discovery, Proceedings of PKDD'99*, Prague, Czech Republic, LNAI, No. 1704 (pp. 518-523). Springer.

Ras, Z., & Wieczorkowska, A. (2000). Action rules: How to increase profit of a company. In D.A. Zighed, J. Komorowski, & J. Zytkow (Eds.), *Principles of data mining and knowledge discovery, Proceedings of PKDD'00*, Lyon, France, LNAI, No. 1910 (pp. 587-592). Springer.

Ras, Z.W., & Dardzinska, A. (2006). Action rules discovery, a new simplified strategy. In F. Esposito et al., *Foundations of Intelligent Systems, Proceedings of ISMIS'06*, Bari, Italy, LNAI, No. 4203 (pp. 445-453). Springer.

Ras, Z.W., Tzacheva, A., & Tsay, L.-S. (2005). Action rules. In J. Wang (Ed.), *Encyclopedia of data warehousing and mining* (pp. 1-5). Idea Group Inc.

Rasiowa, H., & Sikorski, R. (1970). *The mathematics of metamathematics*. Warsaw, Poland: Polish Scientific Publisher.

Rauszer, C. (1974). Semi-Boolean algebras and their applications to intuitionistic logic with dual operators. *Fundamenta Mathematicae, 83*, 219-249.

Reisig, W. (1985). *Petri nets. An introduction*. Berlin: Springer-Verlag.

Roth, M.A., Korth, H.F., & Batory, D.S. (1987). SQL/NF: A query language for non-1NF databases. *Information Systems, 12*, 99-114.

Rudeanu. S. (1974). *Boolean functions and equations*. Amsterdam: North-Holland/American Elsevier.

Russell, B. (1923). Vagueness. *The Australasian Journal of Psychology and Philosophy, 1*, 84-92.

Rzasa, W., & Suraj, Z. (2002). A new method for determining of extensions and restrictions of information systems. *Lecture Notes in Artificial Intelligence, 2475*, 197-204.

Saito, N. (1994). *Local feature extraction and its application using a library of bases*. Unpublished doctoral dissertation.

Sakai, H., & Nakata, M. (2005). Discernibility functions and minimal rules in non-deterministic information systems. In D. Ślęzak G. Wang, M. S. Szczuka, I. Düntsch, & Y.Yao (Eds.), Rough sets, fuzzy sets, data mining, and granular computing, 10th International Conference, Regina, Canada (RSFDGrC 2005). *Lectures Notes in Artificial Intelligence, 3641*, 254-264. Berlin, Heidelberg: Springer-Verlag.

Sant'Anna, A. P. (2004). Rough sets in the probabilistic composition of preferences. In B. de Baets, R. de Caluwe, G. de Tré, J. Fodor, J. Kacprzyk, & S. Sadrosny (Eds.), *Current issues in data and knowledge engineering* (pp. 407-414). Warszawa: EXIT.

Sant'Anna, L. A. F. P., & Sant'Anna, A. P. (2001). Randomization as a stage in criteria combining. In F. S. Fogliato, J. L. D. Ribeiro, & L. B. M. Guimarães (Eds.), *Production and distribution challenges for the 21st century* (pp. 248-256). Salvador: ABEPRO.

Sant'Anna, L. A. F. P., & Sant'Anna, A. P. (in press). A probabilistic approach to evaluate the exploitation of the geographic situation of hydroelectric plants. *Annals of ORMMES 2006*.

Saquer, J., & Deogun, J. (1999). Formal rough concept analysis. In *Proceedings of 7th International Workshop* on *Rough Sets, Fuzzy Sets, Data Mining, and Granular-Soft Computing (RSFDGrC'99)* (pp. 91-99).

Saquer, J., & Deogun, J. (2001). Concept approximations based on rough sets and similarity measures. *International Journal of Applied Mathematics and Computer Science, 11,* 655-674.

Schaffer, J. D. (1984). *Some experiments in machine learning using vector evaluated genetic algorithms.* Unpublished doctoral dissertation, Vanderbilt University, TN.

Selfridge, O. G. (1984). Some themes and primitives in ill-defined systems. In O. G. Selfridge, E. L. Rissland, & M. A. Arbib (Eds.), *Adaptive control of ill-defined systems.* London: Plenum Press.

Selman, B., Kautz, H.A., &. McAllester D.A. (1997). Ten challenges in propositional reasoning and search. In *Proceedings of Fifteenth International Joint Conference on Artificial Intelligence* (pp. 50-54).

Sen, S. (1993). Minimal cost set covering using probabilistic methods. In *SAC '93: Proceedings of the 1993 ACM/SIGAPP Symposium on Applied Computing* (pp. 157-164). New York: ACM Press.

Shafer, G. (1976). *A mathematical theory of evidence.* Princeton: Princeton University Press.

Shafer, G. (1987). Belief functions and possibility measures. In J.C. Bezdek (Ed.), *Analysis of fuzzy information, (Vol. 1) Mathematics and logi* (pp. 51-84). Boca Raton: CRC Press.

Shannon, C.E. (1940). *A symbolic analysis of relay and switching circuits.* MIT, Dept. of Electrical Engineering.

Shao, M.W. & Zhang, W. X. (2005). Approximation in formal concept analysis. In *Proceedings of 10th International Conference on Rough Sets, Fuzzy Sets, Data Mining, and Granular Computing (RSFDGrC'05),* Part I (pp. 43-53).

Shapley, L. (1953). A value for n-person games. In H. Kuhn & A. Tucker, *Contributions to the theory of games II* (pp. 307-317). Princeton: Princeton University Press.

Shen, Q., & Jensen, R. (2004). Selecting informative feature with fuzzy-rough sets and its application for complex systems monitoring. *Pattern Recognition, 37*(7), 1351-1363.

Shi, W., Wang, S., Li, D., & Wang, X. (2003). Uncertainty-based Spatial Data Mining. In *Proceedings of Asia GIS Association,* Wuhan, China (pp. 124-35).

Shi, Y., & Eberhart, R. (1998). A modified particle swarm optimizer. In *Proceedings of the IEEE International Conference on Evolutionary Computation* (pp. 69-73). Anchorage, AK.

Siedlecki, W., & Sklansky, J. (1988). On automatic feature selection. *International Journal of Pattern Recognition and Artificial Intelligence, 2*(2), 197-220.

Siedlecki, W., & Sklansky, J. (1989). A note on genetic algorithms for large-scale feature selection. *Pattern Recognition Letters, 10*(5), 335-347.

Silberschatz, A., & Tuzhilin, A. (1996). What makes patterns interesting in knowledge discovery systems. *IEEE Transactions on Knowledge and Data Engineering, 5*(6).

Silberschatz, A., & Tuzhilin, A., (1995). On subjective measures of interestingness in knowledge discovery. In *Proceedings of KDD'95 Conference.* AAAI Press

Skowron, A. (1993). Boolean reasoning for decision rules generation. In J. Komorowski & Z.W. Raś (Eds.), Seventh International Symposium for Methodologies for Intelligent Systems ISMIS, *Lecture Notes in Artificial Intelligence, 689,* 295-305. Trondheim, Norway. Springer.

Skowron, A. (2001). Toward intelligent systems: Calculi of information granules. *Bulletin of the International Rough Set Society, 5,* 9-30.

Skowron, A. (2003). Approximation spaces in rough neurocomputing. In M. Inuiguchi, S. Hirano, S. Tsumoto (Eds.), *Rough set theory and granular computing* (pp. 13-22). Berlin, Heidelberg, Germany: Springer Verlag.

Skowron, A. (2004). Vague concepts: A rough-set approach. In B. De Beats, R. De Caluwe, G. De Tre, J. Fodor,

J. Kacprzyk, & S. Zadrożny (Eds.), *Data and knowledge engineering: Proceedings of EUROFUSE'2004* (pp. 480-493). Warszawa, Poland: Akademicka Oficyna Wydawnicza EXIT.

Skowron, A. (2005). Rough sets and vague concepts. *Fundamenta Informaticae, 64,* 417-431.

Skowron, A., & Rauszer, C. (1992). The discernibility matrices and functions in information systems. In R. Słowiński (Ed.), *Intelligent decision support – Handbook of applications and advances of the rough sets theory* (pp. 331-362). Dordrecht, The Netherlands: Kluwer Academic Publishers.

Skowron, A., & Stepaniuk, J. (1995). Generalized approximation spaces. In T. Y. Lin & A. M. Wildberger (Eds.), *Soft computing, simulation councils* (pp. 18-21), San Diego.

Skowron, A., & Stepaniuk, J. (1996). Tolerance approximation spaces. *Fundamenta Informaticae, 27,* 245-253.

Skowron, A., & Stepaniuk, J. (2003). Informational granules and rough-neural computing. In S. K. Pal, L. Polkowski, A. Skowron (Eds.), *Rough-neural computing: Techniques for computing with words* (pp. 43-84). Heidelberg, Germany, Springer Verlag.

Skowron, A., & Suraj, Z. (1993). Rough sets and concurrency. *Bulletin of the Polish Academy of Sciences, 41*(3), 237-254.

Skowron, A., Pawlak, Z., Komorowski, J., & Polkowski L. (2002). A rough set perspective on data and knowledge. In W. Kloesgen & J. Żytkow (Eds.), *Handbook of KDD* (pp. 134-149). Oxford: Oxford University Press.

Skowron, A., Stepaniuk, J., & Peters, J.F. (2003). Towards discovery of relevant patterns from parametrized schemes of information granule construction. In M. Inuiguchi, S. Hirano, & S. Tsumoto (Eds.), *Rough set theory and granular computing* (pp. 97-108). Berlin, Heidelberg, Germany: Springer Verlag.

Skowron, A., Stepaniuk, J., Peters, J.F., & Swiniarski, R. (2006). Calculi of approxiamtion spaces. *Fundamenta Informaticae, 72,* 363-378.

Skowron, A., Swiniarski, R., & Synak, P. (2005). Approximation spaces and information granulation. *Transactions on Rough Sets III,* 175-189.

Skowronski B., & Mordecka, Z. (2001). Polish plant for supercritical extraction of hop (in Polish), *Przem. Chem., 80,* 521–523.

Skowronski, B. (2000). Interview by L. Dubiel. Hop extract—Polish at last (in Polish), *Przem. Ferment. i Owocowo -Warzywny, 9,* 30-31.

Ślęzak, D. (1996). Approximate reducts in decision tables. In *Proceedings of the 6th International Conference on Information Processing and Management of Uncertainty in Knowledge-Based Systems (IPMU '96)* (pp. 1159-1164).

Ślęzak, D. (1999). Decomposition and synthesis of decision tables with respect to generalized decision functions. In S.K. Pal, & A. Skowron, (Eds.), *Rough fuzzy hybridization – A new trend in decision making.* Berlin, Heidelberg, Germany: Springer-Verlag.

Slezak, D. (2000). Various approaches to reasoning with frequency based decision reducts: A survey. In L. Polkowski, S. Tsumoto, T. Y. Lin (Eds.), *Rough set methods and applications* (pp. 235-285).

Ślęzak, D. (2002). Approximate entropy reducts. *Fundamenta Informaticae, 53,* 365-387.

Ślęzak, D. (2003). Approximate Markov boundaries and Bayesian Networks: Rough set approach. In M. Inuiguchi, S. Hirano, & S. Tsumoto (Eds.), *Rough set theory and granular computing* (pp. 109-121). Berlin, Heidelberg, Germany: Springer Verlag.

Ślęzak, D. (2005). Rough sets and Bayes factor. *Transactions on Rough Set, III, Journal Subline, Lectures Notes in Computer Science, 3400,* 202-229.

Ślęzak, D., & Wasilewski, P. (in preparation). *Granular sets.* Unpublished manuscript.

Słowiński, R. (Ed.). (1992). *Intelligent decision support, Handbook of applications and advances of the rough set theory.* Dordrecht, Holland: Kluwer Academic Publisher.

Słowiński, R., & Stefanowski, J. (1989). Rough classification in incomplete information systems. *Mathematical and Computer Modelling, 12*(10-11), 1347-1357.

Słowiński, R., & Stefanowski, J. (1994). Handling various types of uncertainty in the rough set approach. In W. Ziarko (Ed.), *Rough Sets, Fuzzy Sets and Knowledge Discovery. Proceedings of the International Workshop on Rough Sets and Knowledge Discovery (RSKD'93)* (pp 366-376). Berlin, Heidelberg: Springer-Verlag.

Słowiński, R., & Stefanowski, J. (1996). Rough set reasoning about uncertain data. *Fundamenta Informaticae, 27*, 229-243.

Słowiński, R., & Vanderpooten, D. (1997). Similarity relation as a basis for rough approximations. In P. P. Wang (Ed.), *Machine intelligence and soft computing, IV* (pp. 17-33). Raleigh, NC: Bookwrights.

Słowiński, R., & Vanderpooten, D. (2000). A generalized definition of rough approximations based on similarity. *IEEE Transactions on Data and Knowledge Engineering, 12*(2), 331-336.

Smith, M.G., & Bull, L. (2003). Feature construction and selection using genetic programming and a genetic algorithm. In *Proceedings of the 6th European Conference on Genetic Programming (EuroGP 2003)* (pp. 229-237).

Smith, N.J.J. (2005). Vagueness as closeness. *Australasian Journal of Philosophy, 83*, 157-183.

Smolinski, T. G. (2004). *Classificatory decomposition for time series classification and clustering.* Unpublished doctoral dissertation, University of Louisville, KY.

Smolinski, T. G., Boratyn, G. M., Milanova, M., Buchanan, R., & Prinz, A. A. (2006). Hybridization of independent component analysis, rough sets, and multi-objective evolutionary algorithms for classificatory decomposition of cortical evoked potentials. *Lecture Notes in Artificial Intelligence, 4146*, 174-183.

Smolinski, T.G., Boratyn, G.M., Milanova, M., Zurada, J.M., & Wrobel, A. (2002). Evolutionary algorithms and rough sets-based hybrid approach to classificatory decomposition of cortical evoked potentials. *Lecture Notes in Artificial Intelligence, 2475*, 621-628.

Smolinski, T.G., Buchanan, R., Boratyn, G.M., Milanova, M., & Prinz, A.A. (2006). Independent component analysis-motivated approach to classificatory decomposition of cortical evoked potentials. *BMC Bioinformatics, 7*(Suppl 2), S8.

Smolinski, T.G., Milanova, M., Boratyn, G.M., Buchanan, R., & Prinz, A.A. (2006). Multi-objective evolutionary algorithms and rough sets for decomposition and analysis of cortical evoked potentials. In Y-Q. Zhang & T.Y. Lin (Ed.), *IEEE International Conference on Granular Computing.* Atlanta, GA: IEEE Press.

Starzyk, J.A., Nelson, D.E., & Sturtz, K. (2000). A mathematical foundation for improved reduct generation in information systems. *Journal of Knowledge and Information Systems, 2*(2), 131-146.

Stefanowski, J. (1998). Handling continuous attributes in discovery of strong decision rules. In L. Polkowski, & A. Skowron, (Eds.), *Rough Sets and Current Trends in Computing, 1st International Conference (RSCTC'98), Warsaw, Poland. Lectures Notes in Artificial Intelligence, 1424* (pp. 394-401). Berlin, Heidelberg, Germany: Springer-Verlag.

Stefanowski, J. (2001). *Algorithms of decision rule induction in data mining.* Poznan, Poland: Poznan, University of Technology Press.

Stefanowski, J., & Tsoukias, A. (2001). Incomplete information tables and rough classification. *Computational Intelligence: An International Journal, 17*(3), 545-566.

Stepaniuk, J. (1998). Approximation spaces, reducts and representatives. In L. Polkowski, A. Skowron (Eds.), Rough sets in knowledge discovery 2. *Studies in Fuzziness and Soft Computing, 19*, 109-126. Heidelberg: Springer-Verlag.

Stepaniuk, J. (1998). Optimizations of rough set model. *Fundamenta Informaticae, 36*(2-3), 265-283.

Stevens, R. (1987). *Hops, An introduction to brewing science and technology.* Series II, Vol. I, p. 23. London: Institute of Brewing.

Stone, M.H. (1963). The theory of representations for Boolean algebras. *Transactions of AMS, 40*, 37–111.

Stone, P. (1936). The theory of representation for Boolean algebras. *Transactions of the American Mathematical Society, 40*(1), 37-111.

Suraj, Z. (1996). Discovery of concurrent data models from experimental tables: A rough set app

Suraj, Z. (1998). The synthesis problem of concurrent systems specified by dynamic information systems. In L. Polkowski & A. Skowron (Eds.), *Rough sets in knowledge discovery 2* (pp. 418-448). Berlin: Physica-Verlag.

Suraj, Z. (2000). Rough set methods for the synthesis and analysis of concurrent processes. In L. Polkowski, S. Tsumoto, & T.Y. Lin (Eds.), *Rough set methods and applications* (pp. 379-488). Berlin: Physica-Verlag.

Suraj, Z. (2001). Some remarks on extensions and restrictions of information systems. *Lecture Notes in Artificial Intelligence, 2005*, 204-211.

Suraj, Z., & Pancerz, K. (2005). Restriction-based concurrent system design using the rough set formalism. *Fundamenta Informaticae, 67*(1-3), 233-247.

Suraj, Z., & Pancerz, K. (2005, November). *The ROSECON system - A computer tool for modelling and analysing of processes*. Paper presented at the International Conference on Computational Intelligence for Modelling, Control and Automation, Vienna, Austria.

Suraj, Z., & Pancerz, K. (2005, September). *Some remarks on computing consistent extensions of dynamic information systems*. Paper presented at the 5th International Conference on Intelligent Systems Design and Applications, Wroclaw, Poland.

Suraj, Z., & Pancerz, K. (2006). Reconstruction of concurrent system models described by decomposed data tables. *Fundamenta Informaticae, 71*(1), 101-119.

Suraj, Z., & Pancerz, K. (2006, September). *Efficient computing consistent and partially consistent extensions of information systems*. Paper presented at the Workshop on Concurrency, Specification and Programming, Wandlitz, Germany.

Suraj, Z., Pancerz, K., & Owsiany, G. (2005). On consistent and partially consistent extensions of information systems. *Lecture Notes in Artificial Intelligence, 3641*, 224-233.

Suraj, Z., Pancerz, K., & Swiniarski R.W. (2005, June). *Prediction of model changes of concurrent systems described by temporal information systems*. Paper presented at the 2005 World Congress in Applied Computing, Las Vegas, Nevada, USA.

Sutton, R.S., & Barto, A.G. (1998). *Reinforcement learning: An introduction.* Cambridge, MA: The MIT Press.

Swiniarski, R. (1999). Rough sets and principal component analysis and their applications in data model building and classification. In S.K. Pal, & A. Skowron (Ed.), *Rough fuzzy hybridization: A new trend in decision-making* (pp. 275-300). Singapore: Springer-Verlag.

Tarski, A. (1935). Zur Grundlegung der Booleschen Algebra, I. *Fundamenta Mathematicae, 24*, 177-198.

Technologic. (2006). *TSUM user's manual*. Retrieved from http://www.embeddedarm.com/

Thiele, H. (1998). *Fuzzy rough sets versus rough fuzzy sets - An interpretation and a comparative study using concepts of modal logics.* (Technical report no. CI-30/98). University of Dortmund.

Tinbergen, N. (1963). On aims and methods of ethology. *Zeitschrift für Tierpsychologie, 20*, 410-433.

Torvalds, L. (2006). *Linux operating system*. Retrieved from http://www.linux.org/

Tsay, L.-S., & Ras, Z.W. (2005). Action rules discovery system DEAR, method and experiments. *Journal of Experimental and Theoretical Artificial Intelligence, 17*(1-2), 119-128.

Tsay, L.-S., & Ras, Z.W. (2006). Action rules discovery system DEAR3. In F. Exposito et al. (Eds.), *Foundations of Intelligent Systems, Proceedings of ISMIS'06*, Bari, Italy, LNAI, No. 4203 (pp. 483-492). Springer.

Tzacheva, A., & Ras, Z.W. (2005). Action rules mining. *International Journal of Intelligent Systems, 20*(7), 719-736.

Vakarelov, D. (1989). Modal logics for knowledge representation systems. In A. R. Meyer, & M. Taitslin (Eds.), *Symposium on Logic Foundations of Computer Science, Pereslav-Zalessky* (pp. 257-277). *Lectures Notes in Computer Science, 363*. Berlin, Heidelberg, Germany: Springer-Verlag.

Vakarelov, D. (1991). A modal logic for similarity relations in Pawlak knowledge representation systems. *Fundamenta Informaticae, 15*, 61-79.

Vakarelov, D. (1998). Information systems, similarity and modal logics. In E. Orłowska (Ed.), *Incomplete information: Rough set analysis* (pp. 492-550). Heidelbelrg, Germany: Physica-Verlag.

Wang, G., Yu, H., & Yang, D. (2002). Decision table reduction based on conditional information entropy, *Chinese Journal of Computers, 25*, 759-766.

Wang, J., & Wang, J. (2001). Reduction algorithms based on discernibility matrix: The ordered attributes method. *Journal of Computer Science and Technology, 16*, 489-504.

Wang, X.Y., Yang, J., Teng, X., Xia, W., & Jensen, R. (2006). Feature selection based on rough sets and particle swarm optimization. Accepted for publication in *Pattern Recognition Letters*.

Wasilewska, A. (1997). Topological rough algebras. In T. Y. Lin, & N. Cercone (Eds.), *Rough sets and data mining* (pp. 411-425). Dodrecht: Kluwer Academic Publisher.

Wasilewska, A., & Banerjee, M. (1995). Rough sets and topological quasi-Boolean algebras. In T. Y. Lin (Ed.), *Workshops on Rough Sets and Data Mining at 23rd Annual Computer Science Conference* (pp. 61-67). Nashville, TN.

Wasilewska, A., & Vigneron, L. (1995). Rough equality algebras. In P. P. Wang, (Ed.), *Proceedings of the International Workshop on Rough Sets and Soft Computing at 2nd Annual Joint Conference on Information Sciences (JCIS'95)* (pp. 26 - 30). Raleigh, NC.

Wasilewski, P. (2004). *On selected similarity relations and their applications into cognitive science* (in Polish). Unpublished doctoral dissertation, Department of Logic, Jagiellonian University, Cracow.

Wasilewski, P. (2005). Concept lattices vs. approximation spaces. In D. Ślęzak G. Wang, M. S. Szczuka, I. Düntsch, & Y.Yao (Eds.), *Rough Sets, Fuzzy Sets, Data Mining, and Granular Computing, 10th International Conference, Regina, Canada (RSFDGrC 2005), Lectures Notes in Artificial Intelligence 3641* (pp. 114-123). Berlin, Heidelberg: Springer-Verlag.

Watkins, C. J. C. H., & Dayan, P. (1992). Technical note: Q-learning. *Machine Learning, 8*, 279-292.

Watkins, C.J.C.H. (1989). *Learning from delayed rewards*. Ph.D. thesis, supervisor: Richard Young, King's College, University of Cambridge, UK.

Wawrzyński. P. (2005). *Intensive reinforcement learning*. PhD dissertation, supervisor: Andrzej Pacut, Institute of Control and Computational Engineering, Warsaw University of Technology.

Wawrzyński. P., & Pacut, A. (2004). Intensive versus nonintensive actor-critic algorithms of reinforcement learning. In *Proceedings of the 7th International Conference on Artificial Intelligence and Soft Computing* (pp. 934-941), Springer 3070,

Wille, R. (1982). Restructuring lattice theory: An approach based on hierarchies of concepts. In I. Rival (Ed.), *Ordered set* (pp. 445-470). Dordecht; Boston: Reidel.

Wiweger, A. (1989). On topological rough sets. *Bulletin of the Polish Academy of Sciences, Mathematics, 37*, 89-93.

Wolski, M. (2003). Galois connections and data analysis. *Fundamenta Informaticae CSP*, 1-15.

Wolski, M. (2005). Formal concept analysis and rough set theory from the perspective of finite topological approximations. *Journal of Transactions on Rough Sets*, III, LNCS 3400, 230-243.

Wong, S.K.M., Wang, L.S., & Yao, Y.Y. (1993). Interval structure: A framework for representing uncertain information. In *Proceedings of 8th Conference on Uncertainty in Artificial Intelligence* (pp. 336-343).

Wong, S.K.M., Wang, L.S., & Yao, Y.Y. (1995). On modeling uncertainty with interval structures. *Computational Intelligence, 11*, 406-426.

Wróblewski, J. (1995). Finding minimal reducts using genetic algorithms. In *Proceedings of the 2nd Annual Joint Conference on Information Sciences* (pp. 186-189).

Wróblewski, J. (1996). Theoretical foundations of order-based genetic algorithms. *Fundamenta Informaticae, 28*(3-4), 423-430.

Wróblewski,. J. (1998). Genetic algorithms in decomposition and classification problem. In L. Polkowski & A. Skowron (Eds.), Rough sets in knowledge discovery 2: Applications, case studies and software systems. *Studies in Fuzziness and Soft Computing, 19*, 471-487. Heidelberg, Germany: Springer.

Wu, Q., Liu, Z.T,. & Li, Y. (2004). Rough formal concepts and their accuracies. In *Proceedings of 2004 International Conference on Services Computing (SCC'04)* (pp. 445-448).

Wu, W.Z., Mi, J.S., & Zhang, W.X. (2003). Generalized fuzzy rough sets. *Information Sciences, 151*, 263-282.

Wu, W.Z., Zhang, M., Li, H.Z., & Mi, Z.H. (2005). Knowledge reduction in random information systems via Dempster-Shafer theory of evidence. *Information Sciences, 174*, 143-164.

Wybraniec-Skardowska, U. (1989). On a generalization of approximation space. *Bulletin of the Polish Academy of Sciences, Mathematics, 37*, 51-61.

Wygralak, M. (1989). Rough sets and fuzzy sets—Some remarks on interrelations. *Fuzzy Sets and Systems, 29*, 241-243.

Xiong, N., & Litz, L. (2002). Reduction of fuzzy control rules by means of premise learning—method and case study. *Fuzzy Sets and Systems, 132*(2), 217-231.

Yao, Y. Y. (1998). A comparative study of fuzzy sets and rough sets. *Information Sciences, 109*(1-4), 21-47.

Yao, Y.Y. (1996). Two views of the theory of rough sets in finite universe. *International Journal of Approximate Reasoning, 15*, 291-317.

Yao, Y.Y. (1998). Constructive and algebraic methods of the theory of rough sets. *Information Sciences, 109*(1-4), 21-47.

Yao, Y.Y. (1998). Generalized rough set models. In L. Polkowski & A. Skowron (Eds.), *Rough sets in knowledge discovery* (pp. 286-318). Heidelberg: Physica-Verlag.

Yao, Y.Y. (1998). On generalizing Pawlak approximation operators. In *Proceedings of 1st International Conference on Rough Sets and Current Trends in Computing (RSCTC'98)* (pp. 298-307).

Yao, Y.Y. (1998). Relational interpretations of neighborhood operators and rough set approximation operator. *Information Sciences 111*(1-4), 239-259.

Yao, Y.Y. (1998). Constructive and algebraic methods of the theory of rough sets. *Information Sciences, 109*, 21-47.

Yao, Y.Y. (2001). Information granulation and rough set approximation. *International Journal of Intelligent Systems, 16*(1), 87-104.

Yao, Y.Y. (2003). On generalizing rough set theory. In *Proceedings of 9th International Conference on Rough Sets, Fuzzy Sets, Data Mining, and Granular Computing (RSFDGrC'03)* (pp. 44-51).

Yao, Y.Y. (2004). A partition model of granular computing. *Transactions on Rough Set, I, Journal Subline, Lectures Notes in Computer Science, 3100*, 232-253.

Yao, Y.Y. (2004). A comparative study of formal concept analysis and rough set theory in data analysis. In W. Zakowski (Ed.), *Proceedings of 3rd International Conference on Rough Sets and Current Trends in Computing (RSCTC'04)* (1983)-68.

Yao, Y.Y. (2004). Concept lattices in rough set theory. In *Proceedings of 23rd International Meeting of the North American Fuzzy Information Processing Society (NAFIPS'04)* (pp. 796-801).

Yao, Y.Y., & Chen, Y.H. (2004). Rough set approximations in formal concept analysis. In *Proceedings of 23rd International Meeting of the North American Fuzzy Information Processing Society (NAFIPS'04)* (pp. 73-78).

Yao, Y.Y., & Chen, Y.H. (2005). Subsystem based generalizations of rough set approximations. In *Proceedings of 15th International Symposium on Methodologies for Intelligent Systems (ISMIS'05)* (pp. 210-218).

Yao, Y.Y., & Chen, Y.H. (2006). Rough set approximations in formal concept analysis. *Journal of Transactions on Rough Sets,* V, LNCS 4100, 285-305.

Yao, Y.Y., & Lin, T.Y. (1996). Generalization of rough sets using modal logic. *International Journal of Intelligent Automation and Soft Computing, 2,* 103-120.

Yao, Y.Y., Zhao, Y., & Wang, J. (2006). On reduct construction algorithms. In *Proceedings of 1st International Conference on Rough Sets and Knowledge Technology (RSKT'06)* (pp. 297-304).

Zadeh, L. (1965). Fuzzy sets. *Information and Control, 8,* 338-353.

Zadeh, L. A. (1975). The concept of a linguistic variable and its application to approximate reasoning. *Information Sciences, 8,* 199-249, 301-357; *9,* 43-80.

Zadeh, L.A. (2003). Fuzzy logic as a basis for a theory of hierarchical definability (THD). In *Proceedings of 33rd International Symposium on Multiple-Valued Logic (ISMVL'03)* (pp. 3-4).

Zaki, M. (1998). Efficient enumeration of frequent sequences. In *Seventh International Conference on Information and Knowledge Management* (pp. 68-75). Washington, D.C.

Zakowski, W. (1983). Approximations in the space (U, Π). *Demonstratio Mathematica,* XVI, 761-769.

Zhang, L., & Malik, S. (2002). The quest for efficient Boolean satisfiability solvers. In *Proceedings of the 18th International Conference on Automated Deduction* (pp. 295-313).

Zhang, M., & Yao, J.T. (2004). A rough sets based approach to feature selection. In *Proceedings of the 23rd International Conference of NAFIPS* (pp. 434-439). Banff, Canada.

Zhang, W.X., Wei, L,. & Qi, J.J. (2005). Attribute reduction in concept lattice based on discernibility matrix. In *Proceedings of 10th International Conference on Rough Sets, Fuzzy Sets, Data Mining, and Granular Computing (RSFDGrC'05),* Part II (pp. 517-165).

Zhang, W.X., Yao, Y.Y., & Liang, Y. (Eds.). (2006). *Rough set and concept lattice.* Xi'An: Xi'An JiaoTong University Press.

Zhong, N., Dong, J., & Ohsuga, S. (2001). Using rough sets with heuristics for feature selection. *Journal of Intelligent Information Systems, 16*(3), 199-214.

Zhu, W., & Wang, F.Y. (2003). Reduction and axiomization of covering generalized rough sets. *Information Sciences, 152,* 217-230.

Zhu, W., & Wang, F.Y. (2006). Covering based granular computing for conflict analysis. In *Proceedings of IEEE International Conference on Intelligence and Security Informatics (ISI'06)* (pp. 566-571).

Ziarko, W. (1993). Variable precision rough set model. *Journal of Computer and Systems Sciences, 46,* 35-59.

Zvieli, A., & Chen, P. (1986). Entity-relationship modeling and fuzzy databases. In *Proceedings of International Conference on Data Engineering* (pp. 320-327). Los Angeles, CA.

About the Contributors

Aboul Ella Hassanien received a BSc (Hons) in 1986 and an MSc in 1993, both from Ain Shams University, Faculty of Science, Pure Mathematics and Computer Science Department, Cairo, Egypt. In September 1998, he received his doctoral degree from the Department of Computer Science, Graduate School of Science & Engineering, Tokyo Institute of Technology, Japan. Currently, he is an associate professor at Cairo University, Faculty of Computer and Information, IT Department. He is a visiting professor at Kuwait University, College of Business Administration, Quantitative and Information System Department. He has served as the review committee chair, program committee member, and reviewer for various international conferences on artificial intelligence, soft computing, image processing, and data mining. He has received the excellence younger researcher award from Kuwait University for the academic year 2003/2004. He is editing several special issues for many international journals. He has directed many funded research projects. He was a member of the interim advisory board committee of the International Rough Set Society. His research interests include rough set theory, wavelet theory, x-ray mammogram analysis, medical image analysis, fuzzy image processing and multimedia data mining.

Zbigniew Suraj received an MSc from the Pedagogical University in Rzeszów (Poland), a PhD from the Warsaw University in mathematics, and a DSc (habilitation) in computer science from the Institute of Computer Science, Polish Academy of Sciences. He is presently a professor at the University of Rzeszów (Poland), head of the Chair of Computer Science at the University. His recent research interests focus on concurrency, rough set theory and its applications, knowledge discovery, data mining, and approximate reasoning. He has published four books and over 150 articles. He is a member of the editorial board of the *Transactions on Rough Sets* (Springer). He was the editor and co-editor of special issues of *Fundamenta Informaticae* (IOS Press). He is a member of the International Rough Set Society, Gesellschaft für Informatik e.V. in Germany, the Polish Information Processing Society, the Polish Mathematical Society, the vice-president of the Scientific Society in Rzeszów.

Dominik Ślęzak received a PhD in computer science in 2002 from Warsaw University, Poland. In an instructional capacity, he has supervised more than 15 graduate students in Canada, Poland, and the UK. He has pursued academic collaborations with Warsaw University, University of Regina, and the Polish-Japanese Institute of Information Technology. Additionally, he serves as an executive member of the International Rough Set Society, editor-in-chief of *Online International Journal on Rough Set Methods*, guest editor and reviewer for a number of international scientific journals, chair of several international scientific conferences, and is a member of

About the Contributors

several IEEE technical committees. He has published over 60 research articles for books, journals, and conference proceedings.

Pawan Lingras is a professor and chairperson of the Department of Mathematics and Computing Science at Saint Mary's University, Halifax, Canada. His undergraduate education from Indian Institute of Technology, Bombay, India was followed by graduate studies at the University of Regina, Canada. He has authored more than 100 research papers in various international journals and conferences. His areas of interests include artificial intelligence, information retrieval, data mining, Web intelligence, and intelligent transportation systems. He has served as the review committee chair, program committee member, and reviewer for various international conferences on artificial intelligence and data mining. He has also served as general workshop chair for IEEE Conference on Data Mining (ICDM), and ACM/IEEE/WIC Conference Web Intelligence, and Intelligent Agent Technologies (WI/IAT). He is an associate editor of *Web Intelligence: An International Journal.*

* * *

Theresa Beaubouef received a PhD in computer science from Tulane University in 1994. Her research interests focus mainly on uncertainty in databases and geographic information systems, but also include other areas such as computation, image processing, and enology. She has worked for the Naval Research Laboratory, Clark Research and Development, and Xavier University, and is currently a professor at Southeastern Louisiana University.

Maciej Borkowski (MSc, Warsaw University, 2000); MSc, University of Manitoba, 2002) is completing his PhD degree. Currently, he is teaching at the University of Winnipeg, a researcher in the Computational Intelligence Laboratory in the ECE Department at the University of Manitoba, a peer reviewer for the *Transactions on Rough Sets Journal* and a number of international conferences. Since 2001, he has published numerous papers, primarily in the rough set theory area and computer vision. His research interests focus on computer vision, design, and implementation of a system used to extract three-dimensional information from two 2D images, digital image processing, geometry, pattern recognition, and parallel computing.

Jerzy W. Grzymala-Busse has been a professor of electrical engineering and computer science at the University of Kansas since August of 1993. His research interests include data mining, machine learning, knowledge discovery, expert systems, reasoning under uncertainty, and rough set theory. He has published three books and over 200 articles. He is a member of editorial boards of the *Foundations of Computing and Decision Science, International Journal of Knowledge-Based Intelligent Engineering Systems, Fundamenta Informaticae, International Journal of Hybrid Intelligent Systems,* and *Transactions on Rough Sets.* He is a vice-president of the International Rough Set Society and a member of the Association for Computing Machinery, American Association for Artificial Intelligence and Upsilon Pi Epsilon.

Cory J. Butz received Bsc, MSc, and PhD degrees in computer science from the University of Regina, Saskatchewan, Canada (1994, 1996, and 2000, respectively). He joined the School of Information Technology and Engineering at the University of Ottawa, Ontario, Canada, as an assistant professor in

2000. In 2001, he joined the Computer Science Department at the University of Regina, Saskatchewan, Canada, in 2001. He was promoted to associate professor in 2003. His research interests include uncertainty reasoning, database systems, information retrieval, and data mining.

Yaohua Chen received a BEng (1997) from Northern East University, People's Republic of China, an MSc (2004) from the University of Regina, Canada. He is currently a PhD candidate in the Department of Computer Science, University of Regina. His research interests include intelligent information systems, granular computing, and cognitive informatics.

Christopher Henry (MSc, 2006) is a full-time PhD student in the Department of Electrical and Computer Engineering (ECE) at the University of Manitoba. Currently, he is the president of the ECE Graduate Students' Association (GSA), and he is also a councilor on the University of Manitoba GSA. Since 2004, he has published six articles (two of which are journal publications) about reinforcement learning, rough sets, and intelligent systems, and has presented papers at two international conferences. His current research interests are in reinforcement learning, rough set theory, approximation spaces, digital image processing, and intelligent systems.

Zdzisław S. Hippe is a full professor of technical sciences at the University of Information Technology and Management in Rzeszów, Poland. His research interests include computer chemistry, expert systems, data mining, machine learning, knowledge discovery, and design of large-scale information systems. He has published 13 books and over 370 articles. He is a vice president of the CODATA Polish National Committee on Data for Science and Technology, and a member of some scientific committees of the Polish Academy of Sciences, such as Computer Science, Scientific Information, Analytical Chemistry, and Chemometrics.

Richard Jensen received a BSc in computer science from Lancaster University, UK, and an MSc and PhD in artificial intelligence from the University of Edinburgh, UK. He is a postdoctoral research fellow with the Department of Computer Science at the University of Wales, Aberystwyth, working in the Advanced Reasoning Group. His research interests include rough and fuzzy set theory, pattern recognition, information retrieval, feature selection, and swarm intelligence. He has published approximately 20 peer-refereed articles in these areas.

Dan Lockery (EIT, BSc, 2003) is completing an MSc in the Department of Electrical and Computer Engineering (ECE) at the University of Manitoba. He is a student member of the IEEE, RAS, and APEGM. His research interests include image processing, robotics and automation, control systems, reinforcement learning, and biologically inspired designs of intelligent systems.

Teresa Mroczek is an assistant of the Department of Expert Systems and Artificial Intelligence at the University of Information Technology and Management, Rzeszów. Her research interests include data mining, especially Bayesian networks, supervised machine learning for classification and identification of objects, reasoning under uncertainty, and rough set theory. She is writing her PhD thesis under the supervision of Prof. Zdzisław Hippe in the Faculty of Computer Science and Management at the Wroclaw University of Technology. She has published about 20 articles. She is a member of the Polish Information Processing Society.

About the Contributors

Hung Son Nguyen is an assistant professor at the Faculty of Mathematics, Informatics and Mechanics, Warsaw University, Poland. He worked as a visiting scientist in Sweden and Northern Ireland. Dr. Hung Son Nguyen is an author and co-author of more than 60 articles published in journals, edited books, and conference proceedings. His main research interests include rough set theory and applications in data mining, hybrid and multilayered learning methods, KDD with applications in text mining, image processing, and bioinformatics.

Krzysztof Pancerz received an MSc in 1998 from the Rzeszów Univeristy of Technology in Electrical Engineering and a PhD in 2006 from the Institute of Computer Science, Polish Academy of Sciences in computer science. He is presently an assistant professor at the University of Information Technology and Management in Rzeszów, Poland. His research interests concern computational intelligence, knowledge discovery, data mining, and data modeling. They especially focus on rough sets and Petri nets. In these areas, he has published over 20 research papers in international journals and conference proceedings. He is a member of the Scientific Society in Rzeszów.

James F. Peters, PhD (1991), is a full professor in the Department of Electrical and Computer Engineering (ECE) at the University of Manitoba. Currently, he is co-editor-in-chief of the *Transactions on Rough Sets Journal* (Springer), co-founder and researcher of the Computational Intelligence Laboratory in the ECE Department (1996-), and current member of the steering committee, International Rough Sets Society. Since 1997, he has published numerous articles about systems that learn in refereed journals, edited volumes, conferences, and workshops. His current research interests are in approximation spaces (near sets), pattern recognition (ethology and image processing), rough set theory, reinforcement learning, biologically inspired designs of intelligent systems (vision systems that learn), and the extension of ethology (study of behavior of biological organisms) to the study of intelligent systems behavior.

Frederick E. Petry received a PhD in computer and information science from The Ohio State University in 1975. He was a professor of computer science at Tulane University, New Orleans, LA, and is currently a computer scientist at the Naval Research Laboratory. His research interests include representation of imprecision via fuzzy sets and rough sets in databases, GIS, and other information systems. Dr. Petry has over 300 scientific publications including 120 journal articles/book chapters and 7 books written or edited. He is an associate editor of *IEEE Transactions on Fuzzy Systems, Neural Processing Letters,* and *Fuzzy Sets and Systems*. He was selected as an IEEE fellow in 1996, and in 2003 was made a fellow of the International Fuzzy Systems Association. In 2002, he was chosen as the outstanding researcher of the year in the Tulane University School of Engineering and in 2004, received the Naval Research Laboratory's Berman Research Publication award.

Astrid A. Prinz received her diploma in physics from the University of Ulm, Germany, in 1996. She proceeded to work on the biophysics of invertebrate neurons with Peter Fromherz at the Max Planck Institute for Biochemistry in Martinsried, Germany, and earned a doctorate in physics from Munich Technical University in 2000. As a postdoctoral fellow with Eve Marder at Brandeis University, Dr. Prinz combined experimental and computational approaches to study pattern generation and robustness in central pattern-generating networks. In 2005, Dr. Prinz was appointed as an assistant professor in the Department of Biology at Emory University in Atlanta, Georgia, where she, and her lab, work on pattern generation, homeostasis, and synchronization phenomena in small neuronal networks.

Zbigniew W. Ras received a PhD from the Warsaw University (Poland) and a DSc (habilitation) from the Polish Academy of Sciences. Since 1987, he is a professor of computer science at the University of North Carolina, Charlotte. He is the KDD laboratory director in the College of Computing and Informatics at UNC-Charlotte, professor at the Polish-Japanese Institute of Information Technology in Warsaw (Poland), editor-in-chief of the *Journal of Intelligent Information Systems* (Springer), deputy editor-in-chief of *Fundamenta Informaticae Journal* (IOS Press), and general chair of ISMIS Symposium. Since 1994, he is a member of the Senate of the Polish-Japanese Institute of Information Technology and affiliate professor in the Institute of Computer Science at the Polish Academy of Sciences.

Edward Roj is a senior researcher of the Fertilizers Research Institute – Puławy, Poland. His research interests include chemical engineering, multiobjective optimization, data mining, and expert systems. He has published over 60 articles. He is a member of the Polish Association of Chemical Engineers.

Annibal Parracho Sant'Anna has graduated in mathematics and economics at the Federal University of Rio de Janeiro (UFRJ). He was conferred an MSc by the Institute of Pure and Applied Mathematics (IMPA-CNPq), in Rio de Janeiro, in 1970, and a PhD in statistics by The University of California, Berkeley, in 1977. He was, for distinct terms, the chairman of the Institute of Mathematics of UFRJ. He is presently full professor at the School of Engineering of Universidade Federal Fluminense, where he has been for the last five years the coordinator of the graduate program in production engineering.

Qiang Shen received BSc and MSc degrees in communications and electronic engineering from the National University of Defence Technology, China, and a PhD in knowledge-based systems from Heriot-Watt University, Edinburgh, UK. He is a professor with the Department of Computer Science at the University of Wales, Aberystwyth. His research interests include fuzzy and imprecise modeling, model-based inference, pattern recognition, and knowledge refinement and reuse. Dr Shen is an associate editor of the *IEEE Transactions on Fuzzy Systems* and of the *IEEE Transactions on Systems, Man, and Cybernetics* (Part B), and an editorial board member of the *Fuzzy Sets and Systems Journal*. He has published around 170 peer-refereed papers in academic journals and conferences on topics within artificial intelligence and related areas.

Bolesław Skowronski is a director of the Fertilizers Research Institute – Puławy, Poland. His research interests include chemical technology and chemical engineering. He has published over 56 articles. He is a member of the Polish Association of Chemical Engineers.

Tomasz G. Smolinski received a BE in computer engineering from the Polish-Japanese Institute of Computer Techniques, Warsaw, Poland, in 1998. After two more years at the same school (although under a new name of the Polish-Japanese Institute of Information Technology), he received an MS in computer science with a double specialization in knowledge discovery in databases as well as artificial intelligence and multimedia. In 2000, he was admitted to the Computer Science and Engineering PhD program and simultaneously offered a research assistantship at the University of Louisville, Kentucky. After successfully defending his doctoral dissertation in 2004, he continued working as a postdoctoral associate at the office of the senior vice president for research at the University of Louisville. In July

About the Contributors

of 2005, he was offered a postdoctoral fellowship in the Biology Department at Emory University, Atlanta, where he continues working on his main area of interest: applications of computer science in neurosciences.

Piotr Wasilewski received his PhD in mathematical logic in 2005 from Jagiellonian University, Krakow, Poland. He is doing his PhD in cognitive psychology at Warsaw University, Poland. He carries out scientific collaborations with Jagiellonian University and University of Regina-Canada. His scientific interest embraces foundations of rough sets, connections between rough set theory and formal concept analysis, as well as philosophical and mathematical foundations of cognitive science. Wasilewski serves as a reviewer for several international scientific journals and conferences. He published a few research articles for journals and conference proceedings. He delivered an invited lecture at the RSFDGrC'2005 conference.

Elzbieta M. Wyrzykowska received a BS from the University of Information Technology and Management, Warsaw, Poland, in 2002. Also, she graduated from Karol Szymanowski's School of Music (First and Second Level), in Warsaw, mastering in violin, viola, and piano. Currently, she is a graduate student at the University of Information Technology and Management. Her research interests include archival science, knowledge discovery, intelligent information systems, and music information retrieval.

Wen Yan received a BSc in electronic engineering from Qingdao University of Science & Technology, Shandong, China, in 2002. He received an MSc in computer science from the University of Regina, Saskatchewan, Canada, in 2006. His research interests include uncertainty reasoning, information retrieval, and rough sets.

Yiyu Yao received a BEng (1983) from Xi'an Jiaotong University, People's Republic of China, an MSc (1988), and a PhD (1991) from the University of Regina, Canada. He was with the Department of Mathematical Sciences, Lakehead University from 1992 to 1998. He joined the Department of Computer Science, University of Regina in 1998, where he is currently a professor of computer science. His research interests include information retrieval, uncertainty management, and intelligent information systems. He serves as an editorial board member for many journals, such as *International Journal of Approximate Reasoning, International Journal of Web Intelligence and Agent Systems, LNCS Transactions on Rough Sets, International Journal of Cognitive Informatics & Natural Intelligence, International Journal of Knowledge and Information Systems, Journal of Computing and Information*, and so forth.

Jacek M. Zurada received MS and PhD degrees (with distinction) in electrical engineering from the Technical University of Gdansk, Poland (1968 and 1975, respectively). Since 1989, he has been a professor, and since 1993 a distinguished Samuel T. Fife professor with the Electrical and Computer Engineering Department at the University of Louisville, Kentucky. He was department chair from 2004 to 2006. He has published 280 journal and conference papers in the areas of neural networks, computational intelligence, data mining, image processing, and VLSI circuits. He also has authored or co-authored three books. From 1998 to 2003 he was the editor-in-chief of *IEEE Transactions on Neural Networks* and president of IEEE Computational Intelligence Society in 2004-05. In 2003 he was conferred the title of national professor by the president of Poland. Since 2005 he has been a member of the Polish Academy of Sciences, and serves as a senior Fulbright specialist in 2006.

Index

A

access point (AP) 191
accuracy of approximation 210, 211, 217
actor critic (AC) 186, 187, 194, 199
  method 187, 195, 197
ant colony optimization (ACO) 97
  framework 98
  model 98
appropriate problem representation 97
approximate information system 6
approximation space 186, 187, 189, 195, 199, 201
association
  analysis 60
  rules 61, 64

B

B-
  boundary region 210
  outside region 210
belief networks 228, 234, 235, 238, 278
binary Boolean algebra 40
Boolean reasoning 38, 44, 61

C

CBF task 205
classificatory decomposition (CD) 204, 206, 208
colored Petri net theory 239, 240, 266
concept lattices 109, 117, 119
concise configuration description 95
concurrent system models 239
cost function definition 95
cross-validation (CV) 222
Cylinder-Bell-Funnel (CBF) 205, 220

D

data 175, 184, 282
  -analysis framework 109
  mining 38, 39, 44, 47, 104, 228, 231, 237, 269, 274
database 175, 176
decision rule construction 53
decision tree 56
  induction 57
  soft 59
degrees of freedom (DOF) 192
directed acyclic graph (DAG) 152, 154
discretization 39, 54
dynamic
  information system 241
  reducts 76

E

elementary-based definition 122
entropy-based
  heuristic 100
  reduction (EBR) 79
equivalence relation 209
evolutionary algorithms (EA) 211, 212

F

feasible solutions, construction of 97
feature selection (FS) 39, 47, 66, 70, 71, 72, 79, 277
filter approach 72
formal concept analysis (FCA) 108, 109, 110, 116, 117
fuzzy rough
  feature selection (FRFS) 85
  measure 90

set dependency measure 99
sets 84
fuzzy set 135, 139, 140

G

genetic algorithms (GAs) 94, 211
granule-based definition 121

H

heuristic desirabilit 97
Heyting algebras 3, 18, 22
hop extraction process 228, 230

I

independent component analysis (ICA) 204, 206, 208, 225, 275
information system 4, 209, 241
  and approximation bases 9
Internet protocol control protocol (IPCP) 191
intuitionistic sets 130, 138, 148, 150

J

Johnson Reducer 81

K

knowledge discovery 129, 149

L

LERS algorithm 180
line-crawling robot 186
link control protocol (LCP) 191

M

MLEM2 algorithm 228
modal-style data operators 111
modified vector evaluated genetic algorithm (M-VEGA) 213
monocular vision 186
multiobjective evolutionary algorithms (MOEA) 205, 206, 208, 211

N

network control protocols (NCPs) 191

O

object-oriented
  concept lattice 117, 118, 123
  database 129, 143, 144

P

Particle swarm optimization (PSO) 100
Petri nets 239
pheromone updating rule 97
point-to-point protocol (PPP) 191
positive region 210
preprocessor 72
preset algorithm 80
primary processes 244
principal component analysis (PCA) 204, 206, 208, 227, 286
probabilistic transition rule 97

Q

quality of classification 210, 211, 217
QuickReduct 74, 86

R

random move generator 95
redundancy 71
reinforcement learning 186, 187, 191, 193, 194, 199
relational
  algebra 136, 141
  database 130, 131, 136
rough-set
  analysis (RSA) 108, 109, 113
  methodology 249
  theory 239, 240, 241, 264
  theory-based schemes 208
rough coverage actor critic (RCAC) 196
rough set 2, 3, 13, 20, 39, 47, 53, 58, 70, 73, 80, 83, 84, 129, 130, 135, 136, 139, 188, 202, 205, 206, 208, 209, 210, 211, 217
  and Boolean reasoning 38
  flow graphs (RSFGs) 152
  method 77
  theory (RST) 1, 70, 73
    based approach 70
rule induction 228, 231, 233, 235
run-and-twiddle (RT) 186, 194, 197, 199

Index

S

satisfiability (SAT) 83
secondary processes 244
significance coefficient 211
simulated annealing (SA) 95
sparse coding with overcomplete bases (SCOB) 215
   methodology 208
spatial data 144, 148, 149
standard rough coverage (SRC) 189
suitable annealing schedule 96
swarm intelligence (SI) 97
Synthetic Control Chart (SCC) 221

T

target tracking 187, 191, 192, 194
temporal difference (TD) 188, 194
tolerance rough sets 22, 28
Traditional Algorithm for RSFG Inference 157
tree-based strategy 175

U

uncertainty management 152

V

vagueness 1, 2, 3, 32, 273
   perspective 1
variable precision rough sets (VPRS) 75
vector evaluated genetic algorithm (VEGA) 213
VEGA algorithm 214

W

wrapper approach 72